Lecture Notes in Physics

Edited by H. Araki, Kyoto, J. Ehlers, München, K. Hepp, Zürich
R. Kippenhahn, München, H. A. Weidenmüller, Heidelberg
and J. Zittartz, Köln

Managing Editor: W. Beiglböck

250

Lie Methods in Optics

Proceedings of the CIFMO-CIO Workshop
Held at León, México, January 7–10, 1985

Edited by J. Sánchez Mondragón and K.B. Wolf

Springer-Verlag
Berlin Heidelberg GmbH

Editors

Javier Sánchez Mondragón
Centro de Investigaciones en Optica A.C.
Apdo. Postal 948, 37000 León, Gto., México

Kurt Bernardo Wolf
Instituto de Investigaciones en Matemáticas Aplicadas y en Sistemas
Universidad Nacional Autónoma de México
Apdo. Postal 20-726, 01000 México D.F., México

ISBN 978-3-662-13579-2 ISBN 978-3-540-39811-0 (eBook)
DOI 10.1007/978-3-540-39811-0

Originally published by Springer-Verlag Berlin Heidelberg New York Tokyo in 1986
Softcover reprint of the hardcover 1st edition 1986

2153/3140-543210

The CIFMO–CIO workshop on

Lie Methods in Optics

LEÓN, GUANAJUATO, JANUARY 7–10, 1985

Informal gatherings of a small number of research workers interested in various aspects of a frontier field, are probably the most efficient way to exchange and consolidate new results and information. With this model in mind, personal invitations were made to the participants of this workshop. The necessary funds were obtained through the **Centro Internacional de Física y Matemáticas Orientadas** (CIFMO) project, given by the Dirección General de Investigación Científica y Superación Académica, Secretaría de Educación Pública. Its director, **Dr. Salvador Malo**, is the promotor of the CIFMO project. These funds were administrated by the **Centro de Investigaciones en Optica** (CIO) of León, Guanajuato, which also undertook the organizational aspects of the workshop. CIO's founding director, **Dr. Daniel Malacara**, is a distinguished scientist working on the design and correction of optical instruments. Several other research workers came from the Instituto Nacional de Astronomía, Optica y Electrónica and from the Instituto de Astronomía, Universidad Nacional Autónoma de México (UNAM), which promote the study of optics. A select group of graduate students attended the sessions; these came from CIO, Facultad de Ciencias, Instituto de Física, and Instituto de Investigaciones en Matemáticas Aplicadas y en Sistemas, UNAM.

The program of the workshop was set up after a short round of talks, in which each participant presented his field of work and area of interest. During the workshop proper, there were lectures and discussions led by the presenter. In many overlap regions a common language was developed.

At the end of the workshop, the participants agreed to contribute to this proceedings volume, with the aim that this may serve as a useful reference book in the subject. Total integration is seldom possible; each author has given his own viewpoint and often his own notation. We would like to thank them for their cooperation, however, in developing the basic material in a careful form, enhancing its readability and lasting value. Editorial work has been done in two appendices, prepared by K.B. Wolf based on standard material, on the symplectic groups and algebra representations. True, traditional optics reads rather differently and Lie methods may seem too mathematical for some practitioners of this old science/art. Also, group-theorists bred with quantum mechanical symmetries, shell-model nuclear physics, and exotic elementary particle multiplets, may find the hamiltonian formulation of optics unfamiliar. Nevertheless, we have seen time and again that it is in blending fields that broader insights are gained and new applications arise. The technological import of image-processing and communication devices based on light is evident, we believe.

JAVIER SÁNCHEZ MONDRAGÓN
KURT BERNARDO WOLF

presenting

The CIFMO Project

Centro Internacional de Física y Matemáticas Orientadas*

CIFMO* is to be an international center where scientists working in *applied* physics and mathematics can meet. It is an academic institution being organized roughly along the lines of the International Centre for Theoretical Physics, in Trieste. The areas of research presently included are: optics and image processing, stochastic models, and plasma physics; other topics are contemplated for the near future. CIFMO will be located in the city of Cuernavaca, Morelos, but its activities may extend freely in collaboration with other institutions in Mexico or abroad.

The objectives of CIFMO are:

- The identification of areas of research in physics and mathematics which are important or promising for the development of a solid scientific and technological basis for the country.
- To contribute to the training of a cadre of specialists of high quality, who may be able to apply this knowledge in an effective way, especially in developing countries.
- To foster the international cooperation between researchers working in these areas in matters of common scientific interest.

CIFMO is intended to be a center of international academic standing, where junior and senior scientists, graduate students and workers in technology can freely exchange their inquiries and results. The activities may take the form of workshops, colloquia, summer and winter schools, short visits by prominent research leaders, repeated or longer stays of up to a year by scientists at every stage of their careers. These activities may take place at the home institutions of the members in Mexico or —when space permits— in Cuernavaca. After the first few years of operation, we expect it to have roughly half of its membership in Mexico and half abroad, in both developed and developing countries.

---

*in formation

The city of León

PLACED in the wide *Bajío* valley in the central Mexican highland, León lies 1786 meters above sea level, 395 km north-west from Mexico city. The region belongs to the *chupícuaro* culture invaded by *guamare* and *pame* indians, and has a preclassic horizon at the two nearby sites of Alfaro and Ibarrillas; it was conquered for the Spanish crown by Nuño de Guzmán in 1530. The founding date of León is January 20, 1576, by order of the viceroy Enríquez de Almanza. The *Villa de León* was then a trading center populated by Spaniards, Tarascans, Mexicans, and Chichimecas; it was needed as defense agains the Chichimecas, Guamares and Cuachichiles. In 1596 it had 180 Spaniards and by 1600 a Franciscan convent was founded.

In 1731 the Jesuits established a school at León, when it had 5000 inhabitants. Nearing a population of 10 000 in 1779, the *Villa* was declared a *Ciudad* —a City. The *Bajío* became the cradle of the independence movement of 1810, and the city of León participated in the wars. Throughout the nineteenth century, León grew to 50 000 inhabitants in 1840, and by 1884 it was the second largest city in the country with 120 000 —while México had 300 000 and Guadalajara 80 000. In 1888 León suffered a devastating flood from which it only slowly recovered. By the turn of the century, however, León was a quite modern town: it had streetcars and railways in 1882, telephone in 1892, electric light in 1898, an airfield in 1911 and in 1914, a cinema.

Presently, León's population has grown to half a million and the city is a manufacturing center with a reputation for its leather goods. The central district was recently remodelled as a pedestrian area. The colonial and moorish buildings around the arcaded *plaza de armas* house a good collection of outdoor cafés, park kiosks, banks, and stores. The Centro de Investigaciones en Optica is in one of the newer sections of the city overlooking the green-and-brown *Bajío* plains.

About this volume

The preparation of the matrix of this Proceedings volume was done at the Instituto de Investigaciones en Matemáticas Aplicadas y en Sistemas, Universidad Nacional Autónoma de México. We have used the TEX text-processing languaje developed by **Donald Knuth** at Stanford, supplemented by the Fácil TEX macro package of **Max Días**. The typography was done by **Alfredo Cortés, Paulino Fermín Ramos, Miguel Navarro Saad**, and one of the editors. This is the second proceedings volume prepared with this system.*

This book also partially covers the commitment by IIMAS stated in Project IVT/EE/NAL/81/1250, "Tipografía Científica Automatizada," presented to the Consejo Nacional de Ciencia y Tecnología in January 1981.

*The first one is: *Nonlinear Phenomena, Proceedings of the CIFMO School and Workshop on–* held at Oaxtepec, México, November 29 – December 17, 1982. Ed. by K.B. Wolf, Lecture Notes in Physics, Vol. 189 (Springer Verlag, 1983).

Participants

ALEX J. DRAGT — Department of Physics, University of Maryland, College Park, MD 20742, USA

JOHN R. KLAUDER — AT&T Bell Laboratories, Murray Hill, NJ 07974, USA

WOLFGANG LASSNER — Naturwissenschaftlich-theoretisches Zentrum, Karl-Marx-Universität, Leipzig, GDR

DANIEL MALACARA — Centro de Investigaciones en Optica, Apdo. Postal 948, 37000 León, Guanajuato

VLADIMIR I. MAN'KO — P.N. Lebedev Institute of Physics, USSR Academy of Sciences, 117924 Moscow, USSR

JORGE OJEDA CASTAÑEDA — Instituto Nacional de Astronomía, Optica y Electrónica, Apdo. Postal 948, Tonanzintla, Puebla

JAVIER SÁNCHEZ MONDRAGÓN — Centro de Investigaciones en Optica, Apdo. Postal 948, 37000 León, Guanajuato

WALTER SCHEMPP — Lehrstuhl für Mathematik, Universität Siegen, D-5900 Siegen, GFR

STANLY STEINBERG — Department of Mathematics and Statistics, The University of New Mexico, Albuquerque, NM 87131, USA

RICARDO WEDER — Instituto de Investigaciones en Matemáticas Aplicadas y en Sistemas, Universidad Nacional Autónoma de México, Apdo. Postal 20-726, 01000 México D.F.

KURT BERNARDO WOLF* — Departamento de Matemáticas, Universidad Autónoma Metropolitana, Iztapalapa, D.F.

Other contributors to this volume

OCTAVIO CASTAÑOS, ENRIQUE LÓPEZ MORENO — Centro de Estudios Nucleares, Facultad de Ciencias, Universidad Nacional Autónoma de México, México D.F.

ETIENNE FOREST — Lawrence Berkeley Laboratory, University of California, Berkeley, CA 94770, USA

HANS RASZILLIER — Physikalisches Institut, Universität Bonn, D-5300 Bonn 1, GFR

*Permanent address: Instituto de Investigaciones en Matemáticas Aplicadas y en Sistemas (Cuernavaca), Universidad Nacional Autónoma de México, Apdo. Postal 20-726, 01000 México D.F.

Research visitors

Luis Raúl Berriel	Instituto Nacional de Astronomía, Optica y Electrónica, Tonanzintla
Alejandro Cornejo	Instituto Nacional de Astronomía, Optica y Electrónica, Tonanzintla
Salvador Cuevas	Instituto de Astronomía, UNAM, México D.F.
Ricardo Flores	Centro de Investigaciones en Optica, León
Enrique Landgrave	Centro de Investigaciones en Optica, León
Zacarías Malacara	Centro de Investigaciones en Optica, León
Arquímedes Morales	Centro de Investigaciones en Optica, León

Graduate students

Víctor Castaño	Instituto de Física, UNAM, México D.F.
Jaime Delgado	Centro de Investigaciones en Optica, León
Enrique López Moreno	Facultad de Ciencias, UNAM, México D.F.
Rogelio Mejía	Instituto de Física, UNAM, México D.F.
Carmela Menchaca	Centro de Investigaciones en Optica, León
Miguel Navarro Saad	IIMAS, UNAM, México D.F.

Contents

Chapter 1. Analog radar signal design and digital signal processing —a Heisenberg nilpotent Lie group approach

by WALTER SCHEMPP

Chapter 2. Fourier optics from the perspective of the Heisenberg group

by HANS RASZILLIER and WALTER SCHEMPP

Chapter 3. Lie series, Lie transformations, and their applications

by STANLY STEINBERG

Chapter 4. Foundations of a Lie algebraic theory of geometrical optics

by ALEX J. DRAGT, ETIENNE FOREST, and KURT BERNARDO WOLF

Chapter 5. Canonical transforms for paraxial wave optics

by Octavio Castaños, Enrique López-Moreno, and Kurt Bernardo Wolf

Chapter 6. Wave theory of imaging systems

by John R. Klauder

Chapter 7. Invariants and coherent states in fiber optics

by VLADIMIR I. MAN'KO

Chapter 8. The influence of spherical aberration on gaussian beam propagation

by VLADIMIR I. MAN'KO and KURT BERNARDO WOLF

Appendix A. The symplectic groups, their parametrization and cover

Appendix B. Representations of the algebra sp(2,$\Re$)

Chapter 1

Analog radar signal design and digital signal processing —a Heisenberg nilpotent Lie group approach

by WALTER SCHEMPP

In every mathematical investigation, the question will arise whether we can apply our mathematical results to the real world. V.I. ARNOLD (1983)

We will cite instances of pure and applied mathematicians doing the same or analogous mathematics, but because of the lack of communication neither knew of the others' work. L. AUSLANDER (1979)

ABSTRACT: The notions of analog and digital radar auto- and cross-ambiguity functions are on the borderline with mathematics, physics, and electrical engineering. This paper presents the solutions of two problems of analog radar signal design: the *synthesis* problem (posed in 1953) and the *invariance* problem for ambiguity surfaces over the symplectic time-frequency plane. Both solutions are achieved via harmonic analysis on the differential principal fiber bundle over the two-dimensional polarized (resp. isotropic) cross-section with structure group isomorphic to the one-dimensional center of the simply connected real Heisenberg nilpotent Lie group. In this way, the linear oscillator representation of the three-dimensional real metaplectic group gives rise to a procedure for generating the energy-preserving linear automorphisms of any given radar ambiguity surface over the time-frequency plane by means of *chirp* waveforms (linear frequency modulated signals).

In the field of digital signal processing, the Whittaker–Shannon–Kotel'nikov sampling theorem also fits the framework of nilpotent harmonic analysis. The basic idea is to realize the linear Schrödinger representation by the linear lattice representation acting in a complex Hilbert space modeled on the compact Heisenberg nilmanifold, to wit, the differential principal fiber bundle over the two-dimensional compact torus $\mathfrak{T}^2$ with structure group isomorphic to the one-dimensional center of the reduced real Heisenberg nilpotent Lie group. In the same vein we look upon the finite Fourier transform, and finally, based on the ambiguity surface conservation principle, the paper deals in a geometric way with the phase discontinuity of Fourier and microwave optics. It follows that analog and digital signal processing, as well as Fourier optics, have a deep geometric common root in nilpotent harmonic analysis. As a mathematical by-product of this research, an identity for Laguerre functions of different orders pops up. Some of its special cases, to wit, a collection of new identities for theta constants have been explicitly calculated and numerically checked.

1.1 Introductory discussion of radar principles

There seems little doubt that radar technology is a permanent and important aspect of research and development in electronics.

L.R. RABINER and B. GOLD (1975)

Let us begin with a brief explanation of the basic principles of radar, in order to point out the role played by the three-dimensional real Heisenberg nilpotent Lie group in the theory of analog radar signal design. To be more precise, we shall be concerned with an application of harmonic analysis on the differential principal fiber bundle over the two-dimensional polarized (resp. isotropic) cross-section with structure group isomorphic to the one-dimensional center of the simply connected real Heisenberg nilpotent Lie group, to the mathematical treatment of the range and velocity measurement on one or more moving targets, by means of the time delay and the Doppler frequency shift which are simultaneously attached to the analog signal return of a transmitted radar signal.

The purpose of *radar* (= **ra**dio **d**etection **a**nd **r**anging) systems is, basically, to survey broad areas of sky in order to detect the presence of a distant object and, at the same time, to gather various kinds of information about the target. In the case of a moving target searched by a radar, the data of main interest are

- the **bearing**,
- the **range** d,
- the radial **velocity** v relative to the radar antenna, and
- the **size** of the object.

Figures 1 below show an elementary form of a conventional radar system using a common stationary antenna (*radar dish*) for both transmission and reception, achieved by means of a duplexer.

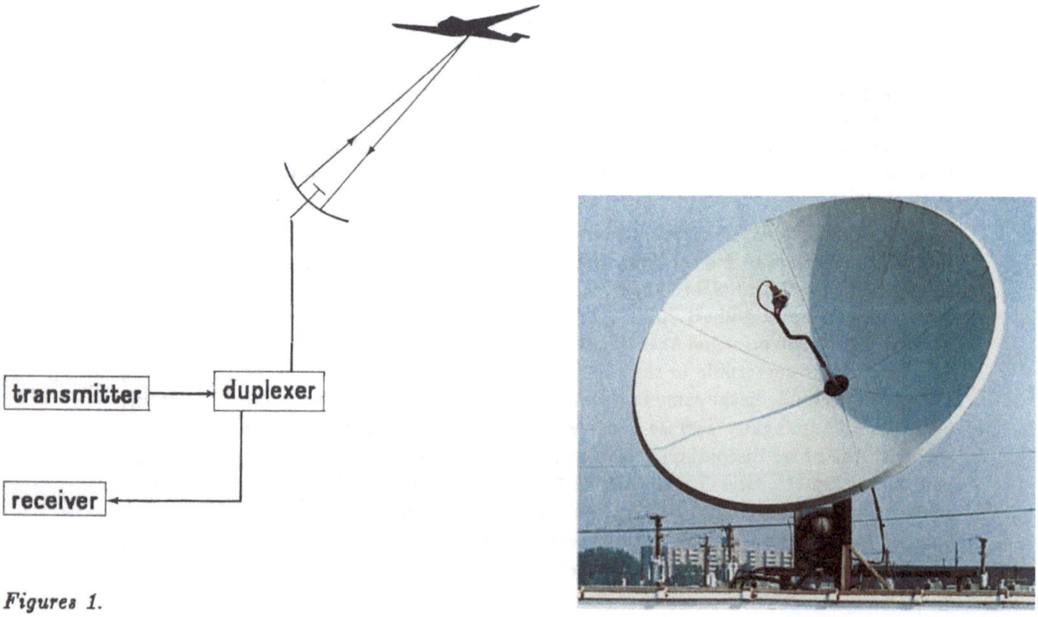

Figures 1.

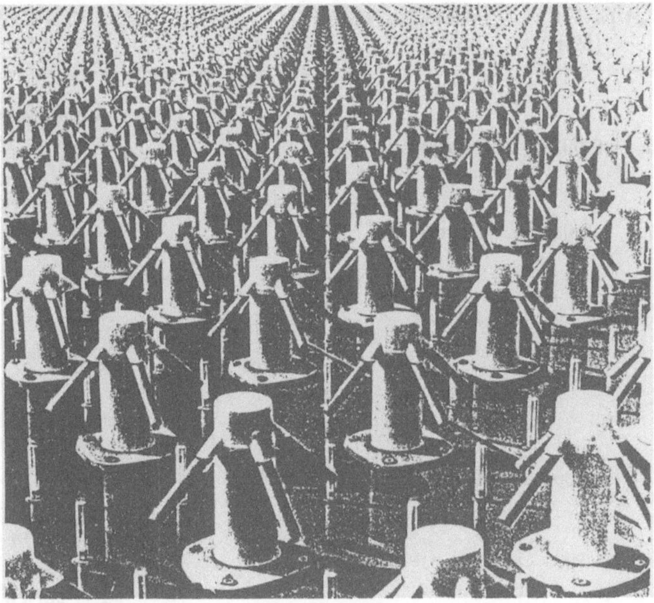

Figure 2.

In a typical air traffic control radar, the stationary antenna rotates mechanically, sweeping out the full 360° by its microwave radiation beam every 4 to 12 s. The azimuth beam resolution is about 1° to 2° and the vertical antenna pattern is a dispersed, fanlike beam, usually having 30° to 45° width. Present airport surveillance radar systems must track up to 50 aircraft within their fields of view (typically 35 to 65 km), and display these tracks to the air traffic controller. Yet in many of the most familiar uses of radar systems such as aviation, air defense, and intelligence, the mechanically steered parabolic antenna of everyday experience is giving way to a new kind of device —see Figure 2 above.

A flat, regular arrangement (*array antenna*) of small, identical antenna elements, each one capable of transmitting and receiving radar signals, takes the place of the physically rotating parabolic reflector. Even as its beam scans expanses of sky, no part of the radar antenna itself moves. Instead, the signal is deflected from target to target electronically, steered througt the physical principle of wave interference.

Today a single radar system of this kind can do what previously might have required a battery of mechanically steered antennas. The new radar device is known as the *electronically steerable* **phased-array** *system*. A military phased-array radar system, for instance, can track several hundred targets scattered through a volume of space spanning about 120° in azimuth and 80° in elevation. In this way, electronically steerable phased array radar systems, used in conjunction with high-speed digital signal processing hardware, have led to great sophistication of radar tracking algorithms. However, the basic principles of radar still remain unchanged when embodied in this advanced technology.

In the transmission mode, the antenna radiates periodically a directed, narrow, pencil-like beam of radio energy in the form of trains (*blocks*) of coherent radar pulses of large amplitude, brief duration, and the same carrier frequency ω. A central oscillator generates the radar signal; transistors or specialized microwave tubes such as traveling-wave tubes amplify it. Most often the electromagnetic energy falls within microwave bands, from 300 MHz to 10 GHz, although some very-long-range radar systems operate instead in the HF and VHF bands (from 3 MHz to 30 MHz and from 30 MHz to 300 MHz, respectively). See the Table, next page.

WAVES		FREQUENCY (MH)	WAVELENGTH (m)		PRESENT APPLICATIONS
ELF	extreme low frequency				submarine communication
		0.003	100 000		
VLF	very low frequency				
		0.03	10 000		
LF	low frequency			long waves	radio
		0.3	1 000		
MF	medium frequency			middle waves	radio
		3	100		
HF	high frequency				radio telegraphy
		30	10	short waves	amateur radio
VHF	very high frequency			TV channels 1–11	FM-radio, TV, police, CB radio
		300	1		
UHF	ultra high frequency			TV channels 14–53	TV, car radio radio navigation
		3 000	0.1		
SHF	super high frequency			microwaves	beacons, radar satellite communication
		30 000	0.01		
EHF	extreme high frequency			millimeter waves	
		300 000	0.001		
IR	infrared			submillimeter waves	optical fibers

Table. Electromagnetic waves and their applications in telecommunication.

If a remote object lies in the path of the propagating beam, a portion of the transmitted signal energy is reflected by the target. Provided that the energy of the transmitted radar pulse, the sensitivity of the radar antenna (operating in the *reception* mode) and the reflective quality of the target are all sufficient, an echo will be detected. The echo signal is then processed in the receiver (the radar installation) to search for the presence of the target and to gauge its parameters for precise tracking.

In the reception mode, the signals received by the radar antenna consist of a high-frequency carrier modulated in amplitude (or phase) by functions f of time, which vary much more slowly than the cycles of the carrier. In radar, the parameters chiefly serving to distinguish or resolve two echo signals are their

arrival times x and the **Doppler shifts** y

of their carrier frequencies from a common reference frequency within the spectral band. The transmitted carrier frequency ω is a natural choice for the reference frequency of a single train of coherent pulses, provided ω is very stable. Thus the analog part of the radar receiver assumes a high degree of accuracy.

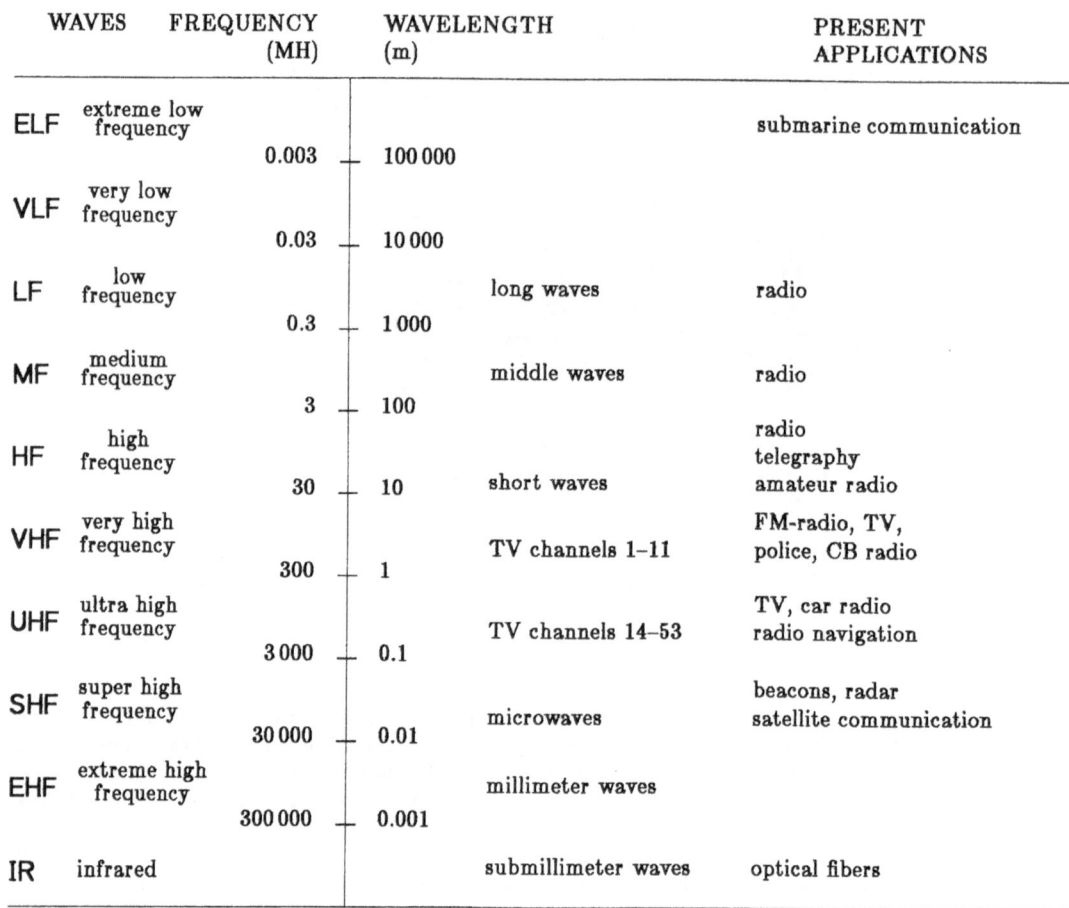

The structure of the receiver and its performance depend upon the symmetrized auto-correlation or analog radar auto-ambiguity function, $(x, y) \rightsquigarrow H(f; x, y)$, associated with the complex envelope f of the monochromatic signal pulse

$$t \rightsquigarrow f(t)\, e^{2\pi i \omega t}$$

that is transmitted by the radar station, and the analog signal return

$$t \rightsquigarrow f(t+x)\, e^{2\pi i \omega (t+y)}$$

that, ideally, is a time-delayed and Doppler frequency-shifted version of the transmitted radar signal. For the purposes of present-day radars we may assume that the target velocity v is small in comparison with the velocity c of electromagnetic radiation. It follows that for the range and frequency shift,

$$d = \frac{1}{2} cx \quad \text{and} \quad v = \frac{1}{2} c \frac{y}{\omega}.$$

We always assume that the signal waveform f belongs to the Schwartz space, $\mathcal{S}(\Re)$, of infinitely differentiable complex-valued functions on the real line $\Re$ that are rapidly decreasing at infinity. By duality we may also include signals having tempered complex distributions $f \in \mathcal{S}'(\Re)$ as waveforms. In the following we shall consider $\mathcal{S}(\Re)$ as an (everywhere dense) vector subspace of the standard complex Hilbert space $\mathcal{L}^2(\Re)$ which is embedded in the natural way into the complex vector space $\mathcal{S}'(\Re)$. The **energy** of the signal is then given by the integral (squared $\mathcal{L}^2$-norm)

$$\|f\|^2 = \int_{\Re} dt\, |f(t)|^2.$$

Thus the $\mathcal{L}^2$-norm on the complex prehilbert space $\mathcal{S}(\Re)$ may be considered as the signal energy norm.

As an example of a signal waveform which has attained practical importance for radar and sonar systems we mention the ***chirp signal*** or linear frequency modulated signal. The envelope of the chirps takes the form

$$t \rightsquigarrow C_u\, e^{-i\pi u t^2},$$

where $u \in \Re$ and $C_u \in \mathbb{C}$ are non-zero constants; see the papers by Klauder, Price, Darlington, and Albersheim [31], and Claasen and Mecklenbräuker [14, 15, 16]. The sound effect caused by the instantaneous signal frequency $\omega - ut$ of the chirp signal explains the name. When trying first to detect a target at long range and then track the target as the range d decreases, a standard procedure is first to transmit a chirp signal followed by a pulse sequence, *i.e.*, a burst yielding precise range and velocity measurements.

The practical realization of any radar system involves an enormous amount of complex electronic equipment. The brains of the radar system is the receiver tracking computer, scheduling the appropriate positions of the mechanically steered dish or the beam positions of the electronically steerable antenna, coordinating the transmitted radar signals via the control path going to the antenna system, and finally running the display system. The transmitted radar signals may vary from simple pulse trains to high bandwidth chirps and bursts of pulses and chirps. In order to estimate the target's size, bursts of extremely short pulses are necessary for an appreciable interval between the echo signals corresponding to the length of the target. It is one of the most important tasks of the receiver tracking computer to control the central signal generator, in order to achieve the best range and velocity measurement.

1.2 Analog radar signal design

The early association of the Heisenberg group with quantum mechanics should not hide the fact that it has been most useful recently as a frame to describe wave systems —optical and radar— where a meaningful phase space and geometric (i.e., classical) limit exist.

M. GARCÍA-BULLÉ, W. LASSNER, and K.B. WOLF (1985)

A major component in the design of the overall radar system is analog signal design, directed toward achieving the best range and velocity measurement on one or more distant targets. At its center is the analog ***radar auto-ambiguity function***

$$(x, y) \rightsquigarrow H(f; x, y)$$

mentioned in Section 1 above, since it represents an idealized mathematical model of a radar system involving the two key variables, arrival time x and Doppler frequency shift y. It takes the symmetrized form (see Ville [54], Woodward [60], and Wilcox [58])

$$H(f; x, y) = \int_{\mathfrak{R}} dt\, f(t + \tfrac{1}{2}x)\overline{f}(t - \tfrac{1}{2}x)\, e^{2\pi i y t}.$$

Since it is immaterial how the signs of time delay and Doppler frequency shift are chosen, there is no unique way of defining the radar auto-ambiguity function. Several essentially equivalent definitions exist in the literature.

While the radar auto-ambiguity function is complex-valued, the ***Wigner distribution function*** used as a phase space technique in quantum mechanics (see Wigner [57] and, for instance, Balazs-Jennings [6], Hillery, O'Connell, Scully, and Wigner [25], and reference [38]),

$$P(f; q, p) = \int_{\mathfrak{R}} dt\, f(q + \tfrac{1}{2}t)\, \overline{f}(q - \tfrac{1}{2}t) e^{2i\pi p t},$$

is real-valued but not always positive on the position–momentum plane $\mathfrak{R} \oplus \mathfrak{R}$. The phase space techniques are motivated by the desire to understand both classical and quantum mechanics in terms of the same basic concepts. Apart of its use in quantum mechanics, the Wigner distribution function has important applications in the analysis of loudspeaker performance (Janse and Kaizer [28], Heyser [23], Gerzon [20]) and optical systems (Bastiaans [7,8], Brenner [11], Brenner and Ojeda-Castañeda [12]). The functions $H(f; \cdot, \cdot)$ and $P(f; \cdot, \cdot)$ are related through a double Fourier transform. The symmetrized cross-correlation or analog ***radar cross-ambiguity function*** $H(f, g; \cdot, \cdot)$ associated with $f \in S(\mathfrak{R})$ and $g \in S(\mathfrak{R})$ is similarly defined via the prescription

$$H(f, g; x, y) = \int_{\mathfrak{R}} dt\, f(t + \tfrac{1}{2}x)\, \overline{g}(t - \tfrac{1}{2}x)\, e^{2\pi i y t},$$

and is of importance in communication theory. Notice that the assignment

$$(f, g) \in S(\mathfrak{R}) \times S(\mathfrak{R}), \quad (f, g) \rightsquigarrow H(f, g; \cdot, \cdot), \quad H(f, g; \cdot, \cdot) \in S(\mathfrak{R} \oplus \mathfrak{R}),$$

defines a sesquilinear mapping which will be seen to be surjective. For $f \in S(\mathfrak{R})$, its restriction $f \rightsquigarrow H(f; \cdot, \cdot) \in S(\mathfrak{R} \oplus \mathfrak{R})$ to the diagonal of $S(\mathfrak{R}) \times S(\mathfrak{R})$, however, is *not* surjective.

The first problem to be solved in analog radar signal design is the *synthesis problem* (see, for instance, Wilcox [58]). It asks for an intrinsic characterization of those functions

$$F \in S(\mathfrak{R} \oplus \mathfrak{R})$$

on the time-frequency plane $\mathfrak{R} \oplus \mathfrak{R}$ (or ***information plane*** in the sense of Gabor [18]) which belong to the range of the mapping $f \rightsquigarrow H(f; \cdot, \cdot)$, *i.e.*, for which there exists a complex-valued envelope $f \in S(\mathfrak{R})$ satisfying the identity

$$F = H(f; \cdot, \cdot).$$

In order words, the problem is to find necessary and sufficient conditions for a given complex-valued smooth function F in the two Fourier dual variables $x \in \Re$ (separation in time) and $y \in \Re$ (separation in frequency) such that F can be realized as an analog radar auto-ambiguity function $(x, y) \rightsquigarrow H(f; x, y)$ with respect to a complex-valued smooth signal waveform $t \rightsquigarrow f(t)$ in one (time) variable $t \in \Re$. In Section 3 we will establish a solution of the radar synthesis problem via harmonic analysis on the differential principal fiber bundle over the two-dimensional polarized (resp. isotropic) cross-section with structure group isomorphic to the one-dimensional center of the simply connected real Heisenberg nilpotent Lie group. Ultimately, our approach is based on the analogy between non-relativistic quantum physics and signal theory, which has been emphasized in the classical study of Ville [54] and particularly in the fundamental works by Gabor [17] and Klauder [30]. However, their point of view is more heuristic than the one adopted in the present paper.

The image $\mathcal{F} = H(f; \Re, \Re)$ of the time frequency plane $\Re \oplus \Re$ under the analog radar auto-ambiguity function $H(f; \cdot, \cdot)$ is called the ***radar ambiguity surface*** over the time-frequency plane generated by the complex envelope $f \in S(\Re)$. For every signal the radar ambiguity surface is peaked at the origin (0,0) of the time-frequency plane $\Re \oplus \Re$, so that certainly not all functions $F \in S(\Re \oplus \Re)$ can be realized as analog radar auto-ambiguity functions with respect to a suitable signal waveform $f \in S(\Re)$. A second signal, arriving with separations x in time and y in frequency that lie under this central peak, will be difficult to distinguish from the first signal. For many types of signals the radar ambiguity surface exhibits additional peaks elsewhere over the time-frequency plane. These sidelobes may conceal weak signals with arrival times and carrier frequencies far from those of the first signal. In a measurement of the arrival time and frequency of a single signal, the subsidiary peaks may lead to gross errors in the result. The taller the sidelobes of the radar ambiguity surface, the greater the probability of such errors in time and Doppler frequency shift. It is desirable, therefore, for the central peak of the radar ambiguity surface to be narrow, and to have as few and low sidelobes as possible.

A transmitted narrow pulse results in good range but poor velocity measurement, while a wide pulse of a single frequency yields good velocity but bad range information. For instance, the chirp signal results in good range measurement but precise measurements of the velocity require additional waveforms such as pulse bursts. By changing the waveform f of a radar signal of given energy, it is possible to change the accuracy of the range and relative radial velocity measurements in such a manner that an increase of the range accuracy results in a decrease of the velocity accuracy, and vice versa (range-velocity coupling).

The basic constraint of analog radar signal design which has serious consequences on radar measurements, namely the fact that a radar signal cannot be designed such that it gives high performance everywhere in the range-velocity plane, constitutes the essence of the *radar uncertainty principle.* If the signal waveform $f \in S(\Re)$ is normalized such that $\|f\| = 1$ holds, the radar uncertainty principle can be expressed in terms of the analog radar auto-ambiguity function by the formula

$$\iint_{\Re \oplus \Re} dx\, dy\, |H(f; x, y)|^2 = 1.$$

This states that the total volume under the normalized radar ambiguity surface $\mathcal{F} = H(f; \Re, \Re)$ over the time-frequency plane equals unity, independently of the signal waveform f. It follows that there are bounds on the achievable resolution performance in range d and radial velocity v, so that radar signal design turns out to be a compromise between range and velocity measurement.

The radar uncertainty principle parallels the Heisenberg uncertainty principle of quantum mechanics although, at first thought, there appears to be no reason why the Heisenberg uncertainty principle should be of any consequence in radar theory. According to quantum mechanics, not all the physical quantities observed in any realizable experiment (even in principle only) can be determined with arbitrarily high accuracy. Even under ideal experimental conditions, an increase in the measurement accuracy of one quantity can be achieved only at the expense of decreasing the measurement accuracy on another ***canonically conjugate*** quantity. The position coordinate q and its momentum p are one example of two such canonically conjugate quantities: It is impossible to determine simultaneously the position q and momentum p of a non-relativistic quantum-mechanical particle (position-momentum coupling). If $f \in S(\mathfrak{R})$ denotes a normalized state vector, the identity

$$\iint_{\mathfrak{R}\oplus\mathfrak{R}} dp\, dq\, |P(f;q,p)|^2 = 1$$

is an expression for the Heisenberg uncertainty principle in terms of the Wigner distribution function.

As everyone knows, to fully understand any mathematical system one has to understand the transformations of the system and, especially, those transformations of the system that leave some particular aspect of the system invariant. In the case of the mathematical theory of analog radar signal design, a close investigation of the radar uncertainty principle leads to a study of the geometry of the radar ambiguity surfaces $\mathcal{F} = H(f;\mathfrak{R},\mathfrak{R})$ over the time-frequency plane, by means of their energy-preserving linear ***automorphisms***. By such an automorphisms of $\mathcal{F}$ we will understand a unitary operator $S : \mathcal{L}^2(\mathfrak{R}) \to \mathcal{L}^2(\mathfrak{R})$ that maps the vector subspace $S(\mathfrak{R})$ onto itself such that, for all waveforms $f \in S(\mathfrak{R})$ and for each pair $(x, y) \in \mathfrak{R} \oplus \mathfrak{R}$, there exists a pair $(x', y') \in \mathfrak{R} \oplus \mathfrak{R}$, depending on S, that satisfies the identity

$$H(f;x,y) = H\big(S(f);x',y'\big).$$

The second problem of analog radar signal design to be solved is the *invariant problem* for radar ambiguity surfaces over the time-frequency plane: calculating explicitly their energy-preserving linear automorphisms. A solution of the invariant problem based on the linear oscillator representation of the metaplectic group $Mp(1,\mathfrak{R}) = Sp(1,\mathfrak{R}) \times \{+1,-1\}$ will be given in Section 4. For every radar ambiguity surface over the time frequency plane, the result exhibits a generating procedure of the energy-preserving linear automorphisms by means of chirp signals. Moreover, it enables us to determine the radially symmetric, *i.e.*, $SO(2,\mathfrak{R})$-invariant radar ambiguity surfaces over the symplectic time-frequency plane $\mathfrak{R} \oplus \mathfrak{R}$, and leads in a natural way to the Laguerre functions. The procedure of replacing the principal fiber bundle over the two-dimensional polarized (resp. isotropic) cross-section with a structure group isomorphic to the one-dimensional center of the simply connected real Heisenberg nilpotent Lie group by the compact Heisenberg nilmanifold, then yields a new identity for Laguerre functions of different orders. Some consequences of this identity for the theta constants will be studied in Section 5 below. Section 6 is concerned with a treatment of cardinal spline interpolation and digital signal processing from the view point of harmonic analysis on the compact Heisenberg nilmanifold. Finally, Section 7 develops the finite Fourier transform via physical systems having finite phase space, whereas Section 8 briefly deals with an application of the ambiguity surface conservation principle which holds in the field of Fourier optics.

1.3 The radar synthesis problem

> *And one morning... I suddenly saw light: Heisenberg's symbolic multiplication was nothing but the matrix calculus, well known to me since my student days... Therefore I was familiar with the fact that matrix multiplication is not commutative.*
> M. BORN (1925)

> *Instead of the commutative law of multiplication, the canonical variables of a system... satisfy the quantum conditions... These equations... are capable of replacing the classical commutative law of multiplication. They appear to be the simplest assumptions one could make which would give a workable theory.*
> P.A.M. DIRAC (1926)

> *...I saw that the noncommutation was really the dominant characteristic of Heisenberg's theory...*
> P.A.M. DIRAC (1971)

D. Gabor (Nobel award 1971) has pointed out in his classic paper [17] that there are two fundamentally distinct approaches to the description of nature: that of time and that of frequency. Both approaches are combined by the notion of ***time-frequency*** (or ***information***) ***plane*** which is of basic importance in information theory. For the purposes of analog radar signal design, rescaling transforms the time-frequency plane into the ***range-velocity plane***. In the following, we look upon these planes as the two-dimensional real vector space $\mathfrak{R} \oplus \mathfrak{R}$ of all pairs $v = (x, y)$. We define the standard *symplectic* (= non-degenerate antisymmetric bilinear) form B on $\mathfrak{R} \oplus \mathfrak{R}$ via the prescription

$$B(v_1, v_2) = \begin{vmatrix} x_1 & y_1 \\ x_2 & y_2 \end{vmatrix}.$$

It is well known that B may be identified with an element of the real vector space of exterior forms, $\bigwedge^2(\mathfrak{R} \times \mathfrak{R}^*)$. A complex-valued function on $\mathfrak{R} \oplus \mathfrak{R}$ is said to be of ***positive type*** on the two-dimensional real ***symplectic*** vector space $(\mathfrak{R} \oplus \mathfrak{R}; B)$ if, for *all* finite sequences of vectors $(v_j)_{1 \leq j \leq N}$ in $\mathfrak{R} \oplus \mathfrak{R}$, the matrix

$$\left(e^{\pi i B(v_j, v_k)} F(v_j - v_k)\right)_{1 \leq j,k \leq N}$$

is a positive definite Hermitian matrix. It should be observed that the notion of positive definiteness on the symplectic plane $(\mathfrak{R} \oplus \mathfrak{R}; B)$ is essentially different from the notion of positive definiteness on the Euclidean plane $\mathfrak{R} \oplus \mathfrak{R}$.

A complex-valued function F on $\mathfrak{R} \oplus \mathfrak{R}$ is called to be of ***pure*** positive type on the symplectic vector space $(\mathfrak{R} \oplus \mathfrak{R}; B)$, if each decomposition $F = F_1 + F_2$ of F into a sum of functions F_1 and F_2 of positive type on $(\mathfrak{R} \oplus \mathfrak{R}; B)$ implies that F_1 and F_2 are proportional to F (*cf.* Souriau [53]).

The ***real Heisenberg group*** $\tilde{A}(\mathfrak{R})$ is the three-dimensional real Lie group with underlying manifold $(\mathfrak{R} \oplus \mathfrak{R}) \times \mathfrak{R}$ and multiplication given by

$$(\mathbf{v}_1, z_1) \cdot (\mathbf{v}_2, z_2) = (\mathbf{v}_1 + \mathbf{v}_2, z_1 + z_2 + \tfrac{1}{2} B(\mathbf{v}_1, \mathbf{v}_2)).$$

Among physicists, the group $\tilde{A}(\mathfrak{R})$ is often called the ***Weyl group***. This terminology, however, would cause too much confusion among mathematicians. Indeed, the Weyl element $\mathbf{J}$ of the real symplectic group $Sp(1, \mathfrak{R}) = SL(2, \mathfrak{R})$ generates a group of order four which acts on $\tilde{A}(\mathfrak{R})$ in a natural way as a group of automorphisms of $\tilde{A}(\mathfrak{R})$. Among mathematicians, *this* group is called the Weyl group. For $\tilde{A}(\mathfrak{R})$, a compromise is the name *Heisenberg-Weyl* group (*cf.* Wolf [59] and García-Bullé, Lassner, and Wolf [19]). Another realization of $\tilde{A}(\mathfrak{R})$ is given by the unipotent matrices

$$\begin{pmatrix} 1 & x & z \\ 0 & 1 & y \\ 0 & 0 & 1 \end{pmatrix},$$

with real entries x, y, z (see, *e.g.*, Auslander [1]). The center $\tilde{C} = \{(0, 0, z) \mid z \in \mathfrak{R}\}$ of $\tilde{A}(\mathfrak{R})$ is isomorphic to the additive group $\mathfrak{R}$, and the descending central series as well as the derived series of $\tilde{A}(\mathfrak{R})$ are given by the descending filtration

$$\tilde{A}(\mathfrak{R}) \hookleftarrow \tilde{C} \hookleftarrow \{1\}.$$

Thus $\tilde{A}(\mathfrak{R})$ is a connected, simply connected, two-step nilpotent, three-dimensional, real Lie group with one-dimensional center; this property characterizes $\tilde{A}(\mathfrak{R})$ within isomorphy. It forms the universal covering group of the *reduced* real Heisenberg nilpotent Lie group $A(\mathfrak{R})$ with center C isomorphic to the one-dimensional compact torus group $\mathfrak{T}$. The reduced real Heisenberg group $A(\mathfrak{R})$ is, in a sense, the simplest possible non-abelian Lie group.

It is not easy to visualize geometrically the action of the real Heisenberg nilpotent Lie group. However, it will be helpful to consider $\tilde{A}(\mathfrak{R})$ as the differential principal fiber bundle over the two-dimensional polarized cross-section $P = \{(\mathbf{v}, 0) \mid \mathbf{v} \in \mathfrak{R} \oplus \mathfrak{R}\}$ to $\tilde{C}$ in $\tilde{A}(\mathfrak{R})$ with structure group isomorphic to $\tilde{C}$. See Figure 3, next page.

$$\begin{array}{ccc} \tilde{C} & \longrightarrow & \tilde{A}(\mathfrak{R}) \\ & & \downarrow \\ & & P \end{array}$$

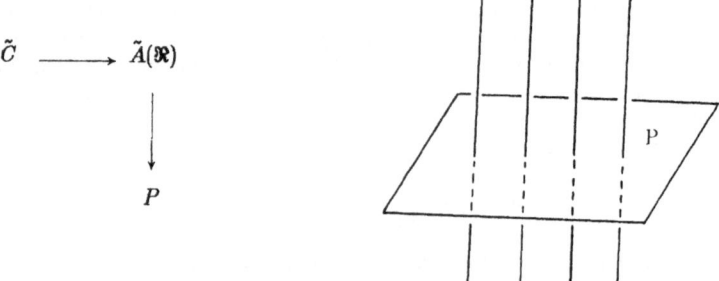

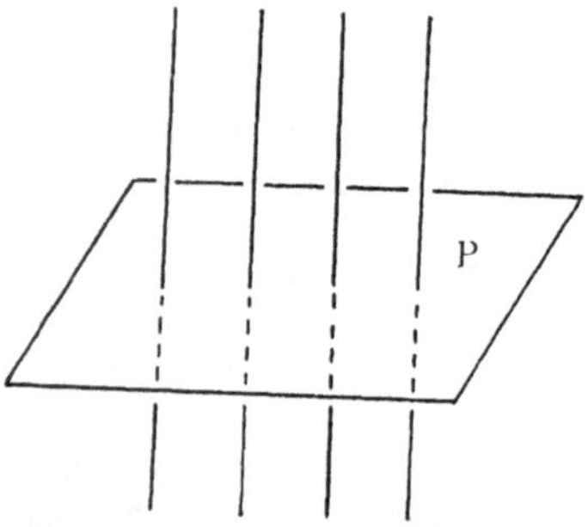

Figure 3.

In order to explain the reason for the central importance of harmonic analysis on the differential principal fiber bundle over P with structure group $\tilde{C}$ for the theory of analog radar signal design, let us describe the unitary dual of $\tilde{A}(\mathfrak{R})$, *i.e.*, the set $\tilde{A}(\mathfrak{R})^\wedge$ of isomorphy classes of topologically irreducible, continuous, unitary, linear representations of $\tilde{A}(\mathfrak{R})$. It is not difficult to prove that for a connected, solvable, locally compact, topological group, all the finite dimensional, topologically irreducible, continuous, unitary, linear representations are necessarily one-dimensional. Thus the unitary dual $\tilde{A}(\mathfrak{R})^\wedge$ admits the following rough classification: It consists of two types of isomorphy classes, namely of

i. continuous unitary characters of $\tilde{A}(\mathfrak{R})$, and

ii. isomorphy classes of infinite dimensional, topologically irreducible, continuous, unitary, linear representations of $\tilde{A}(\mathfrak{R})$. Since $\tilde{A}(\mathfrak{R})$ is simply connected, it forms a monomial real Lie group by the Dixmier-Kirillov theorem. Its Lie algebra $\mathfrak{n}$, the three-dimensional ***real Heisenberg nilpotent Lie algebra***, which is also the Lie algebra of the reduced real Heisenberg nilpotent Lie group $A(\mathfrak{R})$, was first considered by Weyl in his classic reference [56]. It can be realized by the nilpotent matrices

$$\begin{pmatrix} 0 & a & c \\ 0 & 0 & b \\ 0 & 0 & 0 \end{pmatrix}.$$

The matrices

$$\mathbf{X} = \begin{pmatrix} 0 & 1 & 0 \\ 0 & 0 & 0 \\ 0 & 0 & 0 \end{pmatrix}, \qquad \mathbf{Y} = \begin{pmatrix} 0 & 0 & 0 \\ 0 & 0 & 1 \\ 0 & 0 & 0 \end{pmatrix}, \qquad \mathbf{E} = \begin{pmatrix} 0 & 0 & 1 \\ 0 & 0 & 0 \\ 0 & 0 & 0 \end{pmatrix},$$

form a basis of $\mathfrak{n}$ and satisfy the canonical commutation relations of Heisenberg,

$$[\mathbf{X}, \mathbf{Y}] = \mathbf{XY} - \mathbf{YX} = \mathbf{E}, \qquad [\mathbf{X}, \mathbf{E}] = 0, \qquad [\mathbf{Y}, \mathbf{E}] = 0,$$

whence the name of $\mathfrak{n}$, $\tilde{A}(\mathfrak{R})$, and $A(\mathfrak{R})$. The center $\mathfrak{c}$ of $\mathfrak{n}$ is given by $\mathfrak{c} = \mathfrak{R} \cdot \mathbf{E}$. The bracket operation $[\cdot, \cdot]$ induces on every cross-section W to $\mathfrak{c}$ in $\mathfrak{n}$ a natural symplectic form, so that the time-frequency plane of information theory $(\mathfrak{R} \oplus \mathfrak{R}; B)$ and the range-velocity plane of analog radar signal design can be identified with the two-dimensional real symplectic vector space W. The image of W under the exponential map

$$\exp_{\tilde{A}(\mathfrak{R})} : \mathfrak{n} \mapsto \tilde{A}(\mathfrak{R}),$$

gives rise to the two-dimensional isotropic cross-section $\exp_{\tilde{A}(\mathfrak{R})} W = \{(x, y, \frac{1}{2}xy) \mid x, y \in \mathfrak{R}\}$ to $\tilde{C}$ in $\tilde{A}(\mathfrak{R})$ and, therefore, to a differential principal fiber bundle over $\exp_{\tilde{A}(\mathfrak{R})} W$ with structure group isomorphic to $\tilde{C}$. See Figure 4, next page.

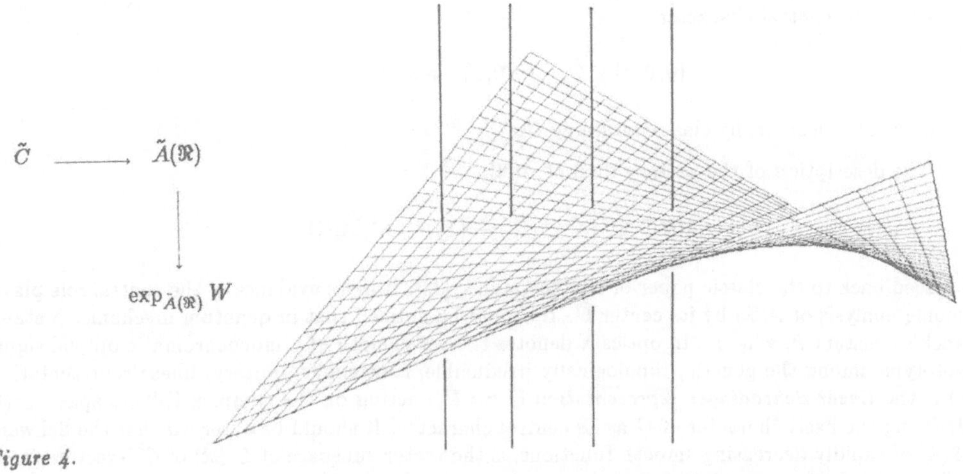

Figure 4.

The Mackey machinery, which is based on a decomposition of $\tilde{A}(\mathfrak{R})$ in a semi-direct product of a properly embedded closed normal abelian subgroup and a closed subgroup, allows us to compute explicitly the unitary dual of $\tilde{A}(\mathfrak{R})$. Adopting a more geometric point of view, the Kirillov coadjoint orbit picture in the dual vector space $\mathfrak{n}^*$ of $\mathfrak{n}$ furnishes a complete classification of the unitary dual $\tilde{A}(\mathfrak{R})^\wedge$ in terms of the coadjoint orbit manifold $\mathfrak{n}^*/\tilde{A}(\mathfrak{R})$. The adjoint action of $\tilde{A}(\mathfrak{R})$ on $\mathfrak{n}$ reads follows:

$$\begin{pmatrix}1 & x & z\\ 0 & 1 & y\\ 0 & 0 & 1\end{pmatrix}\begin{pmatrix}0 & a & c\\ 0 & 0 & b\\ 0 & 0 & 0\end{pmatrix}\begin{pmatrix}1 & x & z\\ 0 & 1 & y\\ 0 & 0 & 1\end{pmatrix}^{-1} = \begin{pmatrix}0 & a & c+bx-ay\\ 0 & 0 & 0\\ 0 & 0 & 0\end{pmatrix}.$$

If $\{\mathbf{X}^*, \mathbf{Y}^*, \mathbf{E}^*\}$ denotes the dual basis of $\{\mathbf{X}, \mathbf{Y}, \mathbf{E}\}$ in $\mathfrak{n}^*$, the coadjoint action of $\tilde{A}(\mathfrak{R})$ in $\mathfrak{n}^*$ is:

$$\begin{pmatrix}1 & x & z\\ 0 & 1 & y\\ 0 & 0 & 1\end{pmatrix}^{-1} \cdot (\alpha\mathbf{X}^* + \beta\mathbf{Y}^* + \lambda\mathbf{E}^*) = (\alpha - \lambda y)\mathbf{X}^* + (\beta + \lambda x)\mathbf{Y}^* + \lambda\mathbf{E}^*.$$

Thus the coadjoint orbit manifold $\mathfrak{n}^*/\tilde{A}(\mathfrak{R})$ falls into two types of orbits (*cf.* Howe [26,27], and [40]; see Figures 5, right):

i. the zero-dimensional (or ***degenerate***) orbits, namely the single points $\{\alpha\mathbf{X}^* + \beta\mathbf{Y}^* \mid \alpha, \beta \in \mathfrak{R}\}$, and

ii. the affine planes (***flat generic orbits***) in $\mathfrak{n}^*$ defined by $\langle \mathbf{E}, \ell \rangle = \lambda \neq 0$.

The Kirillov correspondence $\tilde{A}(\mathfrak{R})^\wedge \to \mathfrak{n}^*/\tilde{A}(\mathfrak{R})$ gives evidence that $\tilde{A}(\mathfrak{R})^\wedge$ consists of two types of isomorphy classes, namely of:

i. continuous unitary characters, *i.e.*, one-dimensional, continuous, unitary, linear representations of $\tilde{A}(\mathfrak{R})$ (***degenerate representations***) parametrized by the set of pairs $(\alpha, \beta) \in \mathfrak{R} \oplus \mathfrak{R}$, and

ii. isomorphy classes of a family (U_λ) of infinite dimensional, topologically irreducible, continuous, unitary, linear representations of $\tilde{A}(\mathfrak{R})$ (***generic representations***), where the parameter λ runs over the set $\mathfrak{R}^\times = \mathfrak{R} - \{0\}$ of non-zero real numbers, and is

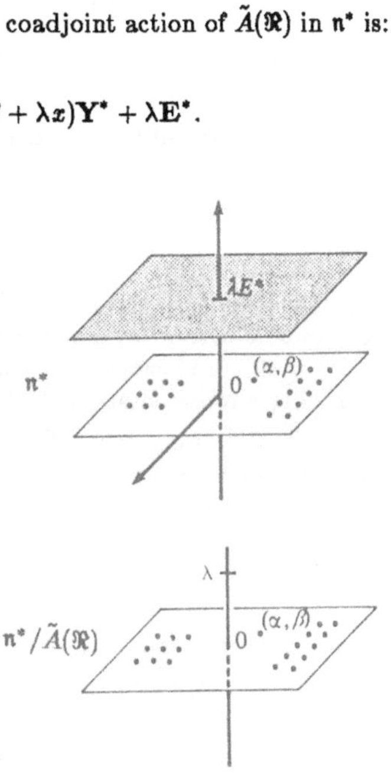

Figures 5.

determined by the central character

$$(0,0,z) \in \tilde{C}, \quad (0,0,z) \rightsquigarrow e^{2\pi i\lambda z} \in \mathfrak{T}$$

associated with the isomorphy class containing U_λ.

The description of the unitary dual of $\tilde{A}(\mathfrak{R})$,

$$\tilde{A}(\mathfrak{R})^\wedge \cong P^\wedge \cup (\tilde{C}^\wedge - \{1\}) \cong \mathfrak{n}^*/\tilde{A}(\mathfrak{R}),$$

can be traced back to the classic paper of von Neumann, [33]. It gives evidence of the *central* role played in harmonic analysis of $\tilde{A}(\mathfrak{R})$ by its center $\tilde{C}$. It should be noticed that in quantum mechanics λ stands for Planck's constant h, whereas in optics λ denotes the wavelength of a monochromatic optical signal. The prototype among the generic, topologically irreducible, continuous, unitary, linear representations of $\tilde{A}(\mathfrak{R})$ is the ***linear Schrödinger representation*** $U := U_1$, acting on the complex Hilbert space $\mathcal{L}^2(\mathfrak{R})$ and admitting the basis character of C as its central character. It should be observed that the Schwartz space $\mathcal{S}(\mathfrak{R})$ of rapidly decreasing smooth functions, is the vector subspace of $\mathcal{L}^2(\mathfrak{R})$ of C^∞-vectors for U, on which it acts via the prescription

$$U(v,z)f(t) = \exp[2\pi i(z + \tfrac{1}{2}xy + yt)]f(t+x), \qquad t \in \mathfrak{R}.$$

Thus the real Heisenberg nilpotent Lie group $\tilde{A}(\mathfrak{R})$ has the key variables, namely the time delay and the Doppler frequency shift built into its structure. Therefore nilpotent harmonic analysis forms the natural foundation of the theory of analog radar signal design. Let the symbol $\otimes$ denote the dyadic tensor product with respect to the prehilbert space structure of $\mathcal{S}(\mathfrak{R})$, and tr_U the trace functional on $\mathcal{S}(\mathfrak{R})$ composed with U. It follows that the analog radar cross-ambiguity function $H(f,g;\cdot,\cdot)$ is equal to the restriction of the ***coefficient function***

$$c_{U,f,g} = \mathrm{tr}_U(f \otimes g)$$

of U onto the two-dimensional polarized cross-section P to $\tilde{C}$ in $\tilde{A}(\mathfrak{R})$. Thus we have the identity

$$H(f,g;x,y) = c_{U,f,g}(x,y,0),$$

for all functions $f,g \in \mathcal{S}(\mathfrak{R})$ and all pairs $(x,y) \in \mathfrak{R} \oplus \mathfrak{R}$. From this basic identity comes evidence of the importance for the theory of analog radar signal design, of harmonic analysis on the differential principal fiber bundle over P with structure group $\tilde{C}$. In view of the flatness of the generic coadjoint orbit $\{\ell \mid \langle \mathbf{E}, \ell\rangle = 1\}$ in $\mathfrak{n}^*$ associated with U under the Kirillov correspondence $\tilde{A}(\mathfrak{R})^\wedge \to \mathfrak{n}^*/\tilde{A}(\mathfrak{R})$, the linear Schrödinger representation U of $\tilde{A}(\mathfrak{R})$ is square integrable modulo $\tilde{C}$ (*cf.* Moore–Wolf [32]). Therefore, the coefficient $c_{U,f,g}$ is a square integrable function on $\tilde{A}(\mathfrak{R})$ modulo $\tilde{C}$ for all elements $f,g \in \mathcal{L}^2(\mathfrak{R})$. The square integrability of U modulo $\tilde{C}$ implies the orthogonality conditions of the coefficient functions $c_{U,f,g}$, which in turn imply the Moyal identities

$$\langle H(f,g;\cdot,\cdot) \mid H(f',g';\cdot,\cdot)\rangle = \langle f \mid f'\rangle\langle g' \mid g\rangle$$

for signal envelopes $f,f',g,g' \in \mathcal{L}^2(\mathfrak{R})$ and, in particular, ***the radar uncertainty principle***

$$\iint_{\mathfrak{R}\oplus\mathfrak{R}} dx\,dy\,|H(f;x,y)|^2 = \|f\|^4$$

in terms of the analog radar auto-ambiguity function. As a consequence we observe:

Theorem 1. *For any given function $F \in S(\mathfrak{R} \oplus \mathfrak{R})$ there exist signal envelopes $f \in S(\mathfrak{R})$ and $g \in S(\mathfrak{R})$ such that the identity*

$$F = H(f, g; \cdot, \cdot)$$

holds.

Thus any smooth function F on the time-frequency plane $\mathfrak{R} \oplus \mathfrak{R}$ can be realized as the analog radar cross-ambiguity function $H(f, g; \cdot, \cdot)$ associated with certain smooth waveforms f and g on $\mathfrak{R}$. More precisely, the sesquilinear mapping

$$(f, g) \in S(\mathfrak{R}) \times S(\mathfrak{R}), \quad (f, g) \rightsquigarrow H(f, g; \cdot, \cdot), \quad H(f, g; \cdot, \cdot) \in S(\mathfrak{R} \oplus \mathfrak{R}),$$

is *surjective*. In the case of the analog radar auto-ambiguity function however [*i.e.*, by restricting the mapping $(f, g) \rightsquigarrow H(f, g; \cdot, \cdot)$ to the diagonal of $S(\mathfrak{R}) \times S(\mathfrak{R})$], the situation becomes quite different. Indeed, a suitable adaptation of the Gelfand–Naĭmark–Segal (GNS) reconstruction to $\tilde{A}(\mathfrak{R})$ and an application of Schur's lemma yields, by fibering over the two-dimensional isotropic cross-section $\exp_{\tilde{A}(\mathfrak{R})} W$ to $\tilde{C}$ in $\tilde{A}(\mathfrak{R})$, the following solution of the radar synthesis problem (*cf.* [42,43]):

Theorem 2. *Let the function $F \in S(\mathfrak{R} \oplus \mathfrak{R})$ be given. There exists a signal envelope $f \in S(\mathfrak{R})$ such that*

$$F = H(f; \cdot, \cdot)$$

holds if and only if F is of pure positive type on the symplectic time-frequency plane $(\mathfrak{R} \oplus \mathfrak{R}; B)$. In this case, the waveforms f which can be synthesized from F are determined uniquely up to a multiplicative complex constant of unit modulus.

The link between the analog radar ambiguity functions and nilpotent harmonic analysis described above in terms of a cross-section to the center $\tilde{C}$ of $\tilde{A}(\mathfrak{R})$, is of central importance for our approach to the mathematical theory of analog radar signal design. Harmonic analysis on the differential principal fiber bundle over the two-dimensional polarized (resp. isotropic) cross-section, with structure group isomorphic to the one-dimensional center $\tilde{C}$ of the simply connected real Heisenberg nilpotent Lie group $\tilde{A}(\mathfrak{R})$, formalizes the analogy between non-relativistic quantum physics and signal theory mentioned in Section 2 above and, in particular, supplies an expression of the *inherent* ambiguity in a simultaneous determination of both the range d and radial velocity v of a moving distant target.

1.4 The radar invariant problem

Now we know (since Felix Klein's Erlangen program) that beneath... lurks one central branch of Mathematics —the linear algebra of modern mathematicians.
J. DIEUDONNÉ (1969)

In his pioneering work on the metaplectic group, Weil [55] reformulated large parts of the theory of quadratic forms using the language of unitary linear group representations. We will follow this use by letting the symplectic group $Sp(\mathfrak{R} \oplus \mathfrak{R}; B) = Sp(1, \mathfrak{R}) = SL(2, \mathfrak{R})$ act in the natural way on the real Heisenberg nilpotent Lie group $\tilde{A}(\mathfrak{R})$, leaving its center $\tilde{C}$ pointwise fixed. Moreover, we let $\tilde{\sigma} \rightsquigarrow T_{\tilde{\sigma}}$ denote the linear oscillator representation of the metaplectic group $Mp(1, \mathfrak{R}) = Sp(1, \mathfrak{R}) \times \mathfrak{Z}/2\mathfrak{Z}$ which doubly covers $Sp(1, \mathfrak{R})$ by means of the covering epimorphism $\tilde{\sigma} \rightsquigarrow \sigma$ (*isomorphism up to a sign*). An application of Segal's metaplectic formula, which on its part is based on the Stone-von Neumann uniqueness theorem [33], then establishes the following solution of the invariant problem for radar ambiguity surfaces over the symplectic time-frequency plane:

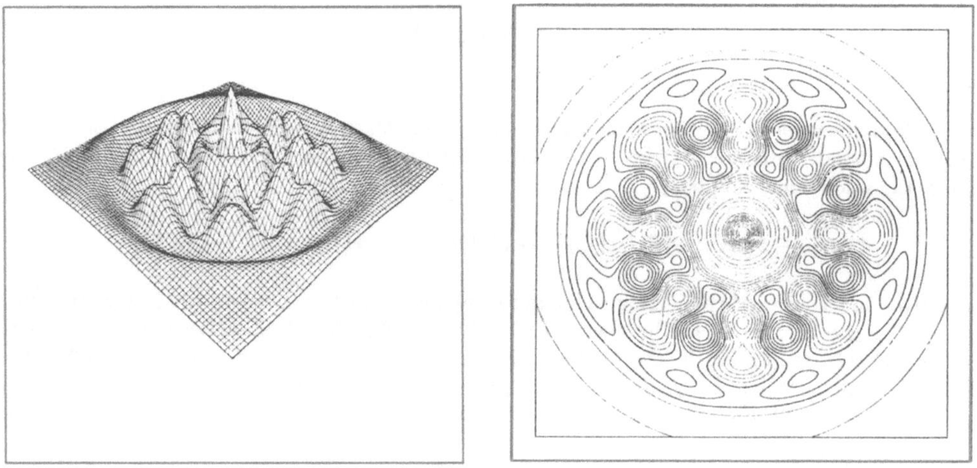

Figures 6.

Theorem 3. *Let the unitary operator*

$$S : \mathcal{L}^2(\mathfrak{R}) \mapsto \mathcal{L}^2(\mathfrak{R})$$

be a linear automorphism of the radar ambiguity surface $\mathcal{F}$ over the symplectic time-frequency plane $(\mathfrak{R} \oplus \mathfrak{R}; B)$. Then, there exists a unique transformation $\sigma \in Sp(1, \mathfrak{R})$ and a complex number $\varsigma_{\tilde{\sigma}}$ of modulus $|\varsigma_{\tilde{\sigma}}| = 1$ such that

$$S = \varsigma_{\tilde{\sigma}} T_{\tilde{\sigma}}$$

holds.

The intertwining operators $T_{\tilde{\sigma}}$ are Fourier transforms with respect to suitable coordinates of the symplectic time-frequency plane $(\mathfrak{R} \oplus \mathfrak{R}; B)$. The symmetries of the radar ambiguity surface displayed in Figures 6 above, from different viewpoints, are computed by the preceding theorem.

It is well known that the matrices

$$\begin{pmatrix} a & 0 \\ 0 & a^{-1} \end{pmatrix}, \quad \begin{pmatrix} 1 & 0 \\ u & 1 \end{pmatrix},$$

where $a \neq 0$ and $u \neq 0$ denote real numbers, together with the Weyl element

$$\mathbf{J} = \begin{pmatrix} 0 & 1 \\ -1 & 0 \end{pmatrix}$$

of $Sp(1, \mathfrak{R})$, give rise to the *Bruhat* decomposition of $Sp(1, \mathfrak{R})$. Based on a suitable polarization (*cf.* [46]) of $\mathfrak{n}$, an explicit computation of the operators $T_{\tilde{\sigma}}$ for the matrices $\sigma \in Sp(1, \mathfrak{R})$ of the Bruhat decomposition, yields the following different solution of the invariant problem for radar ambiguity surfaces over the symplectic time-frequency plane:

Corollary 1. *The energy-preserving linear automorphisms S of the radar ambiguity surface $\mathcal{F} = H(f;\mathfrak{R},\mathfrak{R})$ may be realized by finite sequences of time scalings, pointwise multiplications, and convolutions on the time axis $\mathfrak{R}$ of the signal waveform $f \in S(\mathfrak{R})$ with the chirp signals*

$$t \rightsquigarrow C_u e^{-\pi i u t^2}, \qquad C_u \in \mathfrak{C}, \quad |C_u| = 1,$$

with real parameter $u \neq 0$.

In particular, the preceding result explains the shearing of analog radar ambiguity functions by introducing a quadratic phase in the time and the frequency domain, respectively. For details, see the monograph by Rihaczek [37] and a forthcoming paper devoted to signal geometry. An explicit formula for the phase factor C_u of the chirp signals occurring in Corollary 1 will be given in Section 8 below.

Let $(W_m)_{m\geq 0}$ be the sequence of standardized Hermite functions (harmonic oscillator wave functions) and $(L_n^{(\alpha)})_{n\geq 0}$ the sequence of Laguerre functions of degree n and order $\alpha > -1$. Introducing a complex structure on the two-dimensional cross-section W to $\mathfrak{c}$ in $\mathfrak{n}$ we obtain, by the Bargmann–Fock–Segal model (or complex wave model, *cf.* Ogden–Vági [34]) of the linear Schrödinger representation U of $\tilde{A}(\mathfrak{R})$, the following result:

Corollary 2. *Let $f \in S(\mathfrak{R})$ have energy-norm $\|f\| = 1$. The radar ambiguity surface $\mathcal{F} = H(f;\mathfrak{R},\mathfrak{R})$ over the time-frequency plane $\mathfrak{R} \oplus \mathfrak{R}$ is $SO(2,\mathfrak{R})$-invariant if and only if $f = \varsigma W_m$, for a certain integer $m \geq 0$ and complex number ς of modulus $|\varsigma| = 1$. In this case the analog radar cross-ambiguity functions take the form*

$$H(W_m, W_n; x, y) = \sqrt{\frac{n!}{m!}}\left(\sqrt{\pi}(x+iy)\right)^{m-n} L_n^{(m-n)}\left(\pi(x^2+y^2)\right), \qquad m \geq n \geq 0$$

for all pairs $(x,y) \in \mathfrak{R} \oplus \mathfrak{R}$.

Figures 7,8, and 9 displayed in the following pages, show the graphs of the non-normalized harmonic oscillator wave functions $(W_m)_{0\leq m\leq 5}$ and the associated radially symmetric radar ambiguity surfaces over the time-frequency plane $\mathfrak{R} \oplus \mathfrak{R}$. Ambiguity surfaces of this kind play an important role in the field of loudspeaker design (*cf.* Gerzon [20]).

1.5 The compact Heisenberg nilmanifold

The primary source (Urquell) of all of mathematics are the integers.

H. MINKOWSKI (1864-1909)

A periodic function should always be expanded in a Fourier series.

E. HECKE (1887-1947)

In $\tilde{A}(\mathfrak{R})$ we consider the arithmetic subgroup $A(\mathfrak{Z})$ defined by $x, y, z \in \mathfrak{Z}$. The quotient $A(\mathfrak{Z})\backslash\tilde{A}(\mathfrak{R})$ of right cosets modulo $A(\mathfrak{Z})$ forms a compact two-step nilmanifold, the ***Heisenberg nilmanifold***. It forms a differential principal circle bundle over the two-dimensional compact torus group $\mathfrak{T}^2$. (See Figure 10 on page 19.)

Extend the central character $(0,0,z) \rightsquigarrow e^{2\pi i z}$ from $\tilde{C}$ to the closed normal subgroup $\mathfrak{Z}\times\mathfrak{Z}\times\tilde{C}$ of $\tilde{A}(\mathfrak{R})$ in the trivial way and induce unitarily the associated one-dimensional linear representation from $\mathfrak{Z}\times\mathfrak{Z}\times\tilde{C}$ to $\tilde{A}(\mathfrak{R})$. The monomial representation V of $\tilde{A}(\mathfrak{R})$ obtained in this way is called the linear ***lattice representation*** of $\tilde{A}(\mathfrak{R})$. Construct the ***Weil–Brezin isomorphism*** w, *i.e.*, the periodization

$$f \in S(\mathfrak{R}), \quad w : f \rightsquigarrow \left((x,y,z) \rightsquigarrow e^{2\pi i z}\textstyle\sum_{n\in\mathfrak{Z}} f(n+x)e^{2\pi i n y}\right),$$

with a fundamental domain of $A(\mathfrak{Z})\backslash\tilde{A}(\mathfrak{R})$ (*cf.* Auslander [1,2]). This forms an intertwining operator between the linear Schrödinger representation U and the linear lattice representation V of the real Heisenberg nilpotent Lie group $\tilde{A}(\mathfrak{R})$. Combined with the Plancherel theorem of abelian harmonic analysis, it yields the following identities for the analog radar ambiguity functions (*cf.* [41,44]) which are the basis of the identities for theta constants pointed out below in the Corollary.

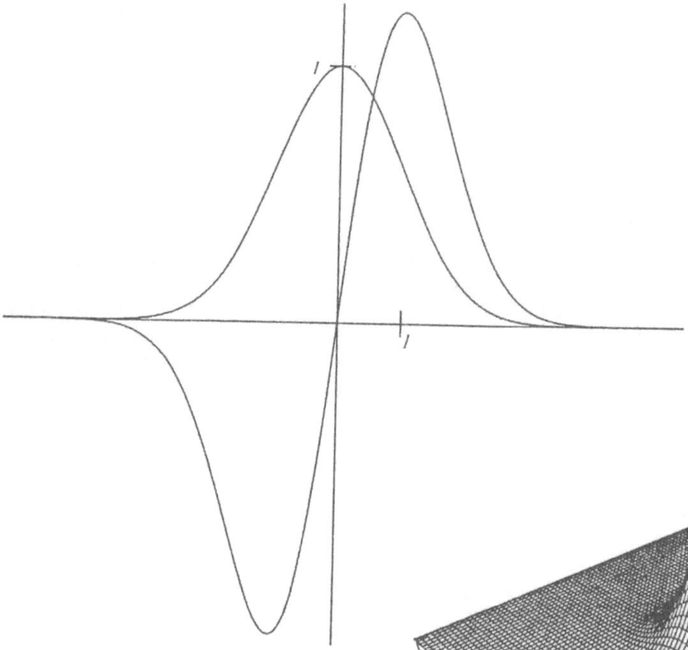

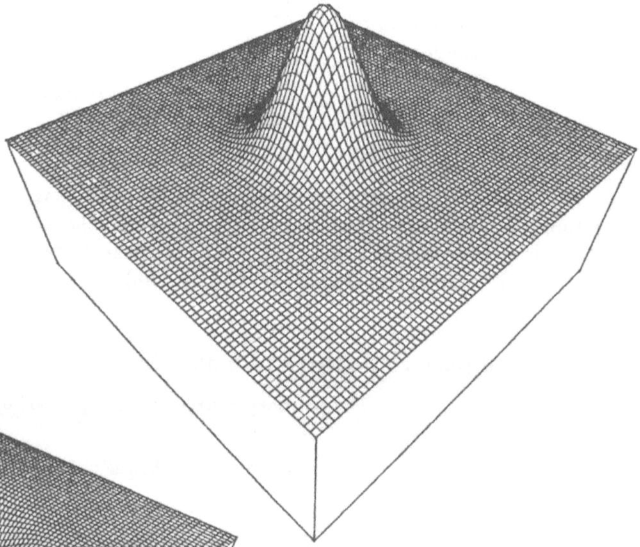

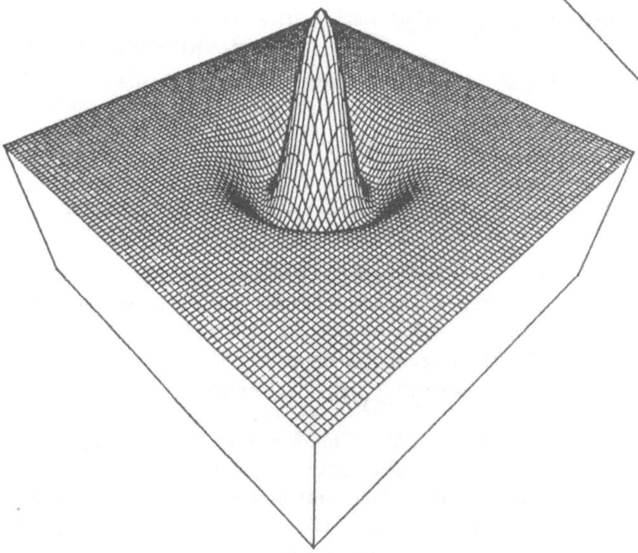

Figures 7. Top: harmonic oscillator wavefunctions for $m = 0, 1$; middle and bottom: $m = 0$ and $m = 1$ ambiguity surfaces.

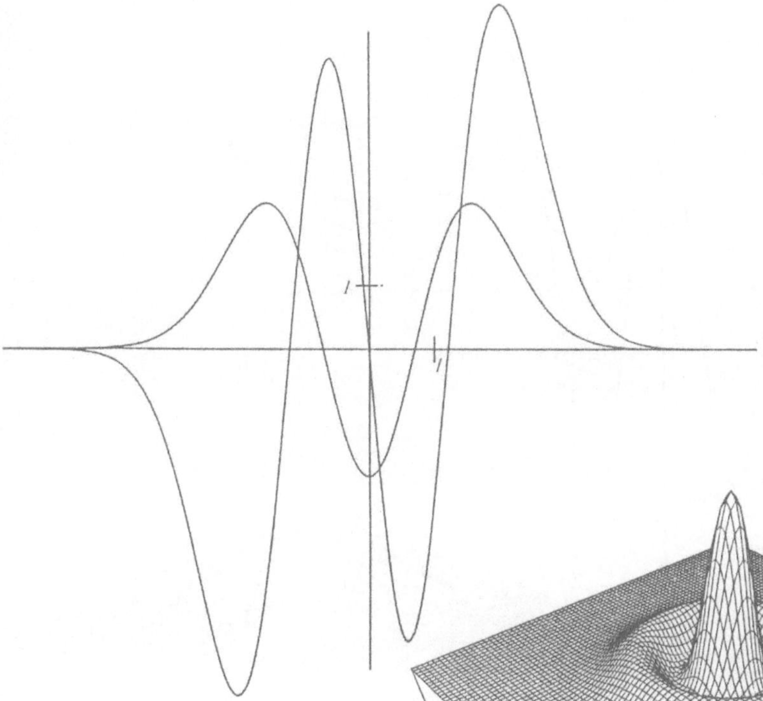

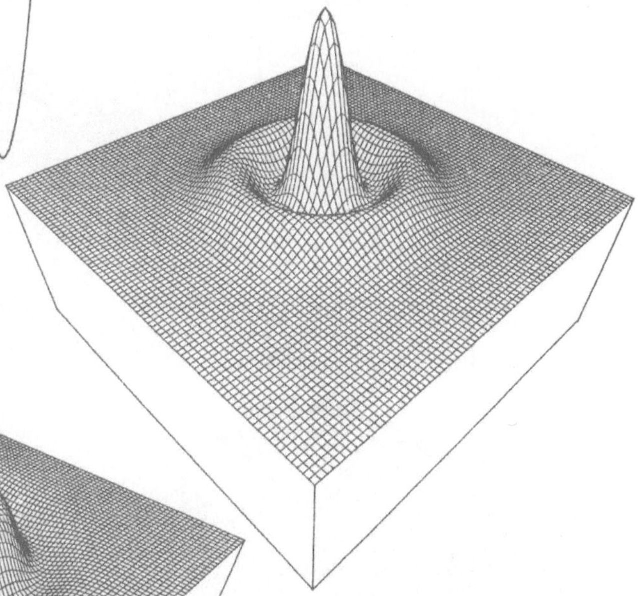

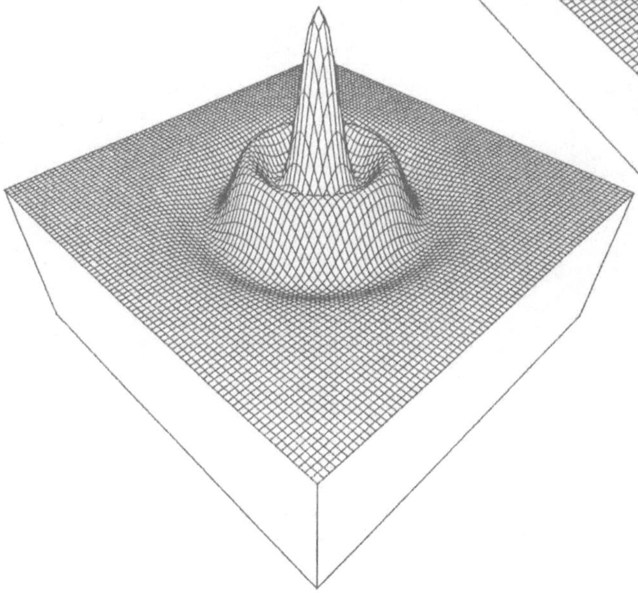

Figures 8. Top: harmonic oscillator wavefunctions for $m = 2, 3$; middle and bottom: $m = 2$ and $m = 3$ ambiguity surfaces.

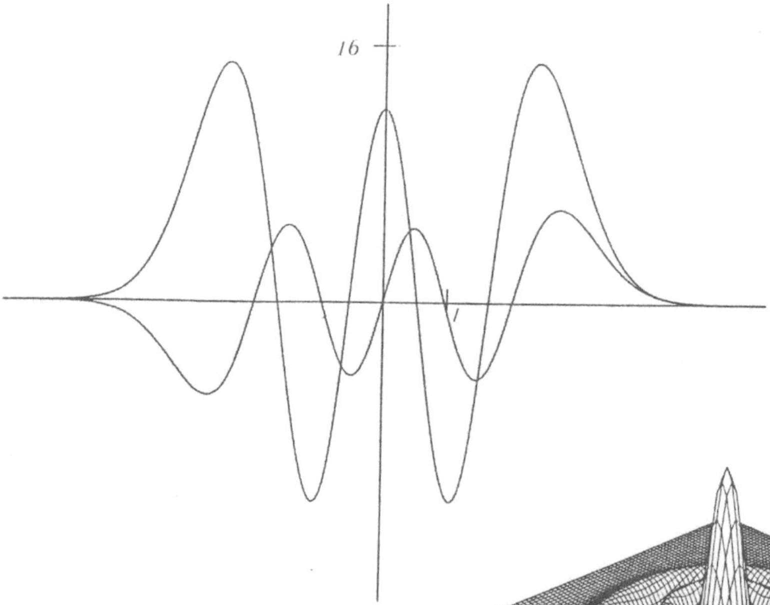

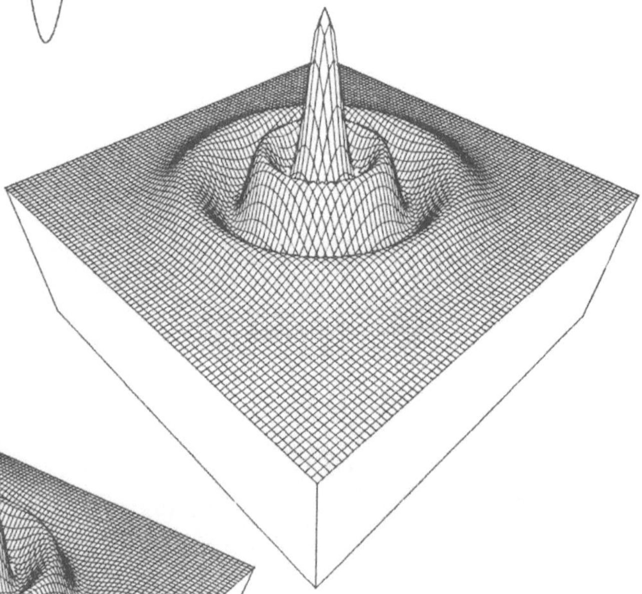

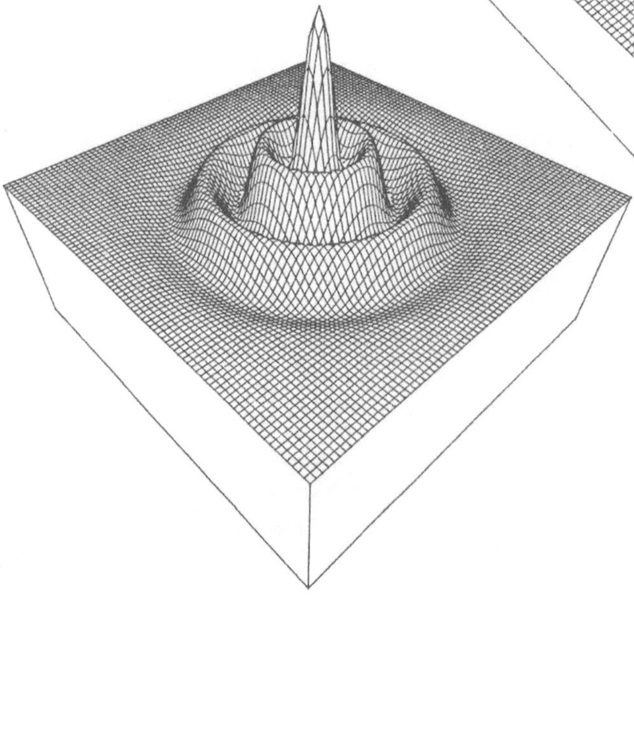

Figures 9. Top: harmonic oscillator wavefunctions for $m = 4, 5$; middle and bottom: $m = 4$ and $m = 5$ ambiguity surfaces.

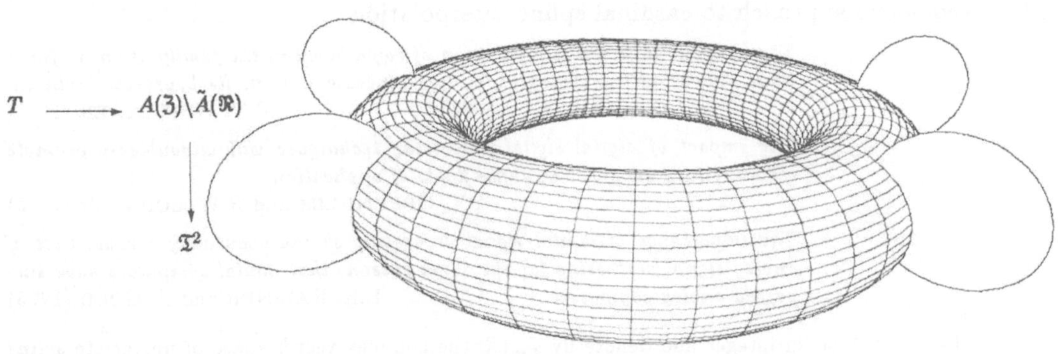

Figure 10.

Theorem 4. *Let m and n denote natural numbers. Then*

$$\sum_{\mu,\nu\in\mathfrak{Z}} H(W_m;\mu,\nu)\,H(W_n;\mu,\nu) = \sum_{\mu,\nu\in\mathfrak{Z}} |H(W_m,W_n;\mu,\nu)|^2$$

holds.

Corollary 3. *For all integers $m \geq n \geq 0$, the following identity for Laguerre functioons of different orders holds:*

$$\sum_{\mu,\nu\in\mathfrak{Z}} L_m^{(0)}(\pi(\mu^2+\nu^2))\,L_n^{(0)}(\pi(\mu^2+\nu^2)) = \frac{n!}{m!}\pi^{m-n} \sum_{(\mu,\nu)\in\mathfrak{Z}\times\mathfrak{Z}} (\mu^2+\nu^2)^{m-n}\Big(L_n^{(m-n)}(\pi(\mu^2+\nu^2))\Big)^2.$$

In the cases $m = n+1 = 1,2,3,4$ we obtain the identities

$$\begin{aligned}
\pi\sum_{\mu\in\mathfrak{Z}}\mu^2 e^{-\pi\mu^2} &= \frac{1}{4}\sum_{\mu\in\mathfrak{Z}} e^{-\pi\mu^2},\\
\pi^3\sum_{\mu\in\mathfrak{Z}}\mu^6 e^{-\pi\mu^2} &= \frac{15}{32}\sum_{\mu\in\mathfrak{Z}}(8\pi^2\mu^4-1)e^{-\pi\mu^2},\\
\pi^5\sum_{\mu\in\mathfrak{Z}}\mu^{10} e^{-\pi\mu^2} &= \frac{45}{64}\sum_{\mu\in\mathfrak{Z}}(16\pi^4\mu^8-140\pi^2\mu^4+21)e^{-\pi\mu^2},\\
\pi^7\sum_{\mu\in\mathfrak{Z}}\mu^{14} e^{-\pi\mu^2} &= \frac{91}{1024}\sum_{\mu\in\mathfrak{Z}}(256\pi^6\mu^{12}-15840\pi^4\mu^8+166320\pi^2\mu^4-25245)e^{-\pi\mu^2}.
\end{aligned}$$

The author is grateful to his student, Martin Schmidt, for help in these calculations and for numerical checking (see Schmidt [47]). A direct proof of the preceding identities for theta constants follows via the integral

$$\int_{\Re} dx\, e^{-x^2}\cos ax = \sqrt{\pi}\, e^{-a^2/4},$$

and the application of the Poisson summation formula (see Grosjean [21]). The first theta identity above is also a consequence of the heat equation for the classical first order Jacobi theta functions (*cf.* Bellman [9]). For this simplest identity see also Borwein [10].

1.6 A geometric approach to cardinal spline interpolation

The cardinal function is a function of royal blood in the family of entire functions, whose distinguished properties seperate it from its bourgeois brethren.
E.T. WHITTAKER (1915)

The impact of digital signal processing techniques will undoubtedly promote revolutionary advances in some fields of application.
A.V. OPPENHEIM and R.W. SCHAFER (1975)

The importance of digital signal processing should eventually surpass that of analog signal processing for the same reasons that digital computers have surpassed analog computers.
L.R. RABINER and B. GOLD (1975)

Let $m \geq 1$ be an integer and denote by $\Phi_m(\mathfrak{K})$ the complex vector space of univariate ***spline functions*** of degree $m-1$ with knot set $\mathfrak{K}$. Thus $S \in \Phi_m(\mathfrak{K})$ if and only if S is a $(m-2)$-times continuously differentiable complex-valued function on $\mathfrak{R}$, and the restrictions of S to the subsequent intervals with end points in $\mathfrak{K}$ are polynomials of degree less or equal to $m-1$, with complex coefficients. In the case of spline knots set $\mathfrak{K} = \mathfrak{Z}$, the ***cardinal spline interpolation problem*** for a given bi-infinite sequence $(y_n)_{n\in\mathfrak{Z}} \in \mathcal{L}^2(\mathfrak{Z})$ reads as follows: Does there exist a cardinal spline function $S \in \Phi_m(\mathfrak{Z})$ such that

$$S(n) = y_n, \quad n \in \mathfrak{Z}$$

holds?

Let $\mathbf{J}$ denote the Weyl element of $Sp(1, \mathfrak{R})$ as in Section 4 and $\overline{\mathcal{F}}_{\mathfrak{R}}$ the Fourier cotransform acting on $\mathcal{L}^2(\mathfrak{R})$ as an automorphism. An application of the Poisson-Weil factorization

$$\overline{\mathcal{F}}_{\mathfrak{R}} = w^{-1} \cdot \mathbf{J} \cdot w$$

of the Fourier cotransform $\overline{\mathcal{F}}_{\mathfrak{R}}$ (*cf.* [39] and Auslander [2]) combined with an argument concerning the inversion of Toeplitz matrices shows that, when m is even, the cardinal spline interpolation problem admits a unique solution. However, when m is odd the knots of the splines must be displaced by $\frac{1}{2}$ to ensure the existence of a unique solution of the cardinal spline interpolation problem (this is the Subbotin–Schoenberg theorem; see Schoenberg [48]). For applications of cardinal spline interpolation to digital signal processing, see Schüßler [50].

Let $PW(\mathfrak{C})$ denote the Paley-Wiener space of all entire functions of exponential type at most π that are square integrable on the real line $\mathfrak{R}$. In view of the Paley-Wiener theorem, the Fourier transform is an isometric isomorphism of the separable complex Hilbert space $PW(\mathfrak{C})$ onto $\mathcal{L}^2(\mathfrak{T})$. The elements $f \in PW(\mathfrak{C})$ that are real valued on the real line $\mathfrak{R}$ are called ***band-limited signals*** or signals with a ***finite spectrum***. Sound signals, radio signals, areal photographs, and holograms are examples of signals of this kind (*cf.* Soroko [52]). Define the function ***sinus cardinalis*** by the usual prescription

$$\operatorname{sinc} z := \begin{cases} \dfrac{\sin \pi z}{\pi z}, & \text{for } z \neq 0, \\ 1, & \text{for } z = 0. \end{cases}$$

In the simplest possible case $m = 1$ the Whittaker-Shannon-Kotel'nikov sampling theorem obtains. For surveys see Jerri [29], Butzer [13], and the recent expository paper by Higgins [24]. Also see the papers [39,45], which emphasize the group theoretical point of view.

Theorem 5. *Each function $f \in PW(\mathfrak{C})$ admits the cardinal series expansion*

$$f(z) = \sum_{\mu\in\mathfrak{Z}} f(\mu)\, \operatorname{sinc}(z-\mu)$$

for all $z \in \mathfrak{C}$. The convergence of the cardinal interpolation series is uniform on the compact subsets of $\mathfrak{C}$.

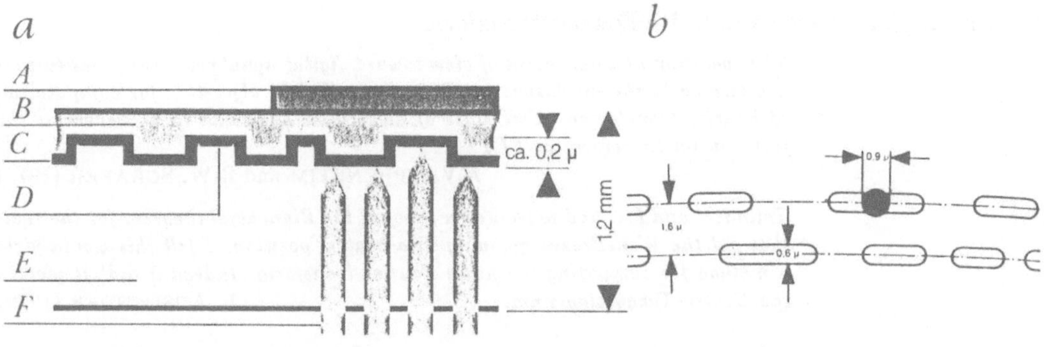

Figures 11. *(a)* Cross section of a CD record: *A* label, *B* protection layer, *C* reflective layer, *D* grove, *E* transparent material, *F* laser beam. *(b)* Information bits along CD record groves; the dark circle represents the ablation spot of a laser beam.

This result may be interpreted in two ways, each of which has found important applications in signal theory.

i. Every signal of finite energy and bandwidth $W = \frac{1}{2}Hz$ may be completely recaptured, in a simple way, from a knowledge of its samples taken at the rate of $2W = 1$ per second (*Nyquist rate*). Moreover —indispensable for any implementation in practice— the recovery is *stable*, in the sense that a small error in reading the sample values produces only a correspondingly small error in the recaptured signal.

ii. Every square-summable sequence of complex numbers may be transmitted at the rate of $2W = 1$ per second over an ideal channel of bandwidth $W = \frac{1}{2}Hz$, by being represented as the samples at the integer points $\mu \in \mathfrak{Z}$ of an easily constructed bandlimited signal of finite energy.

Thus the Whittaker-Shannon-Kotel'nikov sampling theorem as stated above serves as a basis for the interchangeability of analog representations of signals and their representations in digital sequences (*cf.* [45]). Coding of a signal usually consists in its representation in a digital sequence. The digital form of representation rather than the analog representation of signals provides considerably more ways of protecting the signal against various kinds of possible distortion in its storage and transmission. This explains why the digital signal processing, and therefore the Whittaker-Shannon-Kotel'nikov sampling theorem, are so extremely valuable for modern communication systems. For instance, the recently developed CD (= **C**ompact **D**isc) technology forms a very efficient practical application of the digital signal representation. Figures 11 next page, show the structure of the cross-section and the surface of a CD.

The sequence of digital signals located on the surface of the CD are transformed by means of a laser beam into analog electrical signals. Finally, it is the task of the loudspeakers to convert, as electro-acoustical transducers, these electrical signals into corresponding sound waves. The analog auto-ambiguity function and Wigner distribution technique allows the introduction of optimization criteria for both a single transducer and a combination of transducers (Janse and Kaiser [28]). Based on these criteria, loundspeaker systems can be designed which are adapted in a satisfactory way to the high fidelity of CD processing.

Other examples for the application of the Whittaker-Shannon-Kotel'nikov sampling theorem are digital typography, medical computerized tomography, digital holography, and seismic exploration.

1.7 Finite phase space and finite Fourier transform

The evolution of a new point of view toward digital signal processing was further accelerated by the disclosure in 1965 of an efficient algorithm for computation of Fourier transforms. This class of algorithms has come to be known as the fast Fourier transform or FFT.

A.V. OPPENHEIM and R.W. SCHAFER (1975)

Tolimieri and I... had presented a proof of the Plancherel theorem for the reals that put the Weil-Brezin mapping in a central position. I felt this would yield a method for computing the finite Fourier transform. Indeed it did! It yielded the Cooley-Tukey algorithm.

L. AUSLANDER (1979)

Physical systems whose underlying configuration space is a finite set have attracted the attention of physicists for a long time, either from the fundamental point of view or with the aim to get rid of the infinities present in conventional field theories. See, for instance, the paper [51] by Julian Schwinger (Nobel award 1965). On the other hand, it has been known since the beginning of this century that the irreducible linear representations of finite nilpotent groups in complex vector spaces are all monomial. This theorem is the predecessor of the celebrated Dixmier-Kirillov theorem stating the analogous result for all simply connected nilpotent Lie groups. Therefore, the physical as well as the mathematical point of view suggest to investigate the finite Heisenberg nilpotent group $A(\mathfrak{Z}/N\mathfrak{Z}) = \{(x, y, z) \mid x, y, z \in \mathfrak{Z}/N\mathfrak{Z}\}$ where $N \geq 1$ denotes an integer.

Fix the Nth root of unity $\varsigma = e^{2\pi i/N}$ and define the diagonal matrix $\mathbf{D}_N$ according to the prescription

$$\mathbf{D}_N = \begin{pmatrix} \varsigma^0 & & 0 \\ & \ddots & \\ 0 & & \varsigma^{N-1} \end{pmatrix}$$

and the permutation matrix

$$\mathbf{P}_N = \begin{pmatrix} 0 & 1 & 0 & \ddots & 0 & 0 \\ 0 & 0 & 1 & \ddots & 0 & 0 \\ 0 & 0 & 0 & \ddots & 0 & 0 \\ \ddots & \ddots & \ddots & \ddots & \ddots & \ddots \\ 0 & 0 & 0 & \ddots & 0 & 1 \\ 1 & 0 & 0 & \ddots & 0 & 0 \end{pmatrix}.$$

Let $M_0 = \{(x, 0, z) \mid x, z \in \mathfrak{Z}/N\mathfrak{Z}\}$ be maximal abelian subgroup of $A(\mathfrak{Z}/N\mathfrak{Z})$. Inducing unitarily the character $(x, 0, z) \leadsto \varsigma^z$ from M_0 to $A(\mathfrak{Z}/N\mathfrak{Z})$ gives an irreducible, unitary, linear representation U_1 of $A(\mathfrak{Z}/N\mathfrak{Z})$ satisfying

$$U_1(x, 0, 0) = (\mathbf{D}_N)^x,$$
$$U_1(0, y, 0) = (\mathbf{P}_N)^y,$$
$$U_1(0, 0, z) = \varsigma^z.$$

Inducing unitarily $(0, y, z) \leadsto \varsigma^z$ from the maximal abelian subgroup $N_0 = \{(0, y, z) \mid y, z \in \mathfrak{Z}/N\mathfrak{Z}\}$ of $A(\mathfrak{Z}/N\mathfrak{Z})$ gives rise to the irreducible, unitary, linear representation

$$U_2(x, 0, 0) = (\mathbf{P}_N)^x,$$
$$U_2(0, y, 0) = (\mathbf{D}_N)^y,$$
$$U_2(0, 0, z) = \varsigma^z.$$

The finite (or discrete) Fourier cotransform $\overline{\mathcal{F}}_N$, admitting the matrix

$$\frac{1}{\sqrt{N}}(\varsigma^{\alpha\beta})_{0\leq\alpha,\beta\leq N-1}$$

with respect to the canonical basis of $\mathbb{C}^N$ is a unitary isomorphism of U_2 onto U_1. Thus the finite Fourier transform (or DFT) gives the transition between the coordinate representation and the momentum representation. In signal theory, the finite Fourier transform describes the relationship between the time domain and the frequency representation of discrete signals, and therefore plays a central role in the implementation of a variety of digital signal processing algorithms. For large blocklengths N, efficient methods for computing the finite Fourier transform are needed. Several techniques are available, among which the technique that turns a one-dimensional Fourier transform via a *finite* analogue of the Weil-Brezin isomorphism (*cf.* Section 5 above) into a multi-dimensional Fourier transform, is the most powerful one. These techniques are known collectively as the ***fast Fourier transform***, FFT. For a finite Heisenberg nilpotent group approach to the Cooley-Tukey FFT algorithm, see the papers by Auslander and Tolimieri [3] and Auslander, Tolimieri and Winograd [4]. See also the monograph by Schroeder [49] for the connections between FFT and elementary number theory.

1.8 The phase discontinuity of Fourier optics

Je reconnois premièrement... la vérité de ce principe, que la nature agit toujours par les voies les plus courtes. P. DE FERMAT (1891)

... We must modify Fresnel's original description: Not only does the phase of light get multiplied by $e^{id/\lambda}$ as we move along a ray. We get an extra factor $-i = e^{-\pi i/4}\cdot e^{-\pi i/4}$ as the light passes through a conjugate point. As strange as this prescription seems, this fact was experimentally observed by Gouy in 1890. V. GUILLEMIN and S. STERNBERG (1984)

The evolution of a light wave along a coaxial optical system gives information about light intensities. It can be described in paraxial approximation with the aid of a quadratic Hamiltonian. The corresponding ray evolution along the optical *time* axis gives rise to a group of operators acting as canonical transformations in the framework of non-relativistic quantum physics in two dimensions (*cf.* Bacry and Cadilhac [5]). We have seen that the phase space of signal theory is the time-frequency plane $\mathfrak{R}\oplus\mathfrak{R}$ whereas the optical phase space derives from the Fermat principle of *least time* (*cf.* [36]).

In Fourier optics, which is the scalar counterpart of linear geometrical optics, an important theorem states that the ambiguity surface is preserved along the geometrical axis of the optical system (*cf.* Papoulis [35]). We refer to this theorem as the ***ambiguity surface conservation principle***. For the sake of simplicity, let us consider the case when the optical system (composed by ideal lenses) is axially symmetric. In this case the phase factor C_u associated with the matrix of the lens operator

$$\begin{pmatrix}1 & 0\\ u & 1\end{pmatrix}, \qquad u\neq 0$$

(*cf.* Corollary 1 of Theorem 3, above) takes the explicit form

$$C_u = \exp(-i\pi\tfrac{1}{4}\,\mathrm{sign}\,u),$$

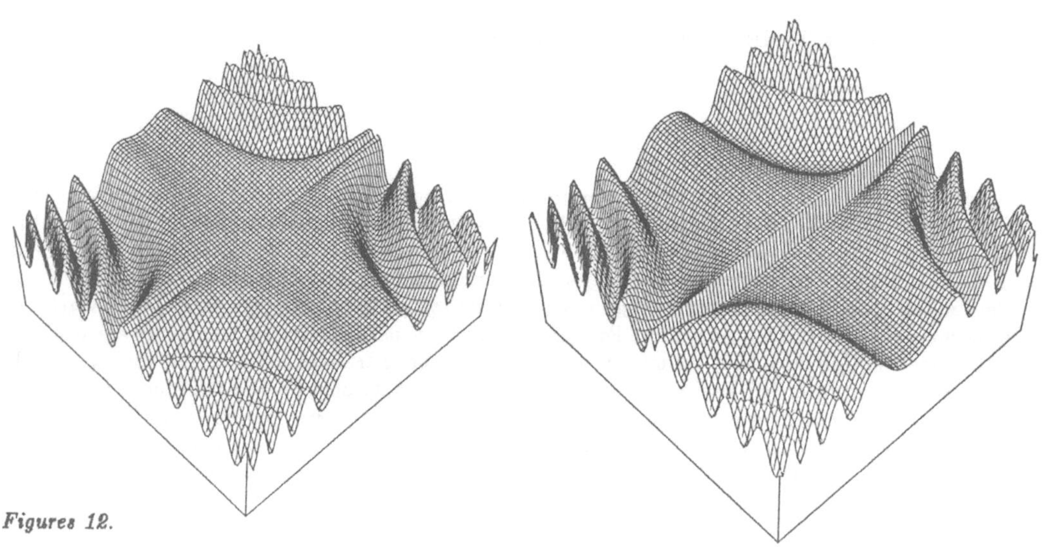

Figures 12.

where $\operatorname{sign} u = u/|u|$ for $u \neq 0$. From this and the ambiguity surface conservation principle mentioned above we may deduce that the light goes through a phase shift of π when passing through a focus. This phenomenon, which is called the ***phase discontinuity*** of Fourier optics, can be checked by elegant experiments in the setting of microwave optics. Figures 12 above (real and imaginary part) may serve as an illustration of phase discontinuity phenomenon by the ambiguity surface conservation principle.

For additional details about the phase discontinuities of the wave field, see the monograph by Guillemin and Sternberg [22] and the paper [36]. In fact, it is the contribution of the group-theoretic Maslov index which properly explains the phase discontinuity of Fourier optics.

1.9 Conclusions

> ***The Heisenberg group is remarkably little known considering its ubiquity.***
> ***The center of the Heisenberg group plays a significant role in what we do.***
> R. HOWE (1980)

Harmonic analysis on the differential principal fiber bundle over the two-dimensional polarized (resp. isotropic) cross-section, with structure group isomorphic to the one-dimensional center of the simply connected real Heisenberg nilpotent Lie group $\tilde{A}(\mathfrak{R})$, plays an important role in non-relativistic quantum physics, Fourier optics, and in analog and digital signal processing. More precisely, the Lie group $\tilde{A}(\mathfrak{R})$ stands at the crossroads of quantum mechanics and analog radar signal design, since the associated phase spaces can be identified with the symplectic tangent space to the non-degenerate coadjoint orbits of $\tilde{A}(\mathfrak{R})$. In particular, within harmonic analysis on the compact Heisenberg nilmanifold a *geometric* proof of an identity for Laguerre functions of different orders pops up and deserves mention also from a purely mathematical point of view because of its deep roots in the theory of the standard theta series, $\theta(z) = \sum_{\mu \in \mathfrak{Z}} e^{\pi i \mu^2 z}$.

Acnowledgements. The author wishes to express his gratitude to Profs. Charles K. Chui and Larry L. Schumaker (College Station, Texas) for inviting him to give a survey lecture on the subject during the Fifth Texas Symposium on Approximation Theory, at the Center of Approximation Theory of Texas A & M University. Moreover, he would like to thank the electrical engineering Prof. Dr.-Ing. Günter Ries and Prof. Dr.-Ing. Rudolf Schwarte (Siegen), and the physicists Prof. Dr. P. Kramer (Tübingen) and Dr. Hans Raszillier (Palaiseau and Bonn) for valuable discussions. A personal note of gratitude goes to Dipl.-Math. Helmut Nienhaus and Dipl.-Math. Karin von Radziewski (Siegen) for their help with the computer plots. Finally, the author's special tbanks are due to Oberbibliotheksrat Ernst Stamm (Siegen) for his expert cooperation over the years and his invaluable assistance with the literature.

References

[1] L. Auslander, *Lecture Notes on Nil-theta Function.* CBMS Regional Conference Series in Mathematics, No. 34. (American Mathematical Society, Providence, R.I., 1977).

[2] L. Auslander, A factorization theorem for the Fourier transform of a separable locally compact abelian group. In: *Special Functions: Group Theoretical Aspects and Applications*, R.A. Askey, T.H. Koornwinder, and W. Schempp, Eds. MIA Series, (Reidel, Dordrecht-Boston-Lancaster, 1984); pp. 261-269.

[3] L. Auslander and R. Tolimieri, Is computing with the finite Fourier transform pure or applied Mathematics? *Bull. Amer. Math. Soc. (New Series)* **1**, 847–897 (1979).

[4] L. Auslander, R. Tolimieri, and S. Winograd, Hecke's theorem in quadratic reciprocity, finite nilpotent groups and the Cooley-Tukey algorithm, *Adv. Math.* **43**, 122–172 (1982).

[5] H. Bacry and M. Cadilhac, The metaplectic group and Fourier optics, *Phys. Rev. A* **23**, 2533–2536 (1981).

[6] N.L. Balazs and B.K. Jennings, Wigner's function and other distribution functions in mock phase space, *Phys. Rep.* **104**, 347–391 (1984).

[7] M.J. Bastiaans, The Wigner distribution function applied to optical signal and systems. *Opt. Comm.* **25**, 26–30 (1978).

[8] M.J. Bastiaans, Wigner distribution functions and its application to first order optics. *J. Opt. Soc. Am.* **69**, 1710–1716 (1979).

[9] R. Bellman, *A Brief Introduction to Theta Functions*, (Holt, Rinehart and Winston, New York, 1961).

[10] J. Borwein, Advanced problem No. 6491, *Amer. Math. Monthly* **92**, 217 (1985).

[11] K. Brenner, *Phasenraumdarstellungen in Optik und Signalverarbeitung.* Dissertation. Universität Erlangen-Nürnberg, 1983.

[12] K. Brenner and J. Ojeda-Castañeda, Ambiguity function and Wigner distribution function applied to partially coherent imagery. *Optica Acta* **31**, 213–223 (1984).

[13] P.L. Butzer, A survey of the Whittaker-Shannon sampling theorem and some of its extensions, *J. Math. Res. Expos. (China)* **3**, 185–212 (1983).

[14] T.A.C.M. Claasen and W.F.G. Mecklenbräuker, The Wigner distribution —a tool for time-frequency signal analysis. Part I. *Philips J. Res.* **35**, 217–250 (1980).

[15] T.A.C.M. Claasen and W.F.G. Mecklenbräuker, The Wigner distribution —a tool for time-frequency signal analysis. Part II. *Philips J. Res.* **35**, 276–300 (1980).

[16] T.A.C.M. Claasen and W.F.G. Mecklenbräuker, The Wigner distribution —a tool for time-frequency signal analysis. Part III. *Philips J. Res.* **35**, 372–389 (1980).

[17] D. Gabor, Theory of communication, *J. Inst. Elec. Eng. (London)* **93**, 429–457 (1946).

[18] D. Gabor, La théorie des communications et la physique. In: *La cybernétique: Théorie du signal et de l'information*, L. de Broglie, Ed. (Éditions de la Revue d'Optique Théorique et Instrumentale, Paris, 1951); pp. 115–149.

[19] M. García-Bullé, W. Lassner, and K.B. Wolf, The metaplectic group within the Heisenberg-Weyl ring. Reporte de Investigación. Departamento de Matemáticas, Universidad Autónoma Metropolitana, Iztapalapa, México 1985; to appear in *J. Math. Phys.*

[20] M.A. Gerzon, Comments on "The delay plane, objective analysis of subjective properties. Part I" *J. Audio Eng. Soc.* **22**, 104–106 (1974).

[21] C.C. Grosjean, Note on two identities mentioned by Dr. W. Schempp near the end of the presentation of his paper, In: *Proceedings of the Laguerre symposium at Bar-le-Duc 1984.* Lecture Notes in Mathematics (Springer Verlag, Heidelberg, 1985); to appear.

[22] V. Guillemin and S. Sternberg, *Symplectic Techniques in Physics*, (Cambridge University Press, 1984).

[23] R.C. Heyser, The delay plane, objective analysis of subjective properties. Part I. *J. Audio Eng. Soc.* **21**, 690–701 (1973).

[24] J.R. Higgins, Five short stories about the cardinal series. *Bull. Amer. Math. Soc. (New Series)* **12**, 45–89 (1985).

[25] M. Hillary, R.F. O'Connell, M.O. Scully, and E.P. Wigner, Distribution functions in physics: Fundamentals. *Phys. Rep.* **106**, 121–167 (1984).

[26] R. Howe, On the role of the Heisenberg group in harmonic analysis, *Bull. Amer. Math. Soc. (New Series)* **3**, 821–843 (1980).

[27] R. Howe, Quantum mechanics and partial differential equations, *J. Funct. Anal.* **38**, 188–254 (1980).

[28] C.P. Janse and A.J.M. Kaizer, Time-frequency distributions of loudspeakers: The application of the Wigner distribution. *J. Audio Eng. Soc.* **31**, 198–223 (1983).

[29] A.J. Jerri, The Shannon sampling theorem —its various extensions and applications: A tutorial review. *Proc. IEEE* **65**, 1565–1596 (1977).

[30] J.R. Klauder, The design of radar signals having both high range resolution and high velocity resolution, *Bell. Syst. Tech. J.* **39**, 809–820 (1960).

[31] J.R. Klauder, A.C. Price, S. Darlington, and W.J. Albersheim, The theory and design of chirp radars, *Bell. Syst. Tech. J.* **39**, 745–808 (1960).

[32] C.C. Moore and J.A. Wolf, Square integrable representations of nilpotent groups, *Trans. Amer. Math. Soc.* **185**, 445–462 (1973).

[33] J. von Neumann, Die Eindeutigkeit der Schrödingerschen Operatoren, *Math. Ann.* **104**, 570–578 (1931).

[34] R.D. Ogden and S. Vági, Harmonic analysis of a nilpotent group and function theory on Siegel domains of type II, *Adv. Math.* **33**, 31–92 (1979).

[35] A. Papoulis, Ambiguity function in Fourier optics, *J. Opt. Soc. Amer.* **64**, 779–788 (1974).

[36] H. Raszillier and W. Schempp, Fourier optics from the perspective of the Heisenberg group. Chapter 2 of this volume.

[37] A.W. Rihaczek, *Principles of High-Resolution Radar*, (McGraw-Hill, New York, 1969).

[38] W. Schempp, On the Wigner quasi-probability distribution function III. *C.R. Math. Rep. Acad. Sci. Canada* **5**, 35–40 (1983).

[39] W. Schempp, Gruppentheoretische Aspekte der Signalübertragung und der kardinalen Interpolationssplines I. *Math. Methods Appl. Sci.* **5**, 195–215 (1983).

[40] W. Schempp, Radar ambiguity functions and the linear Schrödinger representation. In: *Anniversary Volume on Approximation Theory and Functional Analysis*, P.L. Butzer, R.L. Stens, and B. Sz.-Nagy, Eds., ISNM, Vol. 65. (Birkhäuser, Basel, 1984); pp. 481–491.

[41] W. Schempp, Radar ambiguity functions, nilpotent harmonic analysis, and holomorphic theta series. In: *Special functions: Group Theoretical Aspects and Applications*, R.A. Askey, T.H. Koornwinder, and W. Schempp, Eds., MIA Series, (Reidel, Dordrecht, 1984); pp. 217–260.

[42] W Schempp, Radar ambiguity functions of positive type. In: *General Inequalities 4*, W. Walter, Ed. ISNM, Vol. 71, (Birkhäuser, Basel, 1984); pp. 369–380.

[43] W. Schempp, Radar reception and nilpotent harmonic analysis VI. *C.R. Math. Rep. Acad. Sci. Canada* **6**, 179–182 (1984).

[44] W. Schempp, Radar ambiguity functions, the Heisenberg group, and holomorphic theta series. *Proc. Amer. Math. Soc.* **92**, 103–110 (1984).

[45] W. Schempp, On Gabor information cells. In: *Multivariate Approximation Theory III*. W. Schempp and K. Zeller, Eds., ISNM, Vol. 75 (Birkhäuser, Basel, 1985); pp. 349–362.

[46] W. Schempp, *Harmonic Analysis on the Heisenberg Nilpotent Lie Group, with Applications to Signal Theory*, (Pitman, London), to appear.

[47] M. Schmidt, *Die relle Heisenberg-Gruppe und einige ihrer Anwendungen in Radarortung und Physik*. Diplomarbeit. Lehrstuhl für Mathematik I der Universität Siegen, Siegen, 1985.

[48] I.J. Schoenberg, Cardinal spline interpolation. Regional Conference Series in Applied Math., Vol. 12. (SIAM, Philadelphia, 1973).

[49] M.R. Schroeder, *Number Theory in Science and Communication*, (Springer Verlag, Heidelberg, 1984).

[50] H.W. Schüßler, *Digitale Systeme zur Signalverarbeitung*. (Springer Verlag, Heidelberg, 1973).

[51] J. Schwinger, Unitary operator bases. *Proc. Nat. Acad. Sci. USA* **46**, 570–579 (1960).

[52] L.M. Soroko, *Holography and Coherent Optics*, (Plenum Press, New York, 1980).

[53] J.M. Souriau, Géométrie symplectique et physique mathématique, *Gazette des Mathématiciens* **10**, 90–130 (1978).

[54] J. Ville, Théorie et applications de la notion de signal analytique, *Câbles et Transmissions* **2**, 61–74 (1948).

[55] A. Weil, Sur certains groupes d'opérateures unitairs, *Acta Math.* **111**, 143–211 (1964); also in: *Collected papers*, Vol. III, (Springer Verlag, Heidelberg, 1980); pp. 1–69.

[56] H. Weyl, Gruppentheorie und Quantenmechanik, (Wissenschaftliche Buchgesellschaft, Darmstadt, 1981).

[57] E.P. Wigner, Quantum-mechanical distribution functions revisited. In: *Perspectives in Quantum Theory*, W. Yourgrau and A. van der Merwe, Eds., (Dover, New York, 1979); pp. 25–36.

[58] C.H. Wilcox, The synthesis problem for radar ambiguity functions. MRC Tech. Summary Report, N°. 157. (The University of Wisconsin, 1960).

[59] K.B. Wolf, *Integral Transforms in Science and Engineering*, (Plenum Press, New York, 1979).

[60] P.M. Woodward, *Probability and Information Theory, with Applications to Radar*, (Artech House, Dedham MA, 1980).

Chapter 2

Fourier optics from the perspective of the Heisenberg group

by HANS RASZILLIER and WALTER SCHEMPP

ABSTRACT: We discuss (linear) geometrical optics and its scalar wave counterpart, Fourier optics, as a natural outgrowth of the group theoretical machinery emanating from the symplectic structure of (optical) phase space, as it derives from the Fermat principle. We do, in fact, *quantize* geometrical optics. The symplectic structure is preserved by the (symplectic) group of linear canonical transformations. Every element of the symplectic group can be realized as a part of an image system in geometrical optics. On the other hand, one may associate a (nilpotent) Lie algebra to the (symplectic) phase space on which the symplectic group $Sp(4, \mathfrak{R})$ acts in the natural way. Looking further on the irreducible unitary representations of the underlying simply connected Lie group, the Heisenberg group, one finds that the symplectic action on the Heisenberg group induces a group of unitary transformations in the space of each (infinite-dimensional) irreducible representation. This group, called the *metaplectic* group, describes the evolution of the (scalar) wave field (Fourier optics) corresponding to the evolution of rays (geometrical optics), and thereby connects the two views on optics. We put emphasis on the interplay of concepts, and for this reason, we also invoke intuition and terminology from classical and quantum mechanics.

2.1 Introduction

During the last few years much attention has been paid in optics to the construction of wave optics from the knowledge of the system of rays [1–4]. The investigation has concentrated on the optical wave field itself or —for coherent waves, equivalently— on the density matrix of various transforms of it (the autocorrelation or autoambiguity function, and its Fourier transform, the Wigner function).

Since work done so far in this field has described a number of aspects of the problem, we consider it appropriate to present it here in its full natural perspective. Since this perspective is the same as for the *quantization* of classical mechanics, we shall often practice the interplay with (quantum) mechanics, emphasizing both similarities and differences between the two fields.

We are interested in the precise frame of situations in geometrical optics which is called first-order paraxial optics. It is an optical analogue to linear (classical) mechanic systems and can be given a matrix formulation, with matrices belonging to the symplectic group $Sp(4, \mathfrak{R})$ of linear canonical transformations in four-dimensional phase space [4,5].

The symplectic formulation of geometrical optics is, of course, classical [5,6]. The idea of its *quantization* in terms of the metaplectic group (the two-fold covering group of the symplectic group) has been sketched by Guillemin and Sternberg [4]. The concept of the metaplectic group which goes back to A. Weil [7], has also been used by Bacry and Cadilhac [8] while discussing the opposite problem of approaching the *quasigeometric* wave optics from Maxwell's equations (see also [9–11]).

In the present paper we wish to construct first order paraxial wave optics in the framework of group theory, *i.e.*, from the point of view of harmonic analysis on the Heisenberg group. This group is intimately related to quantum mechanics [12] and signal theory [13]. Thereby we will pay special attention to the relation between the group $Sp(2n, \mathfrak{R})$ and its two-fold covering group $Mp(2n, \mathfrak{R})$. We will also discuss in some detail the origin, the location, and the nature of the phase discontinuities of the (scalar) optical wave field. These facts have, in our opinion, not yet been put in their appropriate group theoretical perspective.

We shall first describe the idea of quantization of linear mechanical systems in nontechnical terms. We will be concerned with the relevant mathematical structure, the Heisenberg group, centered around the Heisenberg canonical commutation relations in quantum mechanics. Then we shall apply to it the construction of the Schrödinger wave function, the density operator, as well as the autocorrelation and Wigner functions. By translating the language of quantum mechanics to optics we get the optical wave field propagating along an axis, its density matrix, and its autocorrelation (autoambiguity) function. We locate the phase discontinuities of the wave field by bringing into play concepts related to the group theoretical Maslov index, and finally discuss questions related to symmetries of optical systems.

2.2 The quantization problem

Efforts towards a deeper understanding of the relation between quantum mechanics and classical mechanics concentrate, on the one side, on the proper formulation of the quasiclassical limit ($\hbar \to 0$, $\hbar =$ Planck's constant) of quantum mechanics [14], and on the other side, on the proper formulation of quantum mechanics starting from classical mechanics (*i.e.*, the process of quantization) [15]. These are two opposite starting points aiming at the same result; both are relevant to optics. Their relevance comes from the fact that geometrical optics can be considered as arising from wave optics in the limit $\lambda \to 0$ ($\lambda =$ wave length), and inversely, wave optics may be viewed as a *quantized* geometrical optics. In the optical version of the quantization problem the wave length comes to play the role of Planck's constant.

Quantization of a classical mechanical system becomes transparent (*i.e.*, a purely group theoretical problem) when the Hamiltonian is a (homogeneous) quadratic form of the coordinates of phase space $\mathfrak{R}^{2n}$. Then the equations of motion are linear and homogeneous, the classical time evolution of the system is therefore a group of *linear canonical transformations* or, in mathematical terms, a one-parameter subgroup of the symplectic group $Sp(2n, \mathfrak{R})$. It is in fact this group which leads to quantization.

The road to it can be sketched as follows: quantum mechanics is intimately related to a (nilpotent) Lie group, the Heisenberg group H [9]. This is the simply connected Lie group underlying the Lie algebra of the Heisenberg commutation relations, which are associated (in the spirit of Dirac) to the symplectic phase space; the parameters of H can still be looked upon as classical phase space coordinates, together with a parameter which scales Planck's constant.

The Heisenberg group has essentially a unique infinite-dimensional irreducible unitary representation W for a fixed value, say 1, of the scaling parameter. The group $Sp(2n, \Re)$ of all linear canonical transformations leaves the Heisenberg commutation relations unchanged and can therefore be embedded into the automorphism group of the Heisenberg group.

In view of a theorem by Stone and von Neumann [16], the representation $W(g(h))$, $h \in H$, of the Heisenberg group H, for a fixed element $g \in Sp(2n, \Re)$, is unitarily equivalent to $W(h)$:

$$W(g(h)) = M(g)\, W(h)\, M^{-1}(g). \tag{2.1}$$

It is evident that the operators $M(g)$, $g \in Sp(2n, \Re)$, form a unitary representation *up to a phase factor* (a ***projective*** representation) of the group $Sp(2n, \Re)$. It turns out that one can choose the phase factor such, that the $M(g)$ form a representation up to a sign of this group, or a true representation of a two-fold covering group of $Sp(2n, \Re)$, called *the metaplectic group* $Mp(2n, \Re)$, in the same way as the half-integer spin representations up to a sign of the rotation group $SO(3)$ are true representations of $SU(2)$.

We now take a one-parameter subgroup $\{g(t)\}$, $-\infty < t < \infty$, (of time evolution in classical mechanics) of $Sp(2n, \Re)$ and get a group of unitary operators $M(g(t))$ acting in the representation space of W, the Hilbert space of Schrödinger wave functions. The one-parameter group $\{M(g(t))\}$ describes the quantum mechanical time evolution. Thus, we may conclude that the quantization of a classical mechanical system for any quadratic Hamiltonian, *i.e.*, the construction of the wave mechanics corresponding to it, can be performed as soon as the function $M(g)$ is determined on the group $Sp(2n, \Re)$.

Geometrical optics can be given a canonical (=symplectic) formulation [6], much the same as in classical mechanics. In the so-called paraxial approximation its Hamiltonian is quadratic, the ray evolution along the optical (*time*) axis is therefore a succession of linear canonical transformations belonging to $Sp(4, \Re)$. We can thus apply the procedure just sketched in order to *quantize* explicitly this kind of geometrical optics. We now start to develop the instruments of quantization.

2.3 The Heisenberg group

The canonical commutation relations of quantum mechanics,

$$[p_j, q_k] = -i\hbar\, \delta_{jk}, \qquad j, k = 1, \ldots, n \tag{3.1}$$

($p_j = -i\hbar\partial/\partial x_j$, $q_k = x_k$), may be considered as a realization of the Lie algebra identities:

$$\begin{aligned} [P_j, Q_k] &= -i\delta_{jk} E, \\ [P_j, E] &= 0, \\ [Q_j, E] &= 0, \end{aligned} \tag{3.2}$$

in terms of self-adjoint operators $P_j = -i\hbar\partial/\partial x_j$, $Q_j = x_j$, $E = \hbar$, acting in the Hilbert space of Schrödinger wave functions $\psi(\mathbf{x})$.

The Lie algebra defined by (3.2) is built on the symplectic (phase) space. If $(\cdot,\cdot)$ is the skew symmetric bilinear form of this space and e a vector which is not null, then we define for any two vectors v, v' of the phase space the *Lie bracket* $[v, v']$ by

$$[v, v'] = -i(v, v')e, \tag{3.2'}$$

in terms of this form.

The commutator belongs to the one-dimensional vector space spanned by ie. The vectors $i\lambda e$, $(-\infty < \lambda < \infty)$ of this space are defined to commute among themselves and with any v:

$$[e, v] = 0.$$

The formulation (3.2) already refers to a basis P_j, Q_j, $j = 1, \ldots, n$, with $[P_j, P_k] = 0$, $[Q_j, Q_k] = 0$, $[P_j, Q_k] = -[Q_k, P_j] = \delta_{jk}$ such that

$$v = \sum_{j=1}^{n} (q_j P_j + p_j Q_j) =: (\mathbf{q}, \mathbf{p}) \quad \text{and} \quad v + \lambda e =: (\mathbf{q}, \mathbf{p}; \lambda).$$

We may consider (3.2) as the Lie algebra of a (simply connected) Lie group, the Heisenberg group H[1]. The group again is homeomorphic to the $(2n+1)$–dimensional real vector space

$$H = \{(\mathbf{q}, \mathbf{p}; \lambda) \mid \mathbf{q} \in \Re^n, \mathbf{p} \in \Re^n, \lambda \in \Re\} \tag{3.3}$$

where $\mathbf{q} = (q_1, \ldots, q_n)$, $\mathbf{p} = (p_1, \ldots, p_n)$. The phase space $\{(\mathbf{q}, \mathbf{p})\} = \Re^{2n}$ forms a $2n$-dimensional subspace of H. The form of the composition law of the group depends, of course, on the coordinate system chosen on the Lie group, which is not always convenient to be taken the same as on the Lie algebra. We first mention the so-called basic presentation of the group, $\{(\mathbf{q}, \mathbf{p}; \lambda)_b\}$, because it emphasizes the close connection to the symplectic structure of phase space. It has the composition law

$$\begin{gathered}(\mathbf{q}, \mathbf{p}; \lambda)_b \cdot (\mathbf{q}', \mathbf{p}'; \lambda')_b = \big(\mathbf{q} + \mathbf{q}', \mathbf{p} + \mathbf{p}'; \lambda + \lambda' + \tfrac{1}{2}(\mathbf{q} \cdot \mathbf{p}' - \mathbf{p} \cdot \mathbf{q}')\big)_b, \\ (\mathbf{q} \cdot \mathbf{p}' - \mathbf{p} \cdot \mathbf{q}') =: (v, v').\end{gathered} \tag{3.4a}$$

The *representation* $W_\hbar(\mathbf{q}, \mathbf{p}; \lambda)$ is constructed easiest [16] by the (Mackey) procedure of inducing from a maximal commutative subgroup. Such a subgroup is given *e.g.* by the set $\{(\mathbf{0}, \mathbf{p}; \lambda)\} = \{(\lambda e + \sum_{j=1}^{n} p_j Q_j)\}$.

We consider the Hilbert space $\mathcal{L}^2(\Re^n)$, the $\mathcal{L}^2$-closure of the Schwartz space $S(\Re^n)$ on the set $\{(\mathbf{x}, \mathbf{0}; 0)\} = \{\mathbf{x}\} = \Re^n$. On $S(\Re^n)$ we define, for $(\mathbf{q}, \mathbf{p}; \lambda)_b$,

$$\begin{aligned}(W_\hbar(\mathbf{q}, \mathbf{p}; \lambda)\psi)(\mathbf{x}) &= \exp[i(\lambda + \mathbf{p} \cdot \mathbf{x} + \tfrac{1}{2}\mathbf{q} \cdot \mathbf{p})/\hbar]\psi(\mathbf{x} + \mathbf{q}) \\ &= \exp[i(\lambda + (x, p) + \tfrac{1}{2}(q, p))/\hbar]\, \psi(\mathbf{x} + \mathbf{q}),\end{aligned} \tag{3.5}$$

where

$$x = \sum_{j=1}^{n} x_j P_j, \qquad \mathbf{x} = (x_1, \ldots, x_n),$$

$$q = \sum_{j=1}^{n} q_j P_j, \qquad \mathbf{q} = (q_1, \ldots, q_n),$$

$$p = \sum_{j=1}^{n} p_j Q_j, \qquad \mathbf{p} = (p_1, \ldots, p_n),$$

[1]The Heisenberg group is in this volume also called the *Heisenberg-Weyl* group W_N. (Editor's note.)

and extend it to a unitary representation on $\mathcal{L}^2(\Re^n)$.

Computations are sometimes more conveniently performed in the so-called ***dual pairing*** presentation of the group H, $(\mathbf{a},\mathbf{b};\ell)_d$, with the composition law

$$(\mathbf{a},\mathbf{b};\ell)_d\cdot(\mathbf{a}',\mathbf{b}';\ell')_d=(\mathbf{a}+\mathbf{a}',\mathbf{b}+\mathbf{b}';\ell+\ell'+\mathbf{a}\cdot\mathbf{b}')_d. \tag{3.4b}$$

The relation between the two representations (3.4a) and (3.4b) is given by

$$\begin{aligned}(\mathbf{a},\mathbf{b};\ell)_d=&(\mathbf{q},\mathbf{p};\lambda-\tfrac{1}{2}\mathbf{q}\cdot\mathbf{p})_b,\\(\mathbf{q},\mathbf{p};\lambda)_b=&(\mathbf{a},\mathbf{b};\ell+\tfrac{1}{2}\mathbf{a}\cdot\mathbf{b})_d.\end{aligned} \tag{3.4'}$$

From (3.5) we get

$$\begin{aligned}\mathbf{q}:=&(\mathbf{q},\mathbf{0};0)_b=(\mathbf{q},\mathbf{0};0)_d\mapsto(W_{\hbar}(\mathbf{q})\psi)(\mathbf{x})=\psi(\mathbf{x}+\mathbf{q}),\\\mathbf{p}:=&(\mathbf{0},\mathbf{p};0)_b=(\mathbf{0},\mathbf{p};0)_d\mapsto(W_{\hbar}(\mathbf{p})\psi)(\mathbf{x})=e^{i\mathbf{p}\cdot\mathbf{x}/\hbar}\psi(\mathbf{x}),\\\lambda:=&(\mathbf{0},\mathbf{0};\lambda)_b=(\mathbf{0},\mathbf{0};\lambda)_d\mapsto(W_{\hbar}(\lambda)\psi)(\mathbf{x})=e^{i\lambda/\hbar}\psi(\mathbf{x}),\end{aligned} \tag{3.5'}$$

and because of $(\mathbf{a},\mathbf{b};\ell)_d=(\mathbf{0},\mathbf{0};\ell)(\mathbf{0},\mathbf{b};0)(\mathbf{a},\mathbf{0};0)$, we get in the dual pairing presentation

$$\begin{aligned}(\mathbf{a},\mathbf{b};\ell)_d\mapsto U_{\hbar}(\mathbf{a},\mathbf{b};\ell):=&W_{\hbar}(\ell)W_{\hbar}(\mathbf{b})W_{\hbar}(\mathbf{a}),\\(U_{\hbar}(\mathbf{a},\mathbf{b};\ell)\psi)(\mathbf{x})=&e^{i(\lambda+\mathbf{b}\cdot\mathbf{x})/\hbar}\psi(\mathbf{x}+\mathbf{a}).\end{aligned} \tag{3.6}$$

The theorem of Stone and von Neumann mentioned in section 2 asserts that (3.5)–(3.6) is the *unique* unitary irreducible representation $V(\mathbf{q},\mathbf{p};\lambda)$ of the Heisenberg group with $\big(V(\mathbf{0},\mathbf{0};\lambda)\psi\big)(\mathbf{x})=\exp(i\lambda/\hbar)\psi(\mathbf{x})$, up to unitary equivalence.

The Lie algebra structure (3.2′) is invariant under any linear (symplectic) transformation $v\mapsto gv$ of phase space such that $(gv,gv')=(v,v')$. The set of all these transformations (automorphisms) is the symplectic group $Sp(2n,\Re)$, which may be represented in the coordinates $(\mathbf{q},\mathbf{p})$ by the block matrices

$$g=\begin{pmatrix}A&B\\C&D\end{pmatrix}, \tag{3.7}$$

obeying

$$g^{\top}Jg=J,\qquad J=\begin{pmatrix}0&I_n\\-I_n&0\end{pmatrix}. \tag{3.8}$$

The action of the group $Sp(2n,\Re)$ can be transferred in a natural way to the Heisenberg group H; in the presentation (3.4a) it can, again, be given by the matrices (3.7), *i.e.* by $g(\mathbf{q},\mathbf{p};\lambda)_b=(A\mathbf{q}+B\mathbf{p},C\mathbf{q}+D\mathbf{p};\lambda)_b$.

If we take

$$V(\mathbf{q},\mathbf{p};\lambda)=W_{\hbar}\big(g(\mathbf{q},\mathbf{p};\lambda)\big), \tag{3.9}$$

we get again an irreducible unitary representation of the Heisenberg group. This representation has to be unitarily equivalent to $W_{\hbar}(\mathbf{q},\mathbf{p};\lambda)$, because of $\big(V(\mathbf{0},\mathbf{0};\lambda)\psi\big)(\mathbf{x})=e^{i\lambda/\hbar}\psi(\mathbf{x})$ and the Stone-von Neumann uniqueness theorem:

$$W_{\hbar}\big(g(\mathbf{q},\mathbf{p};\lambda)\big)=M(g)W_{\hbar}(\mathbf{q},\mathbf{p};\lambda)M^{-1}(g). \tag{3.10}$$

The unitary operator $M(g)$ depends on g, the chosen transformation in the group $Sp(2n,\Re)$. Because of the group property, we have, with $g_1,g_2\in Sp(2n,\Re)$, also $g_1g_2\in Sp(2n,\Re)$, and therefore on the one side

$$g_1 g_2 \mapsto M(g_1)M(g_2), \tag{3.11}$$

and on the other

$$g_1 g_2 \mapsto M(g_1 g_2). \tag{3.11'}$$

They have to represent, up to a possible phase factor $c(g_1, g_2)$ ($|c(g_1, g_2)| = 1$) depending on g_1 and g_2, the same transformation:

$$M(g_1 g_2) = c(g_1, g_2) M(g_1) M(g_2). \tag{3.12}$$

The unitary operators $M(g)$ thus give a representation, up to a phase, of the group $Sp\,(2n, \mathfrak{R})$ of linear canonical transformations, in the Hilbert space $\mathcal{L}^2(\mathfrak{R}^n)$ of Schrödinger wave functions.

This is the construction by which we may associate to any linear (classical mechanical) evolution in phase space a (unitary) quantum evolution of wave functions in Hilbert space; or in optical terms: to the evolution of a system of rays along an optical *time* axis, we associate the corresponding evolution of the wave field giving, among other things, information about light intensities.

It has been shown [16] that the phase factor may be chosen in such a way that

$$c^2(g_1, g_2) = \frac{s(g_1)s(g_2)}{s(g_1 g_2)} \tag{3.13}$$

holds, where the function $s(g)$, $|s(g)| = 1$, is well defined on the group $Sp\,(2n, \mathfrak{R})$. The function $s^{\frac{1}{2}}(g)$ can be defined, on the double covering $Mp\,(2n, \mathfrak{R}) = Sp\,(2n, \mathfrak{R}) \times \{\pm 1\}$ of $Sp\,(2n, \mathfrak{R})$ called the metaplectic group, such that

$$U(g) := s^{-\frac{1}{2}}(g) M(g) \tag{3.14}$$

forms a continuous unitary representation of $Mp\,(2n, \mathfrak{R})$.

The operators $U(g)$ may be represented as integral operators in the Hilbert space of Schrödinger wave functions, with distribution kernels $G_g(\mathbf{x}, \mathbf{y})$,

$$(U(g)\psi)(\mathbf{x}) = \int d\mathbf{y}\; G_g(\mathbf{x}, \mathbf{y})\, \psi(\mathbf{y}). \tag{3.15}$$

The kernel is the product of the kernel $M_g(\mathbf{x}, \mathbf{y})$ of $M(g)$ with the function $s^{-\frac{1}{2}}(g)$:

$$G_g(\mathbf{x}, \mathbf{y}) = s^{-\frac{1}{2}}(g)\, M_g(\mathbf{x}, \mathbf{y}). \tag{3.16}$$

There exist explicit formulas, both for $M_g(\mathbf{x}, \mathbf{y})$ and $s(g)$, for any group element $g \in Sp\,(2n, \mathfrak{R})$. The most relevant for our purpose are the group elements $g = \begin{pmatrix} A & B \\ C & D \end{pmatrix}$ with $\det B \neq 0$. For them [16]

$$\begin{aligned} M_g(\mathbf{x}, \mathbf{y}) &= \frac{1}{\sqrt{|\det B|}} \cdot \frac{1}{(2\pi\hbar)^{n/2}} \exp\left[\frac{i}{2\hbar} \sum_{j,k=1}^{n} \left(x_j (DB^{-1})_{jk} x_k + y_j (B^{-1}A)_{jk} y_k - 2x_j (B^{-1})_{jk} y_k \right)\right] \\ s(g) &= \operatorname{sign}(\det B) \cdot e^{\frac{1}{2} in\pi}. \end{aligned} \tag{3.17}$$

For $\det B = 0$, we give, for the sake of illustration only, the formula in the case $n = 1$. Then $g = \begin{pmatrix} a & 0 \\ c & a^{-1} \end{pmatrix}$, and

$$\begin{aligned} M_g(x, y) &= \sqrt{|a|}\, e^{i\pi c a x^2/\hbar}\, \delta(y - ax), \\ s(g) &= \operatorname{sign} a \end{aligned} \tag{3.18}$$

2.4 Description of the wave field

In the operator language of Hilbert space $\mathcal{L}^2(\Re^n)$, a ***wave function*** $\psi(\mathbf{x})$ may be identified (up to an inessential constant phase) with the projection operator P_ψ on its direction

$$(P_\psi \Phi)(\mathbf{x}) = \int d\mathbf{y}\, \psi^*(\mathbf{y})\, \Phi(\mathbf{y}) \cdot \left(\int d\mathbf{y}'\, \psi^*(\mathbf{y}')\, \psi(\mathbf{y}') \right)^{-1} \cdot \psi(\mathbf{x}). \tag{4.1}$$

This operator is a particular kind of ***density*** operator ρ [17], which is a hermitian, positive, trace-class operator:

$$\begin{aligned} &(a) && \rho^+ = \rho, \\ &(b) && \rho > 0, \\ &(c) && \operatorname{Tr} \rho = 1. \end{aligned} \tag{4.2}$$

It represents a pure state (because it is a one-dimensional projector), and therefore

$$\operatorname{Tr} P_\psi^2 = \operatorname{Tr} P_\psi = 1.$$

Generally, for density operators which represent mixed states we have

$$(d) \qquad 0 < \operatorname{Tr} \rho^2 \leq 1. \tag{4.3}$$

A density operator ρ is a Hilbert-Schmidt operator [18] and $\operatorname{Tr} \rho^2$ is its Hilbert-Schmidt norm. In the space $\mathcal{L}^2(\Re^n)$ of Schrödinger wave functions, ρ is, in fact, an integral operator as (4.1) illustrates in the simplest case.

Let us assume that we have

$$(\rho\psi)(\mathbf{x}) = \int d\mathbf{y}\, \rho(\mathbf{x}, \mathbf{y})\, \psi(\mathbf{y}). \tag{4.4}$$

Then the kernel $\rho(\mathbf{x}, \mathbf{y})$ obeys the conditions

$$\begin{aligned} &(a) && \rho^*(\mathbf{x}, \mathbf{y}) = \rho(\mathbf{y}, \mathbf{x}), \qquad (\rho^*(\mathbf{x}, \mathbf{x}) = \rho(\mathbf{x}, \mathbf{x})) \quad (a.e.), \\ &(b) && \int d\mathbf{x}\, d\mathbf{y}\, \psi^*(\mathbf{x})\, \rho(\mathbf{x}, \mathbf{y})\, \psi(\mathbf{y}) > 0 \quad \text{for any} \quad 0 \not\equiv \psi \in \mathcal{L}^2(\Re^n), \quad (\rho(\mathbf{x}, \mathbf{x}) > 0 \quad \text{a.e.}) \\ &(c) && \int d\mathbf{x}\, \rho(\mathbf{x}, \mathbf{x}) = 1, \end{aligned} \tag{4.5}$$

which are equivalent to (4.2), and also

$$(d) \qquad \operatorname{Tr} \rho^2 = \int d\mathbf{x}\, d\mathbf{y}\, \rho(\mathbf{x}, \mathbf{y})\, \rho(\mathbf{y}, \mathbf{x}) = \int d\mathbf{x}\, d\mathbf{y}\, |\rho(\mathbf{x}, \mathbf{y})|^2 \leq 1,$$

which expresses (4.3).

The metaplectic action of $U(g)$ on $\psi(\mathbf{x})$ carries over on ρ as

$$\rho \mapsto \rho_g = U(g) \rho\, U^{-1}(g). \tag{4.6}$$

If $U(g)$ has the (distributional) kernel $G_g(\mathbf{x}, \mathbf{y})$, then $U^{-1}(g)$ [which equals $U(g^{-1})$ locally] has the kernel $G_g^*(\mathbf{y}, \mathbf{x})$, and $\rho(g) = U(g)\rho\, U^{-1}(g)$ has the kernel

$$\rho_g(\mathbf{x}, \mathbf{y}) = \int d\mathbf{x}'\, d\mathbf{y}'\, G_g(\mathbf{x}, \mathbf{x}') \rho(\mathbf{x}'\mathbf{y}')\, G_g^*(\mathbf{y}, \mathbf{y}'). \tag{4.7}$$

We can also use, instead of the density operator ρ [or its kernel $\rho(\mathbf{x},\mathbf{y})$] the ***autocorrelation (autoambiguity) function*** [13,19]

$$\Gamma_\rho(\mathbf{q},\mathbf{p};\lambda) = \mathrm{Tr}\left(W_\hbar(\mathbf{q},\mathbf{p};\lambda)\rho\right), \tag{4.8}$$

which is easily computed to be

$$\Gamma_\rho(\mathbf{q},\mathbf{p};\lambda) = e^{i\lambda/\hbar}\int d\mathbf{x}\, e^{i\mathbf{p}\cdot\mathbf{q}/\hbar}\cdot \rho(\mathbf{x}+\tfrac{1}{2}\mathbf{q},\mathbf{x}-\tfrac{1}{2}\mathbf{q}). \tag{4.9}$$

The symplectically transformed autocorrelation function,

$$\Gamma_{\rho_g}(\mathbf{q},\mathbf{p};\lambda) := \mathrm{Tr}\left(W_\hbar(\mathbf{q},\mathbf{p};\lambda)\rho_g\right),$$

is then

$$\Gamma_{\rho_g}(\mathbf{q},\mathbf{p};\lambda) = \Gamma_\rho(g^{-1}(\mathbf{q},\mathbf{p};\lambda)). \tag{4.10}$$

The function $\Gamma_\rho(\mathbf{q},\mathbf{p};\lambda)$ obeys

$$\frac{1}{(2\pi\hbar)^n}\cdot\int d\mathbf{q}\,d\mathbf{p}\,|\Gamma_\rho(\mathbf{q},\mathbf{p};\lambda)|^2 = \int d\mathbf{x}\,d\mathbf{y}\,|\rho(\mathbf{x},\mathbf{y})|^2 =: \mathrm{Tr}\,\rho^2 \le 1 \tag{4.11}$$

and

$$\sup_{(\mathbf{q},\mathbf{p})\in\Re^{2n}} |\Gamma_\rho(\mathbf{q},\mathbf{p};\lambda)| = \Gamma_\rho(\mathbf{0},\mathbf{0};0) = \int d\mathbf{x}\,\rho(\mathbf{x},\mathbf{x}) = 1. \tag{4.12}$$

These properties tell us that $|\Gamma_\rho(\mathbf{q},\mathbf{p};\lambda)|$ has its maximum at the origin in phase space, and that its $\mathcal{L}^2$-norm is determined by the Hilbert-Schmidt norm of the density operator ρ.

So far we were mainly concerned with considerations related to individual group elements g of $Sp\,(2n,\Re)$. We consider now, in order to come closer to the results we look for, those (continuous) transformations $g(t)\in Sp\,(2n,\Re)$, $-\infty < t < \infty$, in fact one-parameter subgroups, which give the time evolution of a linear mechanical system,

$$g(t_1)\,g(t_2) = g(t_1 t_2), \qquad -\infty < t_1, t_2 < \infty, \tag{4.13}$$

and apply everything we have derived to them. If we do so we get $U(g) = U(g(t))$, the unitary time evolution of the system and its kernel $G_{g(t)}(\mathbf{x},\mathbf{y}) = G(\mathbf{x},\mathbf{y};t)$, the ***Feynman propagator*** [20] in configuration space. In the language of quantum mechanics the explicit result (3.17) is stated as the (known) fact that the Feynman path integral [20] which gives the kernel of the time evolution operator can be evaluated exactly for linear systems.

For the sake of illustration we list a few illustrative simple mechanical and optical examples. Those taken from optics, although extremely simple, already suggest that in optics, in distiction to mechanics, one should not expected the appearence of whole one-parameter subgroups of $Sp\,(2n,\Re)$, but rather *subsets* (in subgroups) of $Sp\,(2n,\Re)$.

2.5 Examples from quantum mechanics

The simplest example is *(a)* **the free particle**, where the one-parameter group is $\{g(t)\}$ [20],

$$g(t) = \begin{pmatrix} I_n & \frac{t}{m}I_n \\ 0 & I_n \end{pmatrix}, \qquad -\infty < t < \infty, \tag{5.1}$$

and for which

$$G(\mathbf{x},\mathbf{y};t) = \left(\frac{m}{2\pi t\hbar}\right)^{n/2} \exp\left(-\frac{in\pi}{4} + \frac{im}{2t\hbar}(\mathbf{x}-\mathbf{y})^2\right), \qquad t > 0, \tag{5.2}$$

with $G(\mathbf{x},\mathbf{y};-t) = G^*(\mathbf{y},\mathbf{x};t)$.

We note also the result for *(b)* **the *n*-dimensional harmonic oscillator**, of mass m and frequencies $\omega_1, \ldots, \omega_n$, for which the subgroup $\{g(t)\}$ is given by [20]

$$g(t) = \begin{pmatrix} \cos\omega_1 t & & 0 & \sin\omega_1 t/m\omega_1 & & 0 \\ & \ddots & & & \ddots & \\ 0 & & \cos\omega_n t & 0 & & \sin\omega_n t/m\omega_n \\ -m\omega_1 \sin\omega_1 t & & 0 & \cos\omega_1 t & & 0 \\ & \ddots & & & \ddots & \\ 0 & & -m\omega_n \sin\omega_n t & 0 & & \cos\omega_n t \end{pmatrix}, \qquad -\infty < t < \infty, \tag{5.3}$$

and

$$G(\mathbf{x},\mathbf{y};t) = \prod_{j=1}^{n} \sqrt{\frac{\omega_j}{2\pi\hbar\,|\sin\omega_j t|}}\ \exp\left[\frac{i\pi}{2}\,\mathrm{Int}\,\frac{\omega_j t}{\hbar} - \frac{in\pi}{4} + \frac{i\omega_j}{2\hbar\sin\omega_j t}\left(2x_j y_j - (x_j^2 + y_j^2)\cos\omega_j t\right)\right], \tag{5.4}$$

for $t > 0$ ($\mathrm{Int}\,x$ is the integer part of x for $x > 0$), with $G(\mathbf{x},\mathbf{y};-t) = G^*(\mathbf{y},\mathbf{x};t)$.

2.6 The phase space of geometrical optics

Geometrical optics may be considered as derived from the Fermat principle of extremal (continuous) optical path $\mathbf{x}(t) : \Re \to \Re^3$, expressed in terms of

$$I = \frac{1}{c}\int_{t_1}^{t_2} dt \left|\frac{d\mathbf{x}}{dt}\right| n(\mathbf{x}(t)), \qquad \mathbf{x}(t_i) = \mathbf{a}_i \quad (i = 1, 2), \tag{6.1}$$

where the (smooth) function $n(\mathbf{x})$ is the (local) refractive index of the optical medium, which is related to the (local) velocity of light propagation $v(\mathbf{x})$ by $v(\mathbf{x}) = c/n(\mathbf{x})$ (c is the velocity of light in vacuum).

We are interested in the nonparametric formulation of this principle and consider only paths $\mathbf{x}(t)$ for which *e.g.* $x_3(t) =: \tau$ is a strictly increasing (continuously differentiable) function of t (propagation along the x_3-axis), and refractive indices $n(\mathbf{x}) =: n(\mathbf{q})$, where $\mathbf{x} = (\mathbf{q}, \tau)$. Then we may reformulate (6.1) as

$$I = \int_{\tau_1}^{\tau_2} L(\mathbf{q}(\tau), \dot{\mathbf{q}}(\tau))d\tau, \qquad \mathbf{q}(\tau_i) = \mathbf{b}_i \qquad (i = 1, 2), \tag{6.2}$$

with the Lagrange function

$$L(\mathbf{q}, \dot{\mathbf{q}}) = n(\mathbf{q})\sqrt{\dot{\mathbf{q}}^2 + 1}, \tag{6.3}$$

and the corresponding Euler equations which give the light rays. The configuration space of geometrical optics is, in this view, $\Re^2 \times \Re^2 = \{\mathbf{q}, \dot{\mathbf{q}}\}$, the tangent bundle of $\Re^2$. The function $L(\mathbf{q}, \dot{\mathbf{q}})$ is convex in $\dot{\mathbf{q}}$; we compute its Legendre transform, the Hamiltonian

$$H(\mathbf{q}, \mathbf{p}) = -\sqrt{n^2(\mathbf{q}) - \mathbf{p}^2}, \tag{6.4}$$

defined by the mapping

$$(\mathbf{q}, \dot{\mathbf{q}}) \mapsto \left(\mathbf{q}, \mathbf{p} = \frac{\partial L}{\partial \dot{\mathbf{q}}} = n(\mathbf{q})\frac{\dot{\mathbf{q}}}{\sqrt{\dot{\mathbf{q}}^2 + 1}}\right), \tag{6.5}$$

on the manifold $M = \{\mathbf{q}, \mathbf{p} \mid \mathbf{p}^2 < n^2(\mathbf{q})\} \subset \Re^2 \times \Re^2$. On M we define a symplectic structure by the skew symmetric bilinear form $(\cdot,\cdot)$ of section 3 on its tangent spaces. Through this form we get from the differential form $dH : dH(\mathbf{q}, \mathbf{p}) = \frac{\partial H}{\partial q_i} dq_i + \frac{\partial H}{\partial p_i} dp_i$ associated to the function $H(\mathbf{q}, \mathbf{p})$ defined on M a (hamiltonian) vector field

$$I\, dH|_{\mathbf{q},\mathbf{p}} = \frac{\partial H}{\partial p_i} P_i - \frac{\partial H}{\partial q_i} Q_i, \tag{6.6}$$

which determines the dynamics of the light rays by the flow

$$\dot{q}_i P_i + \dot{p}_i P_i = I dH|_{\mathbf{q},\mathbf{p}}, \tag{6.7}$$

i.e., by

$$\frac{dq_i}{d\tau} = \frac{\partial H}{\partial p_i}, \qquad \frac{dp_i}{d\tau} = -\frac{\partial H}{\partial q_i}. \tag{6.8}$$

The approximation to this problem corresponds to taking

$$\mathcal{L}(\mathbf{q}, \dot{\mathbf{q}}) = n(\mathbf{q}) + \tfrac{1}{2} n(\mathbf{q}) \dot{\mathbf{q}}^2, \tag{6.9}$$

instead of $L(\mathbf{q}, \dot{\mathbf{q}})$, which is physically reasonable as long as $|\dot{\mathbf{q}}| \ll 1$ (paraxial aproximation). The mapping

$$(\mathbf{q}, \dot{\mathbf{q}}) \mapsto (\mathbf{q}, \mathbf{p} = \partial \mathcal{L} / \partial \dot{\mathbf{q}}) \tag{6.10}$$

leads to the manifold $\mathcal{M} = \Re^2 \times \Re^2$ instead of M, and to the function $\mathcal{H}(\mathbf{q}, \mathbf{p}) = \mathbf{p}^2/2n(\mathbf{q}) - n(\mathbf{q})$ defined on it, instead of H. If we may further take

$$\mathcal{H}_2(\mathbf{q}, \mathbf{p}) = \frac{\mathbf{p}^2}{2n_0} - n_0 - \sum_{i,j} a_{ij} q_i q_j, \tag{6.11}$$

instead of $\mathcal{H}(\mathbf{q}, \mathbf{p})$, then the equations (6.8) for $\mathcal{H}_2(\mathbf{q}, \mathbf{p})$ are linear. Their solution will be

$$\begin{pmatrix} \mathbf{q}(\tau) \\ \mathbf{p}(\tau) \end{pmatrix} = \begin{pmatrix} A(\tau) & B(\tau) \\ C(\tau) & D(\tau) \end{pmatrix} \begin{pmatrix} \mathbf{q}(0) \\ \mathbf{p}(0) \end{pmatrix}, \tag{6.12}$$

with a symplectic 4×4 matrix $g(\tau)$ of the form (3.7). That (6.12) is indeed symplectic results from the fact known in classical mechanics [21], that time evolution is a canonical transformation or, stated in other terms, that a hamiltonian flow preserves the symplectic structure. From (6.12) follows the matrix of free propagation along a distance $\tau = d$ in a medium of constant index of refraction $n = n_0$:

$$g(d) = \begin{pmatrix} I_2 & \frac{d}{n_0} I_2 \\ 0 & I_2 \end{pmatrix}. \tag{6.13}$$

An interesting object of geometric optics is, doubtless, a *lens*. Unfortunately it does not fit directly in the scheme of the classical calculus of variation with continuous (and differentiable) functions. So one should imagine it either as a limiting situation from a rapidly changing index of refraction or consider it in an enlarged frame of the calculus of variation [22]. It will not be surprising, therefore, that the description of thin lenses by elements of $Sp(4,\Re)$ is physically an approximation. The nature of such elements is, however, not hard to describe: A thin lens does not change positions, so $A = I_2$ and $B = 0$. From (3.8) then follows that $D = I_2$ and $C = C^\top$. The real symmetric 2×2 matrix C may be diagonalized by a rotation ($SO(2)$). The eigenvalues λ_i ($i = 1, 2$) are related to the two focal lengths f_i by $\lambda_i = -1/f_i$ and the rotation describes the position with respect to the coordinate axes of the two orthogonal (principal) directions they refer to. If these directions coincide with the axes, then the lens is described by

$$g = \begin{pmatrix} 1 & 0 & 0 & 0 \\ 0 & 1 & 0 & 0 \\ -1/f_1 & 0 & 1 & 0 \\ 0 & -1/f_2 & 0 & 1 \end{pmatrix}. \tag{6.14}$$

For a detailed description of optical devices in terms of $Sp(4,\Re)$ we refer to the literature (*e.g.* [23]).

One remark has still to be added with respect to the interpretation of $\mathbf{p}$ and to the determination of the optical analogue of $\hbar$. From (6.5) we derive, for a constant index of refraction n_0, that $\mathbf{p}^2 < n_0^2$. A (plane) electromagnetic wave has a wave vector $\mathbf{k}$ indicating the direction of propagation and obeying $\mathbf{k}^2 = n_0^2/\bar{\lambda}^2$ ($\bar{\lambda} = \lambda/2\pi$, λ is the wave length in vacuum). So $\mathbf{p}$ is, in fact, for small values, equal to $\bar{\lambda}$ times the projection on the q-plane of the wave vector $\mathbf{k}$. The description of the propagation of this wave along the τ-axis by (3.17) and (6.3) then identifies $\hbar$ as $\bar{\lambda}$.

2.7 Peculiarities of geometrical optics

The representation $U(g)$ is, with the appropriate choice of $\sqrt{s(g)}$, a continuous unitary representation of the metaplectic group $Mp(2n,\Re)$, which implies that

$$U(g')\psi \to U(g)\psi, \qquad \left(\int d\mathbf{x} |(U(g')\psi - U(g)\psi)(\mathbf{x})|^2 \to 0\right), \tag{7.1}$$

whenever the group element g' approaches $g : g' \to g$. If $U(g)$ is represented by an integral operator in the Schrödinger space, its kernel $G_g(\mathbf{x},\mathbf{y})$ is generally a distribution and so could lack continuity properties as a function of g.

In the standard situation of mechanics, the time evolution of a system is a continuous (abelian) group. For quadratic Hamiltonians it is a continuous one-parameter subgroup of $Sp(2n,\Re)$: if $g = g(t), g' = g(t')$ then $g(t') \to g(t)$ when $t \to t'$. In geometrical optics an optical system is not described by a subgroup, not even by a continuous path in the group $Sp(4,\Re)$. The physical reason for this discontinuity lies in the fact that on surfaces of refraction (which lenses are), there is a jump in the momentum of rays of the same kind as in elastic reflection of a particle on a wall, where a singular (*i.e.*, distributional) force acts in order to return it. Thus, strictly speaking, the mechanical analogue of geometrical optics is not *differentiable* classical dynamics, but rather *control theory* [24,25].

Since g in $Sp(4,\Re)$ determines $U(g)$ globally only up to a sign, although it determines it locally (by continuity) in a unique manner, we are *a priori* faced in optics with a *sign ambiguity* of $U(g)$ at every point of discontinuity of the path in $Sp(4,\Re)$, corresponding to the optical system. There is, however, a simple argument of continuity for assigning to a spherical lens a unique $U(g)$, because the group elements g corresponding to these lenses,

$$g(f) = \begin{pmatrix} I_2 & 0 \\ -\frac{1}{f} I_2 & I_2 \end{pmatrix}, \qquad -\infty < \frac{1}{f} < \infty, \tag{7.2}$$

form a continuous (open) path (even a subgroup) in $Sp(4,\Re)$ which passes through the unit element. The orthogonal transmission of an electromagnetic wave through a refracting plane surface (a lens of infinite focal length, $f=\infty$), *i.e.*, the unit group element in $Sp(4,\Re)$ is not accompanied by a phase change in the electromagnetic field, and is therefore represented by the unit element in $Mp(4,\Re)$. Therefore we have to take for these lenses the elements $U(g)$ in $Mp(4,\Re)$, which lie on the path connected to the unit group element.

The problem of how $U(g)$ should be chosen for a group element $g \in Sp(4,\Re)$ corresponding to an arbitrary optical device leads to a detailed analysis of the geometrical and topological structure[2] of $Sp(4,\Re)$. As an example we consider the subgroup of $Sp(4,\Re)$ corresponding to general lenses. Its elements are of the form

$$g_C = \begin{pmatrix} I_2 & 0 \\ C & I_2 \end{pmatrix}, \qquad C^\top = C. \tag{7.3}$$

Since the topology of this subgroup is that of $\Re^3$, every closed path in it can be contracted. They therefore correspond all to the identity element of the fundamental group π_1 of $Sp(4,\Re)$ and their images in $Mp(4,\Re)$ are disconnected. Therefore by continuity $U(g)$ is fixed in its sign by the requirement that to the unit element of $Sp(4,\Re)$ there corresponds precisely the unit element of $Mp(4,\Re)$.

After we have defined precisely $U(g)$ for elements g of $Sp(4,\Re)$ in the context of optics, we consider it as an integral operator and pass to the investigation of its (distributional) kernel.

2.8 Phase discontinuities

So far we have concentrated on the unitary (evolution) operator $U(g)$ of quantum mechanics and (Fourier) optics, particularly its association with the symplectic group elements giving the classical mechanical or geometrical optical evolution. Now we want to discuss its realization as an integral operator in the Hilbert space of Schrödinger wave functions. The kernel of a unitary operator is not square integrable (as it is for Hilbert-Schmidt operators) but rather in general a distribution. We take a look at this distribution $G_g(\mathbf{x},\mathbf{y})$, (3.15), for $U(g)$. Loosely speaking $G_g(\mathbf{x},\mathbf{y})$ represents the wave field at a point $\mathbf{x}$, after an evolution $U(g)$ applied to $\psi(\mathbf{y}'-\mathbf{y})$, *i.e.*, it is the wave field generated by $U(g)$ in $\mathbf{x}$ from a point source situated in $\mathbf{y}$.

If we fix a point source at $\mathbf{q} = \text{constant}\ (=0)$ such that $\{\mathbf{p}\} = \Re^n$, then under the action of the group element g the set $\{\mathbf{q}'\}$ in

$$\begin{pmatrix} \mathbf{q}' \\ \mathbf{p}' \end{pmatrix} = \begin{pmatrix} A & B \\ C & D \end{pmatrix} \begin{pmatrix} \mathbf{q} \\ \mathbf{p} \end{pmatrix}, \tag{8.1}$$

exhausts $\Re^n$, the whole configuration space, except when B is singular ($\det B = 0$). The dimension of the set wherein the particles (rays) collapse when $\det B = 0$, *i.e.*, the dimension of the caustics [6], is given by $n - \dim(\ker B)$. Formula (3.17) thus gives the wave field of a point source *outside* the caustics. *On* the caustics, the kernel has to be given by other formulas.

For the moment we do not raise the question of representing the wave field of a point source on caustics, but rather look at the approach of the caustics through free propagation. Propagation along a distance d in a homogeneous, isotropic medium is given in $Sp(4,\Re)$ by

$$g(d) = \begin{pmatrix} I_2 & \frac{d}{n_0} I_2 \\ 0 & I_2 \end{pmatrix}, \qquad d \geq 0. \tag{8.2}$$

[2] See appendix A. (Editor's note.)

If, after an evolution described by the group element

$$\begin{pmatrix} A & B \\ C & D \end{pmatrix} \in Sp(4, \Re), \qquad \det B \neq 0, \tag{8.3}$$

the system of rays propagates according to (8.2), then its final configuration is determined by

$$\begin{pmatrix} I_2 & \frac{d}{n_0} I_2 \\ 0 & I_2 \end{pmatrix} \begin{pmatrix} A & B \\ C & D \end{pmatrix} = \begin{pmatrix} A + \frac{d}{n_0} C & B + \frac{d}{n_0} D \\ C & D \end{pmatrix}. \tag{8.4}$$

Now,

$$\det(B + \frac{d}{n_0} D) = \det(B) \cdot \det(I_2 + \frac{d}{n_0} DB^{-1}),$$

where $I_2 + \frac{d}{n_0} DB^{-1}$ is a hermitian (real and symmetric) matrix. We denote the (real) eigenvalues of DB^{-1} by λ_1, λ_2, and get

$$\det(I_2 + \frac{d}{n_0} DB^{-1}) = (1 + \frac{d}{n_0} \lambda_1)(1 + \frac{d}{n_0} \lambda_2). \tag{8.5}$$

There are at most two positive values of d : $-n_0/\lambda_1$ and $-n_0/\lambda_2$ (if $\lambda_1, \lambda_2 < 0$), such that $\det(B + \frac{d}{n_0} D) = 0$. As long as $\lambda_1 \neq \lambda_2$, the corresponding caustics are one-dimensional; for $\lambda_1 = \lambda_2 < 0$ it is a point (zero-dimensional).

The geometrical image of a point source at $\mathbf{y}$ through (8.4) is determined by the map $\Re^2 \to \Re^2$ given by

$$\mathbf{p} \mapsto \mathbf{x}, \qquad \mathbf{x} = (B + \frac{d}{n_0} D)\mathbf{p}. \tag{8.6}$$

Along a given trajectory (ray) (8.6), *i.e.*, for fixed $\mathbf{p} \in \Re^2$, the exponent of $G_g(\mathbf{x}, \mathbf{y})$ is continuous, even at the points $d = -n_0/\lambda_1, d = -n_0/\lambda_2$. Since, however, the phase factor in front of the exponential $s(g)$ in (3.17), is discontinuous at $\det(B + \frac{d}{n_0} D) = 0$, the wave of the point source has a phase discontinuity along the rays in the caustics. This phase factor is constructed here from purely group theoretical considerations [16]. Its physical reality is confirmed by experiment [6,26].

2.9 Systems with symmetry

When all elements [in $Sp(4, \Re)$] of an optical system commute with the elements of a certain subgroup of $Sp(4, \Re)$, then the system posesses the *symmetry* of this subgroup. The *maximal* symmetry group for an optical system is the centralizer of the set of its elements in $Sp(4, \Re)$. Of obvious interest are the systems represented by the *diagonal* subgroup $Sp(2, \Re)$ of $Sp(4, \Re)$, with elements of the form

$$\begin{pmatrix} aI_2 & bI_2 \\ cI_2 & dI_2 \end{pmatrix}, \qquad ad - bc = 1. \tag{9.1}$$

The maximal symmetry group (centralized) of this $Sp(2, \Re)$ in $Sp(4, \Re)$ is the group of elements

$$\begin{pmatrix} A & 0 \\ 0 & (A^\top)^{-1} \end{pmatrix}, \qquad A^\top = A^{-1}, \tag{9.2}$$

which is isomorphic to the group of rotations in two dimensions, $O(2)$. This $Sp(2,\Re)$ is the group representing axially symmetric optical systems. There is, further, a ***non-euclidean*** subgroup isomorphic to $Sp(2,\Re)$, with elements

$$\begin{pmatrix} aI_2 & bG_2 \\ cG_2 & dI_2 \end{pmatrix}, \qquad ad - bc = 1, \qquad G_2 := \begin{pmatrix} 1 & 0 \\ 0 & -1 \end{pmatrix}. \tag{9.3}$$

Its centralizer in $Sp(4,\Re)$ is given by the group $O(1,1)$, with elements,

$$\begin{pmatrix} A & 0 \\ 0 & (A^\top)^{-1} \end{pmatrix}, \qquad A^\top G_2 A = G_2, \tag{9.4}$$

We have not looked into the physical realization of such $O(1,1)$-symmetric optical systems.

Generally, if a set $S \subset Sp(4,\Re)$ has a centralizer C, then also the group G_S generated by S has as centralizer the group C. Inversely, G_S lies in the centralizer of C, but it may be a proper subgroup of it. Let us denote the centralizer of C by $G \supset G_S$. The pair (G, C) is called dual [12] in $Sp(4,\Re)$, *i.e.*, the groups are the centralizers of each other. In the examples we have considered the pairs $(Sp(2,\Re), O(2))$ and $(Sp(2,\Re), O(1,1))$ are both dual in $Sp(4,\Re)$. The determination of all optical systems with symmetry thus leads to the determination of all dual pairs (G, C) in $Sp(4,\Re)$.

2.10 Summary and comments

We have sketched the construction of linear geometrical optics and the associated scalar wave optics (Fourier or Fresnel optics) as a problem of group theory. The group which is the backbone of the subject is the Heisenberg group, the Lie group whose Lie algebra is given by the quantum mechanical canonical commutation relations. The set of matrices describing linear geometrical optics represent the symplectic group, the automorphism group of the Heisenberg group leaving its center pointwise fixed. Fourier optics appears as a certain (metaplectic) unitary representation of this symplectic group, which arises naturally via the Stone-von Neumann uniqueness theorem concerning unitary irreducible representations of the Heisenberg group.

The facts we bring together are individually known, in mathematics and also in quantum mechanics. Factorization methods in (geometrical) matrix optics [5,23,27] and operator methods in Fourier optics [1–3] fall under the concepts of (subgroup) decomposition (Iwasawa, Bruhat, ...) of the symplectic group and under those of the metaplectic representation. However, these concepts have been used explicitly in optics only in a limited scope, in more detail in the recent book of Guillemin and Sternberg [4]. There is neccesarily a certain overlap between this paper and Ref. [4], but many of the points stressed by us are not touched in that work. Among other things we insist on a discussion of the *sign ambiguity* in the construction of Fourier optics, in order to make unambiguous statements about phase discontinuities encountered by waves generated by point sources, when they pass through caustics.

References

[1] H. J. Butterweck, Principles of Data Processing In *Progress in Optics*, E. Wolf, Ed. (North-Holland, Amsterdam, 1981).

[2] D. Stoler, Operator methods in physical optics, *J. Opt. Soc. Am.* **71**, 334–341 (1981).

[3] M. Nazarathy and J. Shamir, First order optics —a canonical operator representation: lossless systems, *J. Opt. Soc. Am.* **72**, 356–364 (1982).

[4] V. Guillemin and S. Sternberg, *Symplectic Techniques in Physics* (Cambridge University Press, 1984).

[5] O. N. Stavroudis, *The Optics of Rays, Wavefronts, and Caustics* (Academic Press, London, 1972).

[6] M. Born and E. Wolf, *Principles of Optics* (Pergamon, Oxford, 1975).

[7] A. Weil, Sur certains groupes d'opérateurs unitaires, *Acta Math.* **111**, 143–211 (1964).

[8] H. Bacry and M. Cadilhac, The metaplectic group and Fourier optics, *Phys. Rev.* **A23**, 2533–2536 (1981).

[9] E.C.G. Sudarshan, R. Simon, and N. Mukunda, Paraxial-wave optics and relativistic front description. I. The scalar theory, *Phys. Rev.* **A28**, 2933–2942 (1983).

[10] N. Mukunda, R. Simon, and E.C.G. Sudarshan, Paraxial-wave optics and relativistic front description. II. The vector theory, Phys. Rev. **A28**, 2933–2942 (1983).

[11] H. Bacry, Group theory and paraxial optics. In *Proceedings of the XIII*th *International Colloquium on Group Theoretical Methods in Physics (College Park, Maryland, 21–25 May, 1984)*, W.W. Zachary, Ed. (World Scientific Publ., Singapore, 1984); pp. 215–224.

[12] R. Howe, On the role of the Heisenberg group in harmonic analysis, *Bull. Amer. Math. Soc.* **3**, 821–843 (1980).

[13] W. Schempp, Radar ambiguity functions, the Heisenberg group, and holomorphic theta series, *Proc. Amer. Math. Soc.* **92**, 103–110 (1984).

[14] V.P. Maslov, *Théorie des Perturbations et Méthodes Asymptotiques* (Dunod, Paris, 1972).

[15] D.J. Simms and N.M.J. Woodhouse, *Lectures on Geometric Quantization*, Lecture Notes in Physics, Vol. 53 (Springer Verlag, Berlin, 1976).

[16] G. Lion and M. Vergne, *The Weil Representation, Maslov Index, and Theta series* (Birkhäuser, Basel, 1980).

[17] W. Thirring, *Lehrbuch der Mathematischen Physik, Vol. 4: Quantenmechanik großer Systeme* (Springer Verlag, Wien, 1980).

[18] M. Reed and B. Simon, *Methods of Modern Mathematical Physics, I. Functional Analysis* (Academic Press, New York, 1972).

[19] A. Papoulis, Ambiguity function in Fourier optics, *J. Opt. Soc. Am.* **64**, 779–788 (1974).

[20] R.P. Feynman and A.R. Hibbs, *Quantum Mechanics and Path Integrals* (McGraw Hill, New York, 1965).

[21] V.I. Arnold, *Mathematical Methods of Classical Mechanics* (Springer Verlag, New York, 1978).

[22] C. Carathéodory, *Geometrische Optik* (Springer Verlag, Berlin, 1937).

[23] B. Maculow and H.H. Arsenault, Matrix decompositions for nonsymetrical optical systems, *J. Opt. Soc. Am.* **73**, 1360–1366 (1983).

[24] L.S. Pontryaghin, V.G. Boltyansky, R.V. Gamkrelidse, and E.F. Mishchenko, *Mathematical Theory of Optimal Processes* (in Russian), (Nauka, Moscow, 1969).

[25] A.E. Bryson Jr. and Yu-chi Ho, *Applied Optimal Control* (Blaisdell, Waltham Mass., 1969).

[26] G.W. Farnell, Measured phase distribution in the image space of a microwave lens, *Can. J. Phys.* **36**, 935–943 (1958).

[27] S. Cornbleet, *Microwave Optics* (Academic Press, London, 1976).

Chapter 3

Lie series, Lie transformations, and their applications

by STANLY STEINBERG*

ABSTRACT: This paper is an exposition of the basic properties of Lie series and Lie transformations, which are now finding widespread applications. The applications are of two types: expanding solutions of Hamilton's equations and reducing (simplifying) Hamiltonians to normal form. The expansions are not power series but rather *factored product expansions.* These expansions have the advantage that the approximating systems are also hamiltonian. The normal form procedure has the advantage that it is canonical and explicit. In both cases the methods used are chosen so that they are easy to implement in a general purpose computer symbol manipulator.†

3.1 Introduction

Recently Lie series and Lie transformation have found a wide range of applications, particularly in celestial mechanics, magnetic optics, light optics, neutron transport, plasma physics, and other areas. The purpose of this paper is to provide the reader with the basic tools necessary to apply Lie techniques. Attention will be restricted to a special class of Lie techniques appropriate for the study of hamiltonian systems. However, these ideas can also be applied to general system (see Hlavatý, Wolf, and Steinberg [18], and Steinberg [27, 28]). Applications are not discussed in this paper; its purpose is to motivate and develop the mathematics needed. Extensive references to the applications literature are provided.

3.1.1 Applications of Lie series

The applications are of two types: expanding solutions of Hamilton's equations and simplifying, *i.e.* reducing Hamiltonians to normal form. The expansions are not power series, but rather ***factored product expansions.*** These expansions have the advantage that the approximating systems are also hamiltonian. If the solutions of Hamilton's equations are thought of as a mapping from the initial

*Partially supported by NSF grant # MCS-8102683 and the U.S. Army Research Office.

†This chapter was written by the author using a T-ROFF text processor, and a diskette was sent to the editors, who used the EMACS editor to turn it into a TeX file, half by `replace string` commands, half by hand. Editorial work consisted of proofreading, formatting, and deepening the segmentation. (Editor's note.)

data to the solution at some fixed time, then the factored product expansion writes this mapping as a composition of simpler maps. Naturally, the approximating maps become more complicated as the order of the expansion increases. We first learned about the applications of such expansions from Dragt (see [82]).

The normal form procedure discussed here is advantageous because the transformations that produce the normal form are canonical and explicit, whereas classical methods produce implicit transformations. Solving the implicit transformation for an explicit transformation is generally difficult; solving for the inverse transformation is also difficult. The inverse of a Lie transformation is trivial to compute. As in the factored product expansion calculations, the transformations that produce the canonical forms are computed as a composition of simple transformations. The normal-form procedure is done "term by term"; as the order of the terms increases, so does the complexity of the transformation used to do the simplification. We first read about this procedure in the papers of Dragt [83] and Hori [56, 57]. The transformations that are used in the normal-form procedure form a one-parameter group. This means that the transformations depend on a parameter; if this is thought of as time, the transformations are a solution of a system of autonomous hamiltonian equations. The fact that the transformations form a one-parameter group is what makes Lie's theory ([7, 8]) applicable. Dependence on only one parameter means that, at first, there is only a modest overlap with Lie's theory of finite-dimensional multi-parameter groups (Lie groups).

In both the expansion and the normal-form problem, the mathematics was developed so that the methods are easy to implement in a general purpose computer symbol manipulator. This has been done for the manipulator VAXIMA (MACSYMA) [135].

Lie transformations are written as an exponential of a first-order linear partial-differential operator. It is our opinion that it is hard to overrate the use of this exponential notion. It is this notation that has provided us with much of the insight into how to organize the calculations discussed in this paper. Such notation has been important in general linear analysis of initial value problems for ordinary and partial differential equations (see Steinberg [26] or the Semi-group sections of many of the standard texts on linear analysis). The notation has also been used to study Lie groups and quantum mechanics (see Steinberg [33], Wolf [35], and Wybourne [144]). We first saw this notation applied to a physical problem in the report by Stumpff [79].

3.1.2 The contents of this chapter

In Section 2 it is shown that the solutions of Hamilton's ordinary differential equations can be written using ***Lie series*** and that these series define ***Lie Transformations***. This provides an opportunity to introduce the notions of ***Poisson brackets*** and ***Lie derivatives***.

In Section 3 the elementary properties of Lie transformations are derived. These properties are based on the properties of Lie derivatives which are, in turn, based on the properties of Poisson brackets.

Exponential identities form the backbone of Lie transformation theory. Section 4 provides a discussion of the classical noncommuting exponential identities in a general setting. These identities are analogs of the usual exponential identities for the numerical exponential function.

Section 5 provides some deeper properties of Lie transformations: the ***exponential identities***. Here the results of Section 4 are generalized and applied to Lie transformations. It is these identities which give much of the computational power to Lie transformation theory. Because of the noncommunitivity, there are many forms of the identities.

In Section 6, the discussion of canonical forms is begun. The reduction to canonical form is done with a transformation that is written in factored-product form. The method is perturbative, ***i.e.*** is applied to a finite number of terms in a power series expansion of the Hamiltonian.

Section 7 discusses some detailed aspects of quadratic Hamiltonians which are critical to the normal-form procedure, and then applies the results of the previous section to two examples using a VAXIMA program.

Section 8 discusses the references.

3.1.3 Review of basic concepts

It is assumed that the reader is familiar with the basics of hamiltonian mechanics (Goldstein [140]). To introduce the notation, some of the basic ideas are reviewed here. A hamiltonian system has n degrees of freedom, where n is a positive integer. This means that the system is described using $2n$ variables that are thought of as being in ***phase*** space. The variables are broken into two sets, called ***position*** and ***momentum*** variables. Frequently $\mathbf{q} = (q_1, q_2, \ldots, q_n)$ is chosen for the position variables and $\mathbf{p} = (p_1, p_2, \ldots, p_n)$ is chosen for the momentum variables. The physics of a hamiltonian system will be determined by a hamiltonian function $H = H(\mathbf{q}, \mathbf{p})$. The case when the Hamiltonian is time dependent in not considered (but see Wolf [35]).

This paper is concerned with the local behavior of hamiltonian system, that is, the behavior of the system near a fixed point in phase space. The point is typically a critical point of the system; by translating coordinates, it may be taken to be the origin in phase space. More precisely, we will be interested in high order expansions of solutions of the system when the solutions start near the point. We are interested in the algebraic rather than the analytic aspects of these expansions. The discussion will center on techniques for computing various expansions; the convergence of the resulting expansions will not be discussed.

Lie transformation can be rigorously defined; one definition is based on power series expansions. In addition, most Lie transformation calculations are based on power series expansions, so it is appropriate to assume that all functions are analytic in a neighborhood of the origin. Since the applications require that the functions be real, all functions are assumed to be real analytic. This means that if $f = f(\mathbf{q}, \mathbf{p})$, then f has a power series expansion with real coefficients:

$$f(\mathbf{q}, \mathbf{p}) = \sum_{|l|+|m|\geq 0} f_{lm} \mathbf{q}^l \mathbf{p}^m, \tag{1.1}$$

where f_{lm} is a real number and

$$\begin{aligned} l &= (l_1, \ldots, l_n), & m &= (m_1, \ldots, m_n), \\ \mathbf{q}^l &= (q_1^{l_1}, \ldots, q_n^{l_n}), & \mathbf{p}^m &= (p_1^{m_1}, \ldots, p_n^{m_n}), \\ |l| &= \sum_{j=1}^{n} l_j, & |m| &= \sum_{j=1}^{n} m_j, \end{aligned} \tag{1.2}$$

with l_j and m_j nonnegative integers. The coefficients are given by

$$f_{lm} = \frac{1}{l!\,m!} \partial_q^l \partial_p^m f(\mathbf{q}, \mathbf{p})|_{\mathbf{q}=0,\mathbf{p}=0}\,. \tag{1.3}$$

where

$$l! = l_1! \cdots l_n!, \qquad \partial_q^l = \frac{\partial^{l_1}}{\partial_{q_1}^{l_1}} \cdots \frac{\partial^{l_n}}{\partial_{q_n}^{l_n}}, \tag{1.4}$$

and so forth.

In hamiltonian mechanics, adding a constant to the Hamiltonian does not change the physical interpretation of the system, so choose

$$f(0,0)=0. \tag{1.5}$$

A critical point is a point in phase space where

$$\frac{\partial f}{\partial q_i}(\mathbf{q},\mathbf{p})=0, \quad \frac{\partial f}{\partial p_i}(\mathbf{q},\mathbf{p})=0, \qquad 1\le i\le n. \tag{1.6}$$

As noted above, we will usually assume that the origin in phase space is a critical point, so the power series expansion of f about the origin in phase space begins with quadratic terms:

$$f(\mathbf{q},\mathbf{p})=\sum_{|l|+|m|\ge 2} f_{lm}\mathbf{q}^l\mathbf{p}^m. \tag{1.7}$$

This manuscript is based on several previous papers written by the author, the first of which was written in 1981. Several people have read and commented on the earlier versions and I would like to acknowledge their help. First I would like to thank André Deprit who made many informative comments and suggestions. In addition, John Browning, Alex Dragt, Carl Wulfman, were very helpful. I would also like to thank Kurt Bernardo Wolf for much encouragement over the years.

3.2 Hamiltonian systems

3.2.1 The Hamilton equations of motion

The position of a hamiltonian system at time t will be given by a pair of functions $(\mathbf{Q}(t),\mathbf{P}(t))$. The initial position of the system will be given by $\mathbf{Q}(0)=\mathbf{q}$ and $\mathbf{P}(0)=\mathbf{p}$. In fact, the position of the system is determined by both the time t and the initial position $(\mathbf{q},\mathbf{p})$ so write

$$\mathbf{Q}=\mathbf{Q}(t)=\mathbf{Q}(t,\mathbf{q},\mathbf{p}), \qquad \mathbf{P}=\mathbf{P}(t)=\mathbf{P}(t,\mathbf{q},\mathbf{p}), \tag{2.1}$$

$$\mathbf{Q}(0,\mathbf{q},\mathbf{p})=\mathbf{q}, \qquad \mathbf{P}(0,\mathbf{q},\mathbf{p})=\mathbf{p}. \tag{2.2}$$

The motion of a hamiltonian system is determined by a function $H(\mathbf{q},\mathbf{p})$, which is called the Hamiltonian of the system. The Hamiltonian is determined by the underlying physics. The dynamics of the system are determined by Hamilton's ordinary differential equations:

$$\frac{d\mathbf{Q}}{dt}=\frac{\partial H}{\partial \mathbf{p}}(\mathbf{Q},\mathbf{P}), \qquad \frac{d\mathbf{P}}{dt}=-\frac{\partial H}{\partial \mathbf{q}}(\mathbf{Q},\mathbf{P}). \tag{2.3}$$

Here

$$\frac{\partial H}{\partial \mathbf{q}}=\left(\frac{\partial H}{\partial q_1},\frac{\partial H}{\partial q_2},\dots,\frac{\partial H}{\partial q_n}\right), \qquad \frac{\partial H}{\partial \mathbf{p}}=\left(\frac{\partial H}{\partial p_1},\frac{\partial H}{\partial p_2},\dots,\frac{\partial H}{\partial p_n}\right). \tag{2.4}$$

3.2.2 Example: the harmonic oscillator

A simple example is given by the harmonic oscillator; a system that consists of a mass on the end of a spring where the force due to the spring is linear (obeys Hooke's law) is a harmonic oscillator.

Let Q be the displacement of the mass from equilibrium and P be the momentum of the mass, and choose units so that all constants are simple. This problem has one degree of freedom, so the variables are $(q_1, p_1) = (q, p)$. The Hamiltonian for this system can be chosen to be

$$H(q,p) = \tfrac{1}{2}(q^2 + p^2). \tag{2.5}$$

Hamilton's equations for this system are

$$\frac{dQ}{dt} = P, \qquad \frac{dP}{dt} = -Q. \tag{2.6}$$

The solution of the initial value problem for Hamilton's equations is

$$Q(t) = q\cos t + p\sin t, \qquad P(t) = p\cos t - q\sin t. \tag{2.7}$$

We will return to this example.

3.2.3 Poisson brackets

A concept central to this discussion is that of a ***Poisson bracket***. If $f = f(\mathbf{q},\mathbf{p})$ and $g = g(\mathbf{q},\mathbf{p})$ are given, then the ***Poisson bracket***[1] $[f,g] = [f,g](\mathbf{q},\mathbf{p})$, of f and g is a function given by

$$[f,g] = \sum_{j=1}^{n} \frac{\partial f}{\partial q_j}\frac{\partial g}{\partial p_j} - \frac{\partial f}{\partial p_j}\frac{\partial g}{\partial q_j}. \tag{2.8}$$

The Poisson bracket can also be defined when g is vector valued:

$$\mathbf{g} = \mathbf{g}(\mathbf{q},\mathbf{p}) = (g_1(\mathbf{q},\mathbf{p}), g_2(\mathbf{q},\mathbf{p}), \ldots, g_n(\mathbf{q},\mathbf{p})), \tag{2.9}$$
$$[f,\mathbf{g}] = ([f,g_1],[f,g_2],\ldots,[f,g_n]). \tag{2.10}$$

The Poisson bracket can be used to write Hamilton's system of ordinary differential equations in a particularly symmetric form. First note that

$$[H(\mathbf{q},\mathbf{p}), q_i](\mathbf{q},\mathbf{p}) = -\frac{\partial H}{\partial p_i}(\mathbf{q},\mathbf{p}), \qquad [H(\mathbf{q},\mathbf{p}), p_i](\mathbf{q},\mathbf{p}) = \frac{\partial H}{\partial q_i}(\mathbf{q},\mathbf{p}), \tag{2.11}$$

and consequently Hamilton's equations become

$$\frac{d\mathbf{Q}}{dt} = -[H,\mathbf{q}](\mathbf{Q},\mathbf{P}), \qquad \frac{d\mathbf{P}}{dt} = -[H,\mathbf{p}](\mathbf{Q},\mathbf{P}). \tag{2.12}$$

Not only is it possible to write the differential equation in a symmetric form: higher derivatives of $\mathbf{Q}$ and $\mathbf{P}$ may be written in a similar form. A good way to show this is by induction. Assume that for $k \geq 1$ there are functions $\mathbf{A}_k(\mathbf{p},\mathbf{q})$ and $\mathbf{B}_k(\mathbf{p},\mathbf{q})$ such that

$$\frac{d^k\mathbf{Q}}{dt^k}(t) = \mathbf{A}_k(\mathbf{Q}(t),\mathbf{P}(t)), \qquad \frac{d^k\mathbf{P}}{dt^k}(t) = \mathbf{B}_k(\mathbf{Q}(t),\mathbf{P}(t)). \tag{2.13}$$

[1] The notation in this chapter uses $[\cdot,\cdot]$ for Poisson brackets and commutators; they are isomorphic operations here. In other chapters in this volume, the convention is $\{\cdot,\cdot\}$ for Poisson brackets and $[\cdot,\cdot]$ for commutators.

From (2.12) we know that

$$\mathbf{A}_1 = -[H, \mathbf{q}], \qquad \mathbf{B}_1 = -[H, \mathbf{p}]. \tag{2.14}$$

Use the chain rule to differentiate (2.13) and then use (2.3) to obtain:

$$\begin{aligned}
\frac{d^{k+1}\mathbf{Q}}{dt^{k+1}}(t) &= \sum_{i=1}^{n}\left(\frac{\partial \mathbf{A}_k}{\partial q_i}(\mathbf{Q}(t),\mathbf{P}(t))\frac{dQ_i}{dt}(t) - \frac{\partial \mathbf{A}_k}{\partial p_i}(\mathbf{Q}(t),\mathbf{P}(t))\frac{dP_i}{dt}(t)\right)\\
&= \sum_{i=1}^{n}\left(\frac{\partial \mathbf{A}_k}{\partial q_i}(\mathbf{Q}(t),\mathbf{P}(t))\frac{\partial H}{\partial p_i}(\mathbf{Q}(t),\mathbf{P}(t)) - \frac{\partial \mathbf{A}_k}{\partial p_i}(\mathbf{Q}(t),\mathbf{P}(t))\frac{\partial H}{\partial q_i}(\mathbf{Q}(t),\mathbf{P}(t))\right)\\
&= [\mathbf{A}_k, H](\mathbf{Q}(t),\mathbf{P}(t)).
\end{aligned} \tag{2.15}$$

A similar argument works for $\mathbf{P}$. Thus

$$\begin{aligned}
\mathbf{A}_1 &= -[H, \mathbf{q}], \qquad & \mathbf{A}_{k+1} &= -[H, \mathbf{A}_k],\\
\mathbf{B}_1 &= -[H, \mathbf{p}], \qquad & \mathbf{B}_{k+1} &= -[H, \mathbf{B}_k].
\end{aligned} \tag{2.16}$$

It is clear that the formulas for $\mathbf{A}_k$ and $\mathbf{B}_k$ involve multiple commutators of H with $\mathbf{q}$ and $\mathbf{p}$. To write this nicely we introduce another important concept.

3.2.4 Lie derivatives

A ***Lie derivative*** is an operator that depends on a function and operates on functions. If $H = H(\mathbf{q}, \mathbf{p})$ is a given function and $f = f(\mathbf{q}, \mathbf{p})$ is any other function, the ***Lie derivative*** L_H determined by H is given by

$$L_H f = [H, f]. \tag{2.17}$$

The powers of the Lie derivative are then

$$L_H^0 f = f, \qquad L_H^1 f = [H, f], \qquad L_H^2 f = [H, [H, f]], \tag{2.18}$$

or, in general,

$$L_H^{k+1} f = [H, L_H^k f], k \geq 0. \tag{2.19}$$

The Lie derivative can now be used to write the formulas for the derivatives of $\mathbf{Q}$ and $\mathbf{P}$:

$$\frac{d^k\mathbf{Q}}{dt^k}(t) = -(L_H^k \mathbf{q})(\mathbf{Q}(t),\mathbf{P}(t)), \qquad \frac{d^k Q}{dt^k}(t) = -(L_H^k \mathbf{p})(\mathbf{Q}(t),\mathbf{P}(t)), \tag{2.20}$$

3.2.5 Lie series

It is now possible to motivate the notion of a ***Lie series*** by computing the power series expansion of $\mathbf{Q}$ and $\mathbf{P}$ about $t = 0$:

$$\mathbf{Q}(t) = \sum_{k=0}^{\infty}\frac{(-1)^k}{k!}\frac{d^k\mathbf{Q}}{dt^k}(0)t^k = \sum_{k=0}^{\infty}\frac{(-1)^k}{k!}L_H^k \mathbf{q}\, t^k, \tag{2.21}$$

and

$$\mathbf{P}(t) = \sum_{k=0}^{\infty} \frac{(-1)^k}{k!} \frac{d^k\mathbf{P}}{dt^k}(0)t^k = \sum_{k=0}^{\infty} \frac{(-1)^k}{k!} L_H^k \mathbf{p}\, t^k. \tag{2.22}$$

The power series are nothing but exponential series where the usual real or complex argument has been replaced by a Lie derivative operator:

$$\mathbf{Q}(t) = e^{-L_H}\mathbf{q}, \qquad \mathbf{P}(t) = e^{-L_H}\mathbf{p}. \tag{2.23}$$

The correspondence between $(\mathbf{q}, \mathbf{p})$ and $(\mathbf{Q}, \mathbf{P})$ given by the above formula should be thought of as a transformation, and will be referred to as a *Lie transformation*. Thus when a Lie transformation is expanded as a power series in t, a Lie series is obtained.

The use of exponential notation is not unique to the study of Lie transformations. In the study of systems of linear constant-coefficient ordinary differential equations, the solution of a system can be written as the exponential of the coefficient matrix. Many advanced ordinary differential equations books discuss the notation and several ways of using it to solve and study such problems. Exponential notation is also used in quantum mechanics, there the solution of Schrödinger's equation is written as the exponential of the quantum mechanical Hamiltonian. This is discussed in most quantum mechanics books. This notation, along with Lie algebraic theory, can be used to solve many elementary systems in closed form (see Wolf [35]). The exponential notation used in quantum mechanics is studied by mathematicians under the name "one-parameter semi-groups". This theory is discussed in many books on linear functional analysis. An analog of semi-group theory that is appropriate for applications to Lie transformations has been developed (see Steinberg [26]).

Unfortunately, there are now several popular notations for Lie derivatives. We prefer to write the Lie derivative generated by h operating on f by

$$[h, \circ]f = L_h f = [h, f]. \tag{2.24}$$

This notation is nice because the commutator of two Lie derivatives corresponds to the Poisson bracket of the generating functions (as will be shown in the next section). This, in turn, makes some identities more compact to write. Dragt [87] uses the notation

$$:h\!: f = L_h f = [h, f], \tag{2.25}$$

while Wolf [94] uses

$$\hat{h}(f) = L_h f = [h, f]. \tag{2.26}$$

There are no doubt other notations.[2]

3.3 Lie transformations

One of the advantages of Lie transformations is that they have some nice algebraic properties which help with computations. The properties of Lie transformations are based on the properties of the Lie derivative, which are in turn based on the properties of the Poisson bracket. In the following, let a and b be scalars and let f, g, and h be functions of $(\mathbf{q}, \mathbf{p})$. Recall that the Poisson bracket $[g, h]$ of two functions g and h is defined by

$$[g, h] = \sum_{j=1}^{n} \frac{\partial g}{\partial q_j}\frac{\partial h}{\partial p_j} - \frac{\partial g}{\partial p_j}\frac{\partial h}{\partial q_j}. \tag{3.1}$$

[2] The editors were strongly tempted to use the marvels of the TeX system, to turn all of Prof. Steinberg's $[f, \circ]$'s into Dragt's $: f :$'s, this being the convention in this volume, at the stroke of a button on the computer dashboard. (Editor's note.)

3.3.1 Properties of the Poisson bracket

The Poisson bracket has four important properties.

Skew-symmetry:

$$[g,h] = -[h,g]. \tag{3.2}$$

This property is clear from the definition. Note that it implies the identity $[f,f] = 0$.

Linearity:

$$[f, ag + bh] = a[f,g] + b[f,h], \tag{3.3}$$

where a and b are numbers. This property follows directly from the linearity of differentiation.

Product rule:

$$[f, gh] = [f,g]h + [f,h]g. \tag{3.4}$$

This follows directly from the product rule for differentiation.

Jacobi identity:

$$[f,[g,h]] + [g,[h,f]] + [h,[f,g]] = 0. \tag{3.5}$$

This is a straight-forward calculation. However, in the case of one degree of freedom the expression contains 24 terms, so the calculation is not short. Such things can be easily checked using a computer symbol manipulator.

3.3.2 Properties of Lie derivatives

These properties for the Poisson bracket imply a useful set of properties for the Lie derivative. Recall that the definition of the Lie derivative $[f,\circ]$ generated by f is

$$[f,\circ]g = [f,g]. \tag{3.6}$$

Skew-symmetry:

$$[f,\circ]g = -[g,\circ]f. \tag{3.7}$$

This is the same as the skew-symmetry property for Poisson brackets.

Linearity of the Lie derivative:

$$[f,\circ](ag + bh) = a[f,\circ]g + b[f,\circ]h. \tag{3.8}$$

This follows immediately from the linearity of the Poisson bracket.

Derivation property of the Lie derivative:

$$[f,\circ](gh) = ([f,\circ]g)h + ([f,\circ]h)g. \tag{3.9}$$

A proof is presented to illustrate the use of the Lie derivative notation and to clarify the meaning of the formula.

Proof. This follows from the product rule for the Poisson bracket:

$$\begin{aligned} [f,\circ](gh) &= [f,gh] = [f,g]h + [f,h]g \\ &= ([f,\circ]g)h + ([f,\circ]h)g. \end{aligned}$$

∎

Bracket property:

$$[f,\circ][g,h] = [[f,\circ]g,h] + [g,[f,\circ]h]. \tag{3.10}$$

Again, a proof is given to illustrate the use of the Lie derivative notation.

Proof. This follows from the Jacobi identity for the Poisson bracket:

$$\begin{aligned}[f,\circ][g,h] &= [f,[g,h]]\\ &= -[g,[h,f]] - [h,[f,g]]\\ &= +[g,[f,h]] + [[f,g],h]\\ &= +[g,[f,\circ]h] + [[f,\circ]g,h].\end{aligned}$$

∎

3.3.3 Lie transformations

In the previous section, Lie series and Lie transformations acted on $(\mathbf{q},\mathbf{p})$. Lie transformations also act on functions of these variables. The ***Lie transformation*** generated by f is given by the Lie series

$$e^{t[f,\circ]} = \sum_{k=0}^{\infty} \frac{t^k}{k!}[f,\circ]^k. \tag{3.11}$$

The action of the Lie transformation on a function g is given by

$$e^{t[f,\circ]}g = \sum_{k=0}^{\infty} \frac{t^k}{k!}[f,\circ]^k g. \tag{3.12}$$

Lie transformations also act on vector valued functions; if $\mathbf{g} = (g_1, \ldots, g_k)$ with $k \geq 1$, then

$$e^{t[f,\circ]}\mathbf{g} = \left(e^{t[f,\circ]}g_1, \ldots, e^{t[f,\circ]}g_k\right). \tag{3.13}$$

Theorem. *Lie transformations are well defined.*

This means that if f and g are real analytic functions then $G = G(t,\mathbf{q},\mathbf{p})$, where

$$G = e^{t[f,\circ]}g \tag{3.14}$$

is a real analytic function of $(\mathbf{q},\mathbf{p})$ for t sufficiently small. In fact, G is also a real analytic function of t. Note that the radius of convergence of the power series expansion of G in $(\mathbf{q},\mathbf{p})$ will depend on t. The proof of this fact is a special case of the proof of the Cauchy–Kovalewski theorem on the existence of power series solutions to the initial value problem for a first-order partial differential equation. A discussion of this theorem can be found in many intermediate partial differential equations texts. A more modern proof of this result is given in Theorem 3.1 of Steinberg [26]. Gröbner [16, 17] gives a detailed proof based on direct power series estimates.

3.3.4 Example: the harmonic oscillator

Let us return to the harmonic oscillator example: since

$$H = \tfrac{1}{2}(q^2 + p^2), \tag{3.15}$$

then

$$\begin{aligned}[H,q] &= \frac{\partial H}{\partial q}\frac{\partial q}{\partial p} - \frac{\partial H}{\partial p}\frac{\partial q}{\partial q} = -\frac{\partial H}{\partial p} = -p, \\ [H,p] &= \frac{\partial H}{\partial q}\frac{\partial p}{\partial p} - \frac{\partial H}{\partial p}\frac{\partial p}{\partial q} = \frac{\partial H}{\partial q} = q.\end{aligned} \tag{3.16}$$

Thus,

$$[H,\circ]^2 q = [H,\circ][H,\circ]q = [H,\circ][H,q] = -[H,\circ]p = -[H,p] = -q. \tag{3.17}$$

Repeating this gives

$$\begin{aligned} &[H,\circ]^0 q = +q, && [H,\circ]^0 p = +p, \\ &[H,\circ]^1 q = -p, && [H,\circ]^1 p = +q, \\ &[H,\circ]^2 q = -q, && [H,\circ]^2 p = -p, \\ &[H,\circ]^3 q = +p, && [H,\circ]^3 p = -q, \\ &[H,\circ]^4 q = +q, && [H,\circ]^4 p = +p. \end{aligned}$$

The Lie series definition of the Lie transformation then gives

$$\begin{aligned} e^{t[H,\circ]} q &= q - tp - \tfrac{1}{2}t^2 q + \tfrac{1}{3!}t^3 p + \tfrac{1}{4!}t^4 q - \cdots \\ &= q(1 - \tfrac{1}{2}t^2 + \tfrac{1}{4!}t^4 - \cdots) + p(-t + \tfrac{1}{3!}t^3 - \tfrac{1}{5!}t^5 + \cdots). \end{aligned} \tag{3.18}$$

This gives

$$e^{t[H,\circ]} q = q\cos t - p\sin t; \tag{3.19}$$

replacing t by $-t$ gives the formula derived in the previous section. A similar calculation will produce the value for the Lie transformation operating on p.

3.3.5 Properties of Lie transformations

Lie transformations have four basic properties; other properties will be discussed later. Let f, g, and h be arbitrary Hamiltonians while a and b are real constants.

Linearity:

$$e^{t[f,\circ]}(ag + bh) = ae^{t[f,\circ]}g + be^{t[f,\circ]}h. \tag{3.20}$$

Product preservation:

$$e^{t[f,\circ]}(gh) = (e^{t[f,\circ]}g)(e^{t[g,\circ]}h). \tag{3.21}$$

Poisson bracket preservation:

$$e^{t[f,\circ]}[g,h] = [e^{t[f,\circ]}g, e^{t[f,\circ]}h]. \tag{3.22}$$

Composition:

$$e^{t[f,\circ]}g(\mathbf{q},\mathbf{p}) = g(e^{t[f,\circ]}\mathbf{q}, e^{t[f,\circ]}\mathbf{p}). \tag{3.23}$$

Proofs. The proofs of these properties illustrate how to compute with Lie transformations. The proof of linearity of Lie transformations depends on the linearity of Lie derivatives:

$$[f,\circ](ag + bh) = a[f,\circ]g + b[f,\circ]h.$$

This implies that the square of the Lie derivative is linear:

$$[f,\circ]^2(ag+bh) = [f,\circ]\big(a[f,\circ]g + b[f,\circ]h\big) = \big(a[f,\circ]^2 g + b[f,\circ]^2 h\big).$$

A simple induction gives

$$[f,\circ]^k(ag+bh) = \big(a[f,\circ]^k g + b[f,\circ]^k h\big).$$

The Lie series definition of the Lie transformation then implies the result:

$$\begin{aligned} e^{t[h,\circ]}(af+bg) &= \sum_{k=0}^{\infty} \frac{t^k}{k!}[h,\circ]^k(af+bg) \\ &= a\sum_{k=0}^{\infty} \frac{t^k}{k!}[h,\circ]^k f + b\sum_{k=0}^{\infty} \frac{t^k}{k!}[h,\circ]^k g \\ &= a e^{t[h,\circ]} f + b e^{t[h,\circ]} g. \end{aligned}$$

The proof of the product preservation property is based on the derivation property of the Lie transformation:

$$[f,\circ](gh) = ([f,\circ]g)h + ([f,\circ]h)g.$$

This identity has the same form as the product rule for differentiation, so higher powers of the Lie derivative should satisfy an identity analogous to the Leibnitz rule for derivatives:

$$[f,\circ]^k(gh) = \sum_{i=0}^{k} \binom{k}{i} ([f,\circ]^i g)([f,\circ]^k - ih),$$

where the binomial coefficient is defined in the usual way,

$$\binom{k}{i} = \frac{k!}{i!\,(k-i)!}.$$

This is easily proved using induction. If this is applied to the Lie series definition of a Lie transformation, then

$$\begin{aligned} e^{t[f,\circ]}(gh) &= \sum_{k=0}^{\infty} \frac{t^k}{k!}[f,\circ]^k(gh) \\ &= \sum_{k=0}^{\infty} \frac{t^k}{k!} \sum_{i=0}^{k} \binom{k}{i} ([f,\circ]^i g)([f,\circ]^{k-i} h) \\ &= \sum_{k=0}^{\infty} \sum_{i=0}^{k} \frac{t^{k-i}}{(k-i)!} \frac{t^i}{i!} ([f,\circ]^i g)([f,\circ]^{k-i} h) \\ &= \sum_{j=0}^{\infty} \sum_{i=0}^{\infty} \frac{t^i}{i!} ([f,\circ]^i g) \frac{t^j}{j!} ([f,\circ]^j h) \\ &= (e^{t[f,\circ]} g)(e^{t[f,\circ]} h). \end{aligned}$$

The Poisson bracket preservation property has the same proof as the product preservation property, with gh by $[g,h]$. The composition property is one of the most useful properties of Lie transformations. It is proved by first expanding f in a power series and then using the linearity of the Lie transformation to pass the exponential to each term of the series. Next the product preserving property of Lie transformations is used to pass the exponential onto each variable. Now the series for $\exp[f,\circ]g$ has the same form as the series for g, except that every q_i and p_i has been replaced by $\exp[f,\circ]q_i$ and $\exp[f,\circ]p_i$. Resumming the series gives the result.

3.4 Exponential identities

One of the reasons that Lie transformations are a powerful tool for computation is that they encourage the use of exponential identities; some of these important formulas are derived here. The identities derived in this section do not find as many applications as those that will be derived in the next section. However, the techniques introduced in this section are fundamental to the derivation of all exponential identities.

3.4.1 Commutators

The Lie derivative notation is a bit cumbersome here, so all Lie derivatives are replaced by capital roman characters. Here A, B, C, and so forth are noncommuting variables; they stand for objects that may not commute under multiplication. Thus it is important to remember that

$$AB = BA \quad \text{is **not** true for all } A \text{ and } B. \tag{4.1}$$

However, some of the noncommuting variables will commute under multiplication, so any identities that are derived must reduce to standard identities when the variables commute. The arithmetic for Lie derivatives is the same as for matrices, so another advantage of this notation is that the identities derived are also valid for matrices. It is relatively easy to check some of the identities, or to find counter-examples to incorrect identities, by using 2×2 matrices. Lower case roman variables, a, b, c and so forth will stand for scalars; thus $ab = ba$.

Here is a simple but important identity:

$$(A+B)^2 = A^2 + AB + BA + B^2. \tag{4.2}$$

Another way of writing this is

$$(A+B)^2 = (A^2 + 2AB + B^2) + BA - AB. \tag{4.3}$$

If we define the ***commutator*** of A and B by $[A, B] = AB - BA$, then

$$(A+B)^2 = (A^2 + 2AB + B^2) - [A, B]. \tag{4.4}$$

Formula (4.4) records the identity as a standard identity for commuting variables plus a deviation that involves only ***commutators*** of the variables. Thus it is trivial to check that the identity is true for commuting variables. We will always put identities in such a form.

3.4.2 Properties of the commutator

The commutator satisfies a set of identities similar to the Poisson bracket.

Skew-symmetry:

$$[A, B] = -[B, A]. \tag{4.5}$$

Linearity:

$$[A, bB + cC] = b[A, B] + c[A, C]. \tag{4.6}$$

Product rule:

$$[A, BC] = [A, B]C + B[A, C]. \tag{4.7}$$

Jacobi identity:

$$[A, [B, C]] + [B, [C, A]] + [C, [A, B]] = 0. \tag{4.8}$$

The proofs of these identities are simple computations. These identities have an irritating consequence: expressions involving commutators do not have a unique form. In fact, as the number of commutators in a term increases, so do the possible representations of the expression.

3.4.3 Exponentials and their identities

The exponential of a non-commuting variable is defined by the power series

$$e^A = \sum_{k=0}^{\infty} \frac{A^k}{k!}. \tag{4.9}$$

We make no assumptions on the convergence of the series; consequently all computations in this section are formal. Here formal means that two series are equal if they are equal term by term. Calculations that involve such series are tedious, so we have implemented a simple MACSYMA [135] program (described in an appendix to this section) to do some of these calculations. One important identity which is true for noncommuting variables is

$$e^A e^{-A} = I = e^{-A} e^A. \tag{4.10}$$

This identity is obvious; it only involves one noncommuting variable, so the fact that the variable does not commute with other noncommuting variables plays no role here. Thus the identity can be checked using the power series expansion for commuting variables. This identity shows that the multiplicative inverse B^{-1} of $B = e^A$ is $B^{-1} = e^{-A}$.

Is it true that

$$e^A e^B = e^B e^A? \tag{4.11}$$

This can be checked using the power series expansion:

$$\begin{aligned} e^A e^B &= (I + A + \tfrac{1}{2}A^2 + \cdots)(I + B + \tfrac{1}{2}B^2 + \cdots) \\ &= I + (A + B) + (\tfrac{1}{2}A^2 + AB + \tfrac{1}{2}B^2) + \cdots. \end{aligned} \tag{4.12}$$

Interchanging A and B gives

$$e^B e^A = I + (B + A) + (\tfrac{1}{2}B^2 + BA + \tfrac{1}{2}A^2) + \cdots, \tag{4.13}$$

and then

$$e^A e^B - e^B e^A = AB - BA + \cdots = [A, B] + \cdots. \tag{4.14}$$

Thus, because A and B do not commute, the exponentials do not commute.

Another way to see the same thing and also produce a useful formula is to multiply (4.11) by $\exp(-A)\exp(-B)$, and ask whether it is true that

$$e^{-A} e^{-B} e^A e^B = I? \tag{4.15}$$

This can be checked by expanding the left hand side of the equation:

$$\begin{aligned} e^{-A} e^{-B} e^A e^B &= (I - A + \tfrac{1}{2}A^2 - \cdots)(I - B + \tfrac{1}{2}B^2 - \cdots) \\ &\qquad \times (I + A + \tfrac{1}{2}A^2 + \cdots)(I + B + \tfrac{1}{2}B^2 + \cdots) \\ &= I + AB - BA + \cdots \\ &= I + [A, B] + \cdots. \end{aligned} \tag{4.16}$$

Thus, through quadratic expressions in A and B,

$$e^{-A}e^{-B}e^{A}e^{B} = e^{[A,B]}, \tag{4.17}$$

or

$$e^{A}e^{B} = e^{B}e^{A}e^{[A,B]}. \tag{4.18}$$

If more terms in the series are computed there are more correction. The higher order calculations are very lengthy.

Another important exponential identity is

$$e^{A+B} = e^{A}e^{B}. \tag{4.19}$$

As above, the power series of each side of the above expression do not agree, so this identity cannot be true for noncommuting variables. Another way to see this, and also obtain part of an important formula, is to do a power series computation:

$$e^{-B}e^{-A}e^{A+B} = I - \tfrac{1}{2}[A,B] + \cdots. \tag{4.20}$$

Thus through quadratic expressions in A and B,

$$e^{-B}e^{-A}e^{A+B} = e^{-[A,B]/2}, \tag{4.21}$$

or

$$e^{A+B} = e^{A}e^{B}e^{-[A,B]/2}. \tag{4.22}$$

Again, computing the higher order corrections to this formula is tedious. The use of power series to compute such exponential identities is rather inefficient, so we now turn to another method.

3.4.4 Differentiation of exponential series

A powerful technique for deriving exponential identities involves differentiating noncommuting exponentials. Our first result shows that if the exponent depends in a simple way on t, then the t derivative of the exponential is natural. When the exponent depends in an arbitrary way on t, then the t derivative is more complicated. Let us begin with a simple case.

Proposition. *If $a(t)$ is a scalar function of t and $a'(t) = da(t)/dt$, then*

$$\frac{d}{dt}e^{a(t)A} = a'(t)Ae^{a(t)A}. \tag{4.23}$$

Proof. We have not required the exponential series to be convergent, so this is simply a formal statement; the series must be equal term by term. Of course, if the series converge, then the equality holds for the sum of the series. The proof is a computation:

$$\begin{aligned}\frac{d}{dt}e^{a(t)A} &= \frac{d}{dt}\sum_{k=0}^{\infty}\frac{a^k(t)A^k}{k!} = \sum_{k=0}^{\infty}\frac{d}{dt}\frac{a^k(t)A^k}{k!}\\ &= \sum_{k=1}^{\infty}\frac{a'(t)a^{k-1}(t)A^k}{k-1!} = a'(t)\sum_{j=0}^{\infty}\frac{a^j(t)A^{j+1}}{j!} = a'(t)Ae^{a(t)A}.\end{aligned}$$

∎

Now let us turn to the more interesting case where the exponent $A(t)$ is an arbitrary function of t. First it is important to know that the usual formula for differentiating exponentials is not correct. Let

$$A'(t) = \frac{d}{dt}A(t). \tag{4.24}$$

Then

$$\frac{d}{dt}e^{A(t)} \neq A'(t)e^{A(t)}. \tag{4.25}$$

This can be easily be shown using 2×2 matrices.

We now describe the correct result. We first saw the following argument in a paper by Dragt [81, App. B].

Theorem. *If $A(t)$ is as described above, then*

$$\frac{d}{dt}e^{A(t)} = \int_0^1 e^{\tau A(t)} A'(t) e^{-\tau A(t)} d\tau \, e^{A(t)} \tag{4.26a}$$

$$= e^{A(t)} \int_0^1 e^{-\tau A(t)} A'(t) e^{\tau A(t)} d\tau. \tag{4.26b}$$

Proof. The exponential is defined by a power series:

$$e^{A(t)} = \sum_{n=0}^{\infty} \frac{A^n(t)}{n!}.$$

Thus

$$\frac{d}{dt}e^{A(t)} = \sum_{n=1}^{\infty} \frac{1}{n!} \sum_{m=0}^{n-1} A^m(t)A'(t)A^{n-m-1}(t).$$

Set $A = A(t)$, interchange the order of summation, and do a little algebra, to obtain

$$\begin{aligned}
\frac{d}{dt}e^{A} &= \sum_{m=0}^{\infty} \sum_{n=m+1}^{\infty} \frac{1}{n!} A^m A' A^{n-m-1} \\
&= \sum_{m=0}^{\infty} \sum_{k=0}^{\infty} \frac{1}{(m+k+1)!} A^m A' A^k \\
&= \sum_{m=0}^{\infty} \sum_{k=0}^{\infty} \frac{m!\,k!}{(m+k+1)!} \frac{A^m}{m!} A' \frac{A^k}{k!}.
\end{aligned}$$

However,

$$\frac{m!\,k!}{(m+k+1)!} = \int_0^1 \tau^m (1-\tau)^k d\tau.$$

If this is inserted into the previous expression and the series is summed, then

$$\frac{d}{dt}e^{A(t)} = \int_0^1 \left(\sum_{m=0}^{\infty} \frac{\tau^m A^m}{m!} \right) A' \left(\sum_{k=0}^{\infty} \frac{(1-\tau)^k A^k}{k!} \right)$$

yields the form (4.26a). Changing the variable of integration gives (4.26b). ∎

3.4.5 Adjoint operator action

Another important tool that is useful in its own right, and essential for deriving the exponential identities that are needed, involves something called the ***adjoint action***. This is exactly analogous to the notion of a Lie derivative in hamiltonian mechanics. If A is a given noncommuting variable, the adjoint action, Ad_A, generated by A is given by

$$Ad_A B = [A, B]. \tag{4.27}$$

We will use a notation similar to the notation used for Lie derivatives:

$$[A, \circ]B = Ad_A B = [A, B]. \tag{4.28}$$

The adjoint operator satisfies the same identities as the Lie derivative.

Skew-symmetry:

$$[A, \circ]B = -[B, \circ]A. \tag{4.29}$$

Linearity of the adjoint operator:

$$[A, \circ](bB + cC) = b[A, \circ](B) + c[A, \circ](C). \tag{4.30}$$

Derivation property:

$$[A, \circ](BC) = ([A, \circ]B)C + B[A, \circ]C. \tag{4.31}$$

Bracket property:

$$[A, \circ][B, C] = [[A, \circ]B, C] + [B, [A, \circ]C]. \tag{4.32}$$

The proofs of these identities follow immediately from the definition of the commutator.

The adjoint operator can be exponentiated just as the Lie derivative can be exponentiated:

$$e^{[A,\circ]} = \sum_{i=0}^{\infty} \frac{1}{k!}[A, \circ]^k. \tag{4.33}$$

The case where the adjoint operator is multiplied by the scalar t will be used frequently:

$$e^{t[A,\circ]} = \sum_{i=0}^{\infty} \frac{t^k}{k!}[A, \circ]^k. \tag{4.34}$$

The exponential of the adjoint operator is another operator:

$$\begin{aligned} e^{t[A,\circ]}B &= B + t[A, \circ]B + \tfrac{1}{2}t^2[A, \circ]^2 B + \tfrac{1}{3!}t^3[A, \circ]^3 + \cdots \\ &= B + t[A, B] + \tfrac{1}{2}t^2[A, [A, B]] + \tfrac{1}{3!}t^3[A, [A, [A, B]]] + \cdots. \end{aligned}$$

The next result will be used many times in the following development.

Theorem. *The following equation holds:*

$$e^A B e^{-A} = e^{[A,\circ]}B. \tag{4.35}$$

Proof. Define

$$F(t) = e^{tA}Be^{-tA}.$$

Then $F(0) = B$ and

$$F'(t) = \frac{d}{dt}F(t) = Ae^{tA}Be^{-tA} - e^{tA}Be^{-tA}A = [A, F(t)] = [A, \circ]F(t).$$

Next, define

$$G(t) = e^{t[A,\circ]}B.$$

Then $G(0) = B$ and

$$G'(t) = [A, \circ]G(t).$$

Consequently, $F(t)$ and $G(t)$ satisfy the same initial value problem for a first order ordinary differential equation. Such functions are unique, so $F(t) = G(t)$ and this gives the theorem. ∎

This can also be proved by computing the power series expansions of $F(t)$ and $G(t)$. The differential equations gives a recursion for the coefficients in the power series and then a simple induction gives the result.

3.4.6 Properties of exponentials of adjoint operators

The exponential of an adjoint operator satisfies a set of identities analogous to the identities satisfied by a Lie transform.

Linearity:

$$e^{[A,\circ]}(bB + cC) = be^{[A,\circ]}(B) + ce^{[A,\circ]}(C). \tag{4.36}$$

Product Preservation:

$$e^{[A,\circ]}(BC) = (e^{[A,\circ]}B)(e^{[A,\circ]}C). \tag{4.37}$$

Commutator Preservation:

$$e^{[A,\circ]}[B, C] = [e^{[A,\circ]}B, e^{[A,\circ]}C]. \tag{4.38}$$

Proofs. The linearity is easy. The product preservation follows from the previous theorem:

$$e^{[A,\circ]}(BC) = e^{A}(BC)e^{-A} = e^{A}Be^{-A}e^{A}Ce^{-A} = (e^{[A,\circ]}B)(e^{[A,\circ]}C).$$

The commutator is defined in terms of products, so the commutator preservation follows immediately from the product preservation. ∎

Also, there is a property, analogous to the composition property of Lie series, that will needed near the end of this section. To describe this property it is necessary to know how to apply any function to a noncommuting variable. Let

$$f(z) = \sum_{k=0}^{\infty} a_k z^k. \tag{4.39}$$

Here only convergent series are used, but formal series will do. So far only $f(z) = \exp(z)$ has been considered. Later several functions defined in terms of the exponential will be used. In any case, $f(A)$ is given by the formal series

$$f(A) = \sum_{k=0}^{\infty} a_k A^k. \tag{4.40}$$

The following property is termed similarity rather than composition because it is also analogous to the notion of similarity transformations for matrices.

Similarity:

$$e^{[A,\circ]} f(B) = f(e^{[A,\circ]} B). \tag{4.41}$$

Proof. This is a computation:

$$e^{[A,\circ]} f(B) = e^{[A,\circ]} \sum_{k=0}^{\infty} a_k B^k = \sum_{k=0}^{\infty} a_k e^{[A,\circ]} B^k = \sum_{k=0}^{\infty} a_k (e^{[A,\circ]} B)^k = f(e^{[A,\circ]} B). \qquad \blacksquare$$

The previous two theorems can be combined to give a nice form for the derivative of the exponential of a general t-dependent quantity.

Corollary. *There hold:*

$$\frac{d}{dt} e^{A(t)} = \frac{e^{[A(t),\circ]} - 1}{[A(t),\circ]} A'(t) e^{A(t)}, \tag{4.42a}$$

$$= e^{A(t)} \frac{1 - e^{-[A(t),\circ]}}{[A(t),\circ]} A'(t). \tag{4.42b}$$

Proof. One of the previous formulas for the derivative is

$$\frac{d}{dt} e^{A(t)} = \int_0^1 e^{\tau A(t)} A'(t) e^{-\tau A(t)} d\tau \, e^{A(t)}.$$

The previous theorem allows this to be rewritten as

$$\frac{d}{dt} e^{A(t)} = \int_0^1 e^{\tau [A(t),\circ]} A'(t) d\tau \, e^{A(t)}.$$

The antiderivative with respect to τ is simple:

$$\int e^{\tau [A(t),\circ]} d\tau = \frac{e^{\tau [A(t),\circ]}}{[A(t),\circ]}.$$

Evaluating the antiderivative at the limits of integration gives the first part of the corollary, (4.42a), and a similar calculation gives the second part, (4.42b). ∎

3.4.7 Baker–Campbell–Hausdorff formulas

The results of the next two theorems are two important exponential identities that are frequently called the Baker–Campbell–Hausdorff (BCH) formulas (or identities). The proofs of these identities are base on computing logarithmic derivatives and then using some of the previous results. If $a(t)$ is a scalar function then the logarithmic derivative of $a(t)$ is given by

$$\frac{d}{dt}\ln(a(t)) = \frac{a'(t)}{a(t)}. \tag{4.43}$$

The logarithmic derivative of

$$F(t) = e^{tA}e^{tB}, \tag{4.44}$$

which is frequently used, has two possible forms:

$$F'(t)F^{-1}(t) \qquad \text{and} \qquad F^{-1}(t)F'(t). \tag{4.45}$$

It is important to note whether the inverse is multiplied on the left or right. The derivative is given by the product rule;

$$F'(t) = Ae^{tA}e^{tB} + e^{tA}Be^{tB} = Ae^{tA}e^{tB} + e^{tA}Be^{-tA}e^{tA}e^{tB}. \tag{4.46}$$

The inverse F^{-1} of F is given by

$$F^{-1}(t) = e^{-tB}e^{-tA}, \tag{4.47}$$

and consequently

$$F'(t)F^{-1}(t) = A + e^{tA}Be^{-tA} = A + e^{t[A,\circ]}B. \tag{4.48}$$

Theorem. *The following holds:*

$$e^{A+B} = e^{A}e^{B}e^{C_2}e^{C_3}e^{C_4}\cdots, \tag{4.49}$$

and

$$e^{A+B} = \cdots e^{-C_4}e^{C_3}e^{-C_2}e^{B}e^{A}, \tag{4.50}$$

where C_k is a linear combination of k-fold commutators of A and B. In particular,

$$C_2 = -\tfrac{1}{2}[A,B], \qquad C_3 = \tfrac{1}{6}[A,[A,B]] + \tfrac{1}{3}[B,[A,B]]. \tag{4.51}$$

Proof. We first prove that the second identity follows from the first. Note that

$$C_k = C_k(A,B),$$

and because C_k is a k-fold commutator,

$$C_k(-A,-B) = (-1)^k C_k(A,B).$$

Replacing A and B by $-A$ and $-B$ in the first identity gives

$$e^{-A-B} = e^{-A}e^{-B}e^{C_2}e^{-C_3}e^{C_4}\cdots.$$

Taking the inverse of both sides of this identity gives the second formula in the theorem.

Let now

$$F(t) = e^{-tA}e^{-tB}e^{t(A+B)}.$$

Then

$$F'(t)F^{-1}(t) = -B - e^{-tB}Ae^{tB} + e^{-tB}e^{-tA}(A+B)e^{tA}e^{tB},$$

or

$$F'(t)F^{-1}(t) = -B - e^{-t[B,\circ]}A + e^{-t[B,\circ]}e^{-t[A,\circ]}(A+B).$$

Next define

$$G(t) = e^{t^2C_2}e^{t^3C_3}e^{t^4C_4}\cdots.$$

Then

$$G'(t)G^{-1}(t) = 2tC_2 + 3t^2e^{t^2[C_2,\circ]}C_3 + 4t^3e^{t^2[C_2,\circ]}e^{t^3[C_3,\circ]}C_4 + \cdots.$$

Expanding the exponentials of the adjoint operators yields:

$$\begin{aligned} F'(t)F^{-1}(t) &= -B - A + t[B,A] - \tfrac{1}{2}t^2[B,[B,A]] + \cdots \\ &\quad + (A+B) - t[A,B] + \tfrac{1}{2}t^2[A,[A,B]] + \cdots \\ &\quad - t[B,A] + t^2[B,[A,B]] + \cdots + \tfrac{1}{2}t^2[B,[B,A]] + \cdots \\ &= -t[A,B] + \tfrac{1}{2}t^2([A,[A,B]] + 2[B,[A,B]]) + \cdots \\ G'(t)G^{-1}(t) &= 2tC_2 + 3t^2C_3 + \cdots. \end{aligned}$$

Collecting coefficients of powers of t gives the formulas for C_2 and C_3. Any number of the C_k may be computed by carrying the calculations though the k-th power of t. An inspection of the computation shows that the coefficients of k-th power of t must involve only k-fold commutators of A and B. ∎

This theorem shows how to break up an exponential of a sum into a product of exponentials. The next theorem is a converse; it shows how to combine a product of exponentials into an exponential of a sum.

Theorem. *The following equation holds:*

$$e^Ae^B = e^{A+B+D_2+D_3+\cdots}, \tag{4.52}$$

where D_k is a linear combination of k-fold commutators of A and B. In particular,

$$D_2 = \tfrac{1}{2}[A,B], \qquad D_3 = \tfrac{1}{12}[A,[A,B]] - \tfrac{1}{12}[B,[A,B]]. \tag{4.53}$$

Proof. Set

$$\begin{aligned} F(t) &= e^{tA}e^{tB}, \\ D(t) &= t(A+B) + t^2D_2 + t^3D_3 + \cdots, \\ G(t) &= e^{D(t)}. \end{aligned}$$

Then

$$\begin{aligned} F'(t)F^{-1}(t) &= A + e^{t[A,\circ]}B \\ &= A + B + t[A,B] + \tfrac{1}{2}t^2[A,[A,B]] + \cdots, \end{aligned}$$

and

$$\begin{aligned} G'(t)G^{-1}(t) &= \int_0^1 e^{\tau[D,\circ]}D'd\tau \\ &= \int_0^1 (D' + \tau[D,D'] + \cdots)d\tau \\ &= D' + \tfrac{1}{2}[D,D'] + \cdots. \end{aligned}$$

Now

$$D' = (A+B) + 2tD_2 + 3t^2D_3 + \cdots,$$

and

$$[D,D'] = t^2[A+B,D_2] + \cdots.$$

Thus

$$G'(t)G^{-1}(t) = (A+B) + 2tD_2 + t^2(3D_3 + \tfrac{1}{2}[A+B,D_2]) + \cdots.$$

Comparing coefficients of powers of t gives the values gives the formulas for D_2 and D_3. ∎

The sum in the exponent in the previous theorem can be rearranged, and then some of the terms can be summed in closed form. We first saw this result in Dragt and Finn [81] and Dragt [82,84], where it is applied to mirror machine problems.

Theorem. *The following equation holds:*

$$e^Ae^B = e^{C(A,B)}, \tag{4.54}$$

where

$$C(A,tB) = \sum_{k=0}^{\infty} t^k C_k(A,B). \tag{4.55}$$

Note that C_k contains all terms that contain exactly k factors equal to B. In particular,

$$C_0(A,B) = A, \tag{4.56}$$

$$C_1(A,B) = \frac{[A,\circ]}{1-e^{-[A,\circ]}}B = B + \frac{1}{2}[A,B] + \frac{1}{12}[A,[A,B]] + \cdots, \tag{4.57}$$

$$C_2(A,B) = -\frac{1}{2}\frac{[A,\circ]}{1-e^{-[A,\circ]}}\int_0^1 [S_1(\tau),S_2(\tau)]d\tau = -\frac{1}{12}[B,[A,B]] + \cdots. \tag{4.58}$$

where

$$S_1(\tau) = \frac{1-e^{-\tau[A,\circ]}}{[A,\circ]}C_1, \qquad S_2(\tau) = e^{-\tau[A,\circ]}C_1. \tag{4.59}$$

Proof. It is clear that the sum can be rearranged into the form given. Setting $t = 0$ gives $C_0(A,B) = A$. Next, set

$$F(t) = e^Ae^{tB}, G(t) = e^{C(A,tB)}.$$

Then

$$F^{-1}(t)F'(t) = B,$$

and

$$G^{-1}(t)G'(t) = \int_0^1 e^{-\tau C} C' e^{\tau C} d\tau = \frac{1 - e^{-[C,\circ]}}{[C,\circ]} C'.$$

Combining these two equations gives:

$$B = \int_0^1 e^{-\tau C} C' e^{\tau C} d\tau = \frac{1 - e^{-[C,\circ]}}{[C,\circ]} C'.$$

Setting $t = 0$ gives:

$$B = \frac{1 - e^{-[A,\circ]}}{[A,\circ]} C_1(A, B),$$

or

$$C_1(A, B) = \frac{[A,\circ]}{1 - e^{-[A,\circ]}} B.$$

Recall that if

$$f(z) = \frac{z}{1 - e^{-z}} = 1 + \frac{z}{2} + \frac{z^2}{12} - \frac{z^4}{720} + \cdots,$$

then

$$f([A,\circ]) = \frac{[A,\circ]}{1 - e^{-[A,\circ]}} = 1 + \frac{[A,\circ]}{2} + \frac{[A,\circ]^2}{12} - \frac{[A,\circ]^4}{720} + \cdots.$$

This formula gives the expanded form for C_1.

The computation of C_2 begins by differentiating $B = \int_0^1 e^{-\tau C} C' e^{\tau C} d\tau$;

$$0 = \int_0^1 \left(e^{-\tau C} C'' e^{\tau C} + \frac{e^{-\tau[C,\circ]} - 1}{-\tau[C,\circ]} (-\tau C') e^{-\tau C} C' e^{\tau C} + e^{-\tau C} C' e^{\tau C} \frac{1 - e^{-\tau[C,\circ]}}{\tau[C,\circ]} (\tau C') \right) d\tau.$$

Set

$$R_1(\tau) = \frac{e^{-\tau[C,\circ]} - 1}{[C,\circ]} C',$$
$$R_2(\tau) = -e^{-\tau C} C' e^{\tau C} = e^{-\tau[C,\circ]} C'.$$

Then

$$0 = \int_0^1 \left(e^{-\tau[C,\circ]} C'' + R_1(\tau)R_2(\tau) - R_2(\tau)R_1(\tau)\right) d\tau.$$

Integrating the left-most term gives

$$\frac{1 - e^{-[C,\circ]}}{[C,\circ]} C'' = -\int_0^1 [R_1(\tau), R_2(\tau)] d\tau$$

or

$$C'' = -\frac{[C,\circ]}{1 - e^{-[C,\circ]}} \int_0^1 [R_1(\tau), R_2(\tau)] d\tau.$$

Evaluating the previous formulas at $t = 0$ gives $C'' = 2C_2$, and

$$C_2 = -\frac{1}{2}\frac{[A,\circ]}{1-e^{-[A,\circ]}}\int_0^1 [S_1(\tau), S_2(\tau)]d\tau,$$

where

$$S_1(\tau) = \frac{e^{-\tau[A,\circ]}-1}{[A,\circ]}C_1, \qquad S_2(\tau) = e^{-\tau[A,\circ]}C_1.$$

These quantities are now computed through third order commutators:

$$\begin{aligned}
C_1 &= B + \tfrac{1}{2}[A,B] + \tfrac{1}{12}[A,[A,B]] + \cdots,\\
S_1 &= -\tau B + \left(\frac{\tau^2}{2} - \frac{\tau}{2}\right)[A,B] + \left(-\frac{\tau^3}{6} + \frac{\tau^2}{4} - \frac{\tau}{12}\right)[A,[A,B]] + \cdots,\\
S_2 &= B + \left(-\tau + \frac{1}{2}\right)[A,B] + \left(\frac{\tau^2}{2} - \frac{\tau}{2} + \frac{1}{12}\right)[A,[A,B]] + \cdots,\\
[S_1, S_2] &= \tfrac{1}{6}[B,[A,B]] + \cdots,\\
\int_0^1 [S_1, S_2]d\tau &= -\tfrac{1}{12}[B,[A,B]] + \cdots.
\end{aligned}$$

We do not see any method of writing the next term in a simple form. ∎

3.4.8 The MACSYMA programs

MACSYMA "knows about" both commuting and noncommuting variables. The product for commuting variables is "*" while the product for noncommuting variables is ".". To use the following program, one must be careful to multiply exponential series using the noncommuting product ".". The commuting variables should be declared "scalar" while the noncommuting variables should be declared "nonscalar". The program is a straight-forward implementation of the formulas in this section.

Before presenting the programs a comment is in order about why more was not done with MACSYMA. We found it impossible to implement a commutator operator, $[A, B]$, which is nonassociative, skew-symmetric and bilinear. Consequently, the terms in the BCH formulas could not be genereated; only the hand computations could be checked. Moreover, the MACSYMA noncommuting product and the Taylor series package are not completely compatible, so expressions could not be computed efficiently.

Program Listings

```
/* Set some flags that control the ''.'' product. */
    dotscrules : true;
    declare([r, s, t], scalar);
    declare([a, b, c], nonscalar);


/* The variable ''degree'' will be used
       to control the truncation of series. */
    degree : 2;


/* Define the commutator of two noncommuting variables. */
    comm(a, b) := a . b - b . a ;
```

```
/* Define the exponential series. */
    ncexp(a) := block( [term, sum],
        term : 1,
        sum : 1,
        for i thru degree do (
            term : term . a / i,
            sum : sum + term,
        end+i+do ),
        sum );


/* Define the adjoint action of a on b. */
    ad(a, b) := block( [term, sum],
        term : b,
        sum : term,
        for i thru degree do (
            term : comm(a, term)/i,
            sum : sum + term,
        end+i+do ),
        sum );


/* Here are some functions that can be used to
        check some of the calculations done in the section. */


/* Check the commutation of exponentials. */
    check1(t) :=
        taylor(expand(
            ncexp( - t * a) .
            ncexp( - t * b) .
            ncexp( + t * a) .
            ncexp( + t * b)
                ), t, 0, degree);


/* Check the exponential of a sum. */
    check2(t) :=
        taylor(expand(
            ncexp( - t * a) .
            ncexp( - t * b) .
            ncexp( + t * (a + b))
                ), t, 0, degree);


/* The first BCH formula. */
    check3(t) := block(
        degree : 3,
        declare([c2, c3], nonscalar),
        c2 : - comm(a, b)/2,
        c3 : comm(a, comm(a, b))/6 + comm(b, comm(a, b))/3,

        temp1 : taylor(ncexp(t * (a + b)), t, 0, degree),
        temp2 : taylor(ncexp(t * a), t, 0, degree),
```

```
        temp2 : taylor(expand(temp2 . ncexp(t * b)), t, 0, degree),
        temp3 : taylor(ncexp(t↑2 * c2), t, 0, degree),
        temp2 : taylor(expand(temp2 . temp3), t, 0, degree),
        temp3 : taylor(ncexp(t↑3 * c3), t, 0, degree),
        temp2 : taylor(expand(temp2 . temp3), t, 0, degree),
        expand(temp1 - temp2)
    );


/* The second BCH formula. */
    check4(t) := block(
        degree : 3,
        declare([d2, d3], nonscalar),
        d2 : comm(a, b)/2,
        d3 : comm(a, comm(a, b))/12 - comm(b, comm(a, b))/12,

        temp : t * a + t * b + t↑2 * d2 + t↑3 * d3,
        term : 1,
        temp1 : 1,

        term : term . temp,
        temp1 : temp1 + term,
        term : taylor(term . temp/2, t, 0, degree),
        term : expand(term),
        temp1 : temp1 + term,
        term : taylor(term . temp/3, t, 0, degree),
        term : expand(term),
        temp1 : temp1 + term,

        temp2 : taylor(expand(ncexp(t * a) . ncexp(t * b)), t, 0, degree),
        expand(temp1 - temp2)
    );
```

3.5 Factored products

The goal of this section is to expand solutions of Hamilton's equations. The solutions are written using Lie transformations, hence these must be expanded. The applications dictate that these expansions be based on the degree of the polynomials occurring in the Hamiltonian. The expansions generated in the previous section are based on the number of commutators in an expression. As it turns out, these two ideas are not completely compatible, so most of the results of the previous section cannot be applied directly here. However, the techniques used previously can be applied. Some of the results of the previous section need to be translated to the hamiltonian setting. This is done by replacing operators (*e.g.*, A) by Lie derivatives (*e.g.*, $[f, \circ]$).

The first thing to note is that the Poisson bracket and the commutator bracket are compatible.

Proposition. *It holds that*

$$[[f,\circ],[g,\circ]] = [[f,g],\circ]. \tag{5.1}$$

This proposition says that the commutator of two Lie derivatives is generated by the Poisson bracket of the functions generating the Lie derivatives. This identity is the justification behind using square brackets for both commutator and Poisson brackets.

Proof. The outer-most brackets on the left-hand side of the previous equation are commutator brackets; all other brackets are Poisson brackets. The objects in this identity are operators, so to check the identity we compute (the Jacobi identity is applied)

$$\begin{aligned}[[f,\circ],[g,\circ]]h &= [f,\circ][g,\circ]h - [g,\circ][f,\circ]h \\ &= [f,\circ][g,h] - [g,\circ][f,h] \\ &= [f,[g,h]] - [g,[f,h]] \\ &= -[h,[f,g]] \\ &= [[f,g],h] \\ &= [[f,g],\circ]h.\end{aligned}$$

∎

3.5.1 Similarity for Lie transformations

As in the previous section, similarity operations play an important role in the computations. Similarity for Lie transformations may be written as

$$e^{t[f,\circ]}ge^{-t[f,\circ]} = (e^{t[f,\circ]}g), \tag{5.2}$$

$$e^{t[f,\circ]}[g,\circ]e^{-t[f,\circ]} = [e^{t[f,\circ]}g,\circ]. \tag{5.3}$$

Proof. These are operator identities which follow from the similarity property given in the previous section. This can be seen by choosing $A = [f,\circ]$ and $B = g$ to prove the first identity. Here B is the operator that multiplies functions by g. The second identity can be derived by choosing A as before and $B = [g,\circ]$. The proofs are similar to the proofs of the previous proposition. Because of the concrete representation of the operators, there are even simpler proofs.

The first identity means

$$e^{t[f,\circ]}\big(g(e^{-t[f,\circ]}h)\big) = (e^{t[f,\circ]}g)h.$$

The product-preservation property for Lie transformations given in Section 3 yields

$$e^{t[f,\circ]}\big(g(e^{-t[f,\circ]}h)\big) = (e^{t[f,\circ]}g)(e^{t[f,\circ]}e^{-t[f,\circ]}h) = (e^{t[f,\circ]}g)h.$$

The second identity means

$$e^{t[f,\circ]}\big([g,\circ](e^{-t[f,\circ]}h)\big) = [e^{t[f,\circ]}g,\circ]h.$$

The bracket preservation property for Lie transformations given in Section 3 yields

$$\begin{aligned}e^{t[f,\circ]}\big([g,\circ](e^{-t[f,\circ]}h)\big) &= e^{t[f,\circ]}[g,e^{-t[f,\circ]}h] \\ &= [e^{t[f,\circ]}g, e^{t[f,\circ]}e^{-t[f,\circ]}h] \\ &= [e^{t[f,\circ]}g,h] = [e^{t[f,\circ]}g,\circ]h.\end{aligned}$$

∎

As in the BCH identities, the formula for differentiating exponentials with time-dependent exponents plays a central role in the calculations.

Theorem. *The following equations hold:*

$$\frac{d}{dt}e^{[f(t),\circ]} = \left[\frac{e^{[f(t),\circ]}-1}{[f(t),\circ]}f'(t),\circ\right]e^{[f(t),\circ]}, \tag{5.4}$$

and

$$\frac{d}{dt}e^{[f(t),\circ]} = e^{[f(t),\circ]}\left[\frac{1-e^{-[f(t),\circ]}}{[f(t),\circ]}f'(t),\circ\right]. \tag{5.5}$$

Proof. The formula for differentiating an exponential derived in the previous section yields

$$\frac{d}{dt}e^{[f,\circ]} = \int_0^1 e^{\tau[f,\circ]}[f',\circ]e^{-\tau[f,\circ]}\,d\tau\, e^{[f,\circ]}.$$

Linearity and the similarity formulas yield

$$\frac{d}{dt}e^{[f,\circ]} = \left[\int_0^1 e^{\tau[f,\circ]}f',\circ\right]d\tau\, e^{[f,\circ]}.$$

Completing the integration gives

$$\frac{d}{dt}e^{[f,\circ]} = \left[\frac{e^{[f,\circ]}-1}{[f,\circ]}f',\circ\right]e^{[f,\circ]}.$$

The second form of the result can be obtained from the above by noting that

$$\frac{d}{dt}e^{[f,\circ]} = e^{[f,\circ]}e^{-[f,\circ]}\left(\left[\frac{e^{[f,\circ]}-1}{[f,\circ]}f',\circ\right]e^{[f,\circ]}\right).$$

The product and bracket-preservation properties yield

$$\frac{d}{dt}e^{[f,\circ]} = e^{[f,\circ]}\left[e^{-[f,\circ]}\frac{e^{[f,\circ]}-1}{[f,\circ]}f',\circ\right] = e^{[f,\circ]}\left[\frac{1-e^{-[f,\circ]}}{[f,\circ]}f',\circ\right].$$

∎

Recall that Hamilton's equations were discussed in Section 2. In the next theorem, the first result implies that the solutions of Hamilton's equations can be written using a Lie transformation, while the second result implies that the solutions of Liouville's equation can be written using the same Lie transformation. In this result, replacing t by $-t$ is the same as replacing H by $-H$, so signs are changed to reduce the number of minus signs in the formulas.

Theorem. *Let* $H = H(\mathbf{q},\mathbf{p})$, $\mathbf{Q}(t) = \mathbf{Q}(t,\mathbf{q},\mathbf{p})$, *and* $\mathbf{P}(t) = \mathbf{P}(t,\mathbf{q},\mathbf{p})$. *If*

$$\mathbf{Q}(t) = e^{t[H,\circ]}\mathbf{q}, \qquad \mathbf{P}(t) = e^{t[H,\circ]}\mathbf{p}, \tag{5.6}$$

then

$$\mathbf{Q}'(t) = [H,\mathbf{Q}(t)], \qquad \mathbf{P}'(t) = [H,\mathbf{P}(t)], \tag{5.7a}$$

and

$$\mathbf{Q}(0) = q, \qquad \mathbf{P}(0) = p. \tag{5.7b}$$

Moreover, if $f = f(\mathbf{q},\mathbf{p})$ *and* $F(t) = F(t,\mathbf{q},\mathbf{p})$, *then*

$$F(t) = e^{t[H,\circ]}f \tag{5.8}$$

satisfies

$$F'(t) = [H,F(t)], \qquad F(0) = f. \tag{5.9}$$

Proof. If $\mathbf{Q}(t) = e^{t[H,\circ]}\mathbf{q}$, then

$$\mathbf{Q}'(t) = e^{t[H,\circ]}[H,\circ]\mathbf{q} = e^{t[H,\circ]}[H,\mathbf{q}].$$

The bracket-preservation property gives

$$\mathbf{Q}'(t) = [e^{t[H,\circ]}H, e^{t[H,\circ]}\mathbf{q}] = [H,\mathbf{Q}(t)].$$

The proof for $\mathbf{P}$ is essentially the same.

In the case of Liouville's equation, $F(t) = e^{t[H,\circ]}f$, so

$$F'(t) = e^{t[H,\circ]}[H,\circ]f = e^{t[H,\circ]}[H,f].$$

Again, the bracket-preservation property gives

$$F'(t) = [e^{t[H,\circ]}H, e^{t[H,\circ]}f] = [H,F(t)].$$

∎

3.5.2 Degree of homogeneity

The Poisson bracket of polynomials plays a critical role in the computations that will be done in this section, so some results about such polynomials will now be described. A polynomial f in the variables $(\mathbf{q},\mathbf{p})$ is homogeneous of degree k if

$$f(\epsilon\mathbf{q},\epsilon\mathbf{p}) = \epsilon^k f(\mathbf{q},\mathbf{p}). \tag{5.10}$$

Let X_k be the space of all polynomials in the $2n$ variables $(\mathbf{q},\mathbf{p})$ which are homogeneous of degree k.

Proposition. *X_k is a real linear space of dimension*

$$\binom{2n-1+k}{k} = \frac{(2n-1+k)!}{(2n-1)!\,k!}. \tag{5.11}$$

The linearity of X_k is clear, while the dimension can be calculated by induction on the number of variables in the polynomial.

If f is a function of $(\mathbf{q},\mathbf{p})$, it was assumed (in the introduction) that f has a convergent power series expansion that starts with the terms of degree two. If all of the terms that are homogeneous of degree k are collected together, they form a polynomial $f_k(\mathbf{q},\mathbf{p})$ that is in X_k, and f is the sum of all such terms. Consequently any $f = f(\mathbf{q},\mathbf{p})$ has the form

$$f = f(\mathbf{q},\mathbf{p}) = \sum_{k=2}^{\infty} f_k(\mathbf{q},\mathbf{p}), \qquad f_k \in X_k, \tag{5.12}$$

where the series is convergent. Note that

$$f(\epsilon\mathbf{q},\epsilon\mathbf{p}) = \sum_{k=2}^{\infty} \epsilon^k f_k(\mathbf{q},\mathbf{p}). \tag{5.13}$$

Proposition. *If $f_i \in X_i$ and $f_j \in X_j$ then*

$$[f_i, f_j] \in X_{i+j-2}. \tag{5.14}$$

Proof. First note that if $f_i \in X_i$, then

$$\frac{\partial f_i}{\partial q_j} \in X_{i-1}, \qquad \frac{\partial f_i}{\partial p_j} \in X_{i-1}.$$

Also, if $f_i \in X_i$ and $f_j \in X_j$ then $f_i f_j \in X_{i+j}$. Combining this with the formula for the Poisson bracket,

$$[f, g] = \sum_{i=1}^{n} \frac{\partial f}{\partial q_i}\frac{\partial g}{\partial p_i} - \frac{\partial f}{\partial p_i}\frac{\partial g}{\partial q_i},$$

gives the result. ∎

We also define

$$X = \bigcup_{k=2}^{\infty} X_k. \tag{5.15}$$

The above results say that X is a graded Lie algebra under the Poisson bracket. This means that X is a linear space and, on X, the Poisson bracket is skew-symmetric, bilinear, and satisfies the Jacobi identity. Moreover, X must be the union of disjoint finite dimensional subspaces X_k, and

$$[X_i, X_j] \subset X_{i+j-2}. \tag{5.16}$$

Note that $[X_2, X_j] \subset X_j$ and in all other cases the commutator belongs to a subspace with an index strictly greater than the index of either of the subspaces being commuted. The index 2 has special significance in the following computations.

Proposition. *If $f \in X_2$, the action of $\exp(t[f, \circ])$ can be computed as a matrix exponential or, equivalently, by solving a system of linear constant-coefficient ordinary differential equations. Moreover, $\exp(t[f, \circ])$ maps X_k into X_k for all $k \geq 2$.*

Proof. The last statement follows from the Lie series for the Lie transformation and the fact that $[f, \circ]$ maps X_k into X_k. If $f \in X_2$, then f is quadratic:

$$f(\mathbf{q}, \mathbf{p}) = \sum_{i=1}^{n}\sum_{j=1}^{n} b_{ij} q_i q_j + \sum_{i=1}^{n}\sum_{j=1}^{n} c_{ij} q_i p_j + \sum_{i=1}^{n}\sum_{j=1}^{n} d_{ij} p_i p_j.$$

Here $B = (b_{ij})$, $C = (c_{ij})$, and $D = (d_{ij})$ are $n \times n$ symmetric matrices. Also:

$$[f, q_i] = -\sum_{j=1}^{n} c_{ij} q_j - \sum_{j=1}^{n} d_{ij} p_j, \; [f, p_i] = \sum_{j=1}^{n} b_{ij} q_j + \sum_{j=1}^{n} a_{ij} p_j.$$

A column vector notation is now needed, so that Hamilton's equations can be written as a single system of ordinary differential equations. Let

$$\mathbf{z} = (\mathbf{q}, \mathbf{p})^{\mathsf{T}}, \qquad \mathbf{Z}(t) = \big(\mathbf{Q}(t), \mathbf{P}(t)\big)^{\mathsf{T}}$$

(where the superscript ^{T} means transpose), and let A be the $2n \times 2n$ matrix given by

$$A = \begin{pmatrix} -C & -B \\ D & C \end{pmatrix}.$$

The previous formulas yield

$$[f, \circ]\mathbf{z} = [f, \mathbf{z}] = ([f, q_1], \ldots, [f, p_1], \ldots)^{\mathsf{T}} = A\mathbf{z}.$$

Consequently,

$$e^{t[f,\circ]}\mathbf{z} = \sum_{k=0}^{\infty} \frac{t^k}{k!}[f, \circ]^k \mathbf{z} = \sum_{k=0}^{\infty} \frac{t^k}{k!} A^k \mathbf{z} = e^{tA}\mathbf{z}.$$

However, if

$$M(t) = e^{tA},$$

then

$$M'(t) = Ae^{tA} = AM(t), \qquad M(0) = I,$$

that is, M is the fundamental solution matrix of the system of ordinary differential equations determined by A. The matrices $M(t)$ are called ***symplectic*** (see Laub [6]). In fact, the quadratic Hamiltonians form a finite-dimensional Lie algebra, so the classical theory of Lie algebras (see Wybourne [144]) can be used to calculate f_2. ∎

Once the action of $[f, \circ]$ on $\mathbf{q}$ and $\mathbf{p}$ has been determined, the action of $[f, \circ]$ on any function $f(\mathbf{q}, \mathbf{p})$ can be computed using the composition property. This also makes it clear that $[f, \circ]$ is a linear mapping of X_k into X_k.

3.5.3 Factored product expansions

We call the following the ***factored product expansion theorem.*** It shows how to write an exponential of a sum as a product of exponentials.

Theorem. *Let $k \geq 2$. If $h_k \in X_k$ then there exist $g_k \in X_k$ such that*

$$e^{[h_2+h_3+h_4+\cdots,\circ]} = e^{[g_2,\circ]}e^{[g_3,\circ]}e^{[g_4,\circ]}\cdots. \tag{5.17}$$

Formulas for computing the g_k recursively in terms of h_k are displayed below. In particular:

$$\begin{aligned}
g_2 &= h_2, \\
g_3 &= \frac{1 - e^{-[h_2,\circ]}}{[h_2, \circ]} h_3, \\
g_4 &= \int_0^1 e^{-\tau[h_2,\circ]}\left(h_4 - \frac{1}{2}\left[\frac{e^{\tau[h_2,\circ]} - 1}{[h_2, \circ]} h_3, h_3\right]\right) d\tau.
\end{aligned}$$

Proof. Assume that

$$e^{t[h_2+h_3+h_4+\cdots,\circ]} = e^{[f_2(t),\circ]}e^{[f_3(t),\circ]}e^{[f_4(t),\circ]}\cdots,$$

and then compute the logarithmic derivative of both sides of this expression:

$$\begin{aligned}[h_3 + h_3 + h_4 + \cdots, \circ] &= \left[\frac{e^{[f_2,\circ]}-1}{[f_2,\circ]}f_2', \circ\right] \\ &\quad + e^{[f_2,\circ]}\left[\frac{e^{[f_3,\circ]}-1}{[f_3,\circ]}f_3', \circ\right]e^{-[f_2,\circ]} + \cdots.\end{aligned}$$

In the next step more terms for this calculation will be recorded. The similarity properties can be used to move all terms inside the brackets. The Lie derivatives are equal if and only if the functions determining them are equal, so

$$\begin{aligned}h_2 + h_3 + h_4 + \cdots &= \frac{e^{[f_2,\circ]}-1}{[f_2,\circ]}f_2' \\ &\quad + e^{[f_2,\circ]}\frac{e^{[f_3,\circ]}-1}{[f_3,\circ]}f_3' \\ &\quad + e^{[f_2,\circ]}e^{[f_3,\circ]}\frac{e^{[f_4,\circ]}-1}{[f_4,\circ]}f_4' + \cdots.\end{aligned}$$

The pattern for computing all of the terms is now clear.

Comparing the terms of degree 2 gives

$$h_2 = \frac{e^{[f_2,\circ]}-1}{[f_2,\circ]}f_2'. \tag{5.18}$$

An obvious solution is $f_2 = t\,h_2$.

Now multiply the determining equation by $\exp(-t[h_2,\circ])$ to obtain:

$$\begin{aligned}e^{-t[h_2,\circ]}(h_3 + h_4 + \cdots) &= \frac{e^{[f_3,\circ]}-1}{[f_3,\circ]}f_3' \\ &\quad + e^{[f_3,\circ]}\frac{e^{[f_4,\circ]}-1}{[f_4,\circ]}f_4' \\ &\quad + e^{[f_3,\circ]}e^{[f_4,\circ]}\frac{e^{[f_5,\circ]}-1}{[f_5,\circ]}f_5' + \cdots.\end{aligned}$$

Again, the pattern for computing all of the terms is now clear.

Expanding the exponentials as power series and comparing degrees of homogeneity yields:

$$\begin{aligned}e^{-t[h_2,\circ]}h_3 &= f_3', \\ e^{-t[h_2,\circ]}h_4 &= f_4' + \tfrac{1}{2}[f_3, f_3'], \\ e^{-t[h_2,\circ]}h_5 &= f_5' + [f_3, f_4'] + \tfrac{1}{6}[f_3,[f_3, f_3']], \\ e^{-t[h_2,\circ]}h_6 &= f_6' + [f_3, f_5'] + \tfrac{1}{2}[f_4, f_4'] + \tfrac{1}{2}[f_3,[f_3, f_4']] + \tfrac{1}{24}[f_3,[f_3,[f_3, f_3']]].\end{aligned}$$

Notice that these equations can be used to determine f_k' in terms of h_k and f_i with $i < k$. Integrating f_k' will give f_k. Also, setting $g_k = f_k(1)$ gives the main part of the theorem.

The equation for the terms of degree three can be integrated:

$$\frac{1-e^{-t[h_2,\circ]}}{[h_2,\circ]}h_3 = f_3(t). \tag{5.19}$$

A simple substitution, a little algebra with exponentials, and an integration give the solution to the equation involving terms of degree four. ∎

3.5.4 Inverse factored product expansions

Attention will now be focussed on a process that can be thought of as inverse to the factored-product expansion; *combining* products of exponentials. This result is a bit more difficult than the above. The difficulties arise from the terms of degree two. If these terms are absent then the above result generalizes to one which can be used both to combine and expand exponentials.

Corollary. *Let $k \geq 3$. If $h_k \in X_k$ then there exist $g_k \in X_k$ such that*

$$e^{[h_3+h_4+h_5+\cdots,\circ]} = e^{[g_3,\circ]}e^{[g_4,\circ]}e^{[g_5,\circ]}\ldots. \tag{5.20}$$

Conversely, if $g_k \in X_k$ then there exist $h_k \in X_k$ such that the above formula holds. In particular:

$$\begin{aligned} g_3 &= h_3, \\ g_4 &= h_4, \\ g_5 &= h_5 - \tfrac{1}{2}[h_3, h_4], \\ g_6 &= h_6 - \tfrac{1}{2}[h_3, h_5] + \tfrac{1}{6}[h_3,[h_3,h_4]]. \end{aligned}$$

Clearly these equations can be solved for h_k.

Proof. This is an elementary calculation using the formulas generated in the above theorem. ∎

Theorem. *Let $k \geq 2$. If $g_k \in X_k$ then there exist $h_k \in X_k$ such that*

$$e^{[g_2,\circ]}e^{[g_3,\circ]}e^{[g_4,\circ]}\ldots = e^{[h_2+h_3+h_4\cdots,\circ]}. \tag{5.21}$$

In particular:

$$\begin{aligned} h_2 &= g_2, \\ h_3 &= \frac{[g_2,\circ]}{1-e^{-[g_2,\circ]}}g_3, \\ h_4 &= \frac{[g_2,\circ]}{1-e^{-[g_2,\circ]}}\left(g_4 - \frac{1}{2}\int_0^1 e^{-\tau[g_2,\circ]}\left[\frac{e^{\tau[g_2,\circ]}-1}{[g_2,\circ]}f_3, f_3\right]d\tau\right), \end{aligned}$$

where

$$f_3 = \frac{[g_2,\circ]}{1-e^{-[g_2,\circ]}}g_3.$$

Proof. The formula for g_3 given in the previous theorem can be solved for h_3. We see no way of doing this for higher k. Use the previous corollary to write

$$e^{[g_2,\circ]}e^{[g_3,\circ]}e^{[g_4,\circ]}\cdots = e^{[g_2,\circ]}e^{[g_3+g_4+g_5+[g_3,g_4]/2+\cdots,\circ]}$$

From the last theorem of the previous section we have,

$$e^A e^B = e^{A+C_2+C_3+\cdots}.$$

Choose

$$A = [g_2, \circ], \qquad B = [g_3 + g_4 + g_5 + \tfrac{1}{2}[g_3, g_4] + \cdots, \circ].$$

All of the operators in that theorem can be now be written as Lie derivatives:

$$C_1 = [c_1, \circ] = \frac{[[g_2, \circ], \circ]}{1 - e^{[-[g_2,\circ],\circ]}} B.$$

Consequently,

$$c_1 = \frac{[g_2, \circ]}{1 - e^{-[g_2,\circ]}}(g_3 + g_4 + \cdots).$$

Also:

$$S_1(\tau) = [s_1(\tau), \circ],$$

where

$$s_1(\tau) = \frac{1 - e^{-\tau[g_2,\circ]}}{[g_2, \circ]} c_1,$$

and

$$S_2(\tau) = [s_2(\tau), \circ],$$

where

$$s_2(\tau) = e^{-\tau[g_2,\circ]} c_1.$$

Also

$$C_2(\tau) = [c_2(\tau), \circ],$$

where

$$c_2 = -\frac{1}{2}\frac{[g_2, \circ]}{1 - e^{-[g_2,\circ]}} \int_0^1 [s_1(\tau), s_2(\tau)]\, d\tau,$$

or

$$c_2 = -\frac{1}{2}\frac{[g_2, \circ]}{1 - e^{-[g_2,\circ]}} \int_0^1 \left[\frac{1 - e^{-\tau[g_2,\circ]}}{[g_2, \circ]} c_1, e^{-\tau[g_2,\circ]} c_1\right] d\tau.$$

Factoring out an exponential in the integrand yields:

$$c_2 = -\frac{1}{2}\frac{[g_2, \circ]}{1 - e^{-[g_2,\circ]}} \int_0^1 e^{-\tau[g_2,\circ]}\left[\frac{e^{\tau[g_2,\circ]} - 1}{[g_2, \circ]} c_1, c_1\right] d\tau.$$

Collecting terms with the same degree of homogeneity yields the result. ∎

3.5.5 Products of Lie transformations

In applications, it is common to encounter products of Lie transformations. The next result allows for the simplification of such products.

Theorem. *Let $k \geq 2$. If $g_k, h_k \in X_k$ then there exist $f_k \in X_k$ such that*

$$e^{[g_2,\circ]}e^{[g_3,\circ]}e^{[g_4,\circ]}\cdots e^{[h_2,\circ]}e^{[h_3,\circ]}e^{[h_4,\circ]}\cdots = e^{[f_2,\circ]}e^{[f_3,\circ]}e^{[f_4,\circ]}\cdots, \tag{5.22}$$

where

$$f_3 = e^{-[h_2,\circ]}g_3 + h_3,$$
$$f_4 = e^{-[h_2,\circ]}g_4 + h_4 + \tfrac{1}{2}[e^{-[h_2,\circ]}g_3, h_3],$$

and so forth. Also, f_2 can be calculated using classical finite-dimensional Lie theory.

Proof. The similarity property gives

$$e^{-[h_2,\circ]}[g_k,\circ]e^{[h_2,\circ]} = [e^{-[h_2,\circ]}g_k,\circ],$$

and consequently

$$e^{-[h_2,\circ]}e^{[g_k,\circ]}e^{[h_2,\circ]} = e^{[e^{-[h_2,\circ]}g_k,\circ]}.$$

This can be rearranged as

$$e^{[g_k,\circ]}e^{[h_2,\circ]} = e^{[h_2,\circ]}e^{-[h_2,\circ]}e^{[g_k,\circ]}e^{[h_2,\circ]} = e^{[h_2,\circ]}e^{[e^{[h_2,\circ]}g_k,\circ]}.$$

Apply this to (5.22) to move the $\exp([h_2,\circ])$ all the way to the left, which gives:

$$E = e^{[g_2,\circ]}e^{[h_2,\circ]}e^{[\hat{g}_3,\circ]}e^{[\hat{g}_4,\circ]}\cdots e^{[h_3,\circ]}e^{[h_4,\circ]}\cdots,$$

where

$$\hat{g}_k = e^{-[h_2,\circ]}g_k.$$

∎

The previous corollary can be used to compute the nonquadratic terms. For example, if term up to degree 4 are computed, then the expression becomes

$$E \approx e^{[g_2,\circ]}e^{[h_2,\circ]}e^{[\hat{g}_3,\circ]}e^{[\hat{g}_4,\circ]}e^{[h_3,\circ]}e^{[h_4,\circ]}$$

where all terms of degree greater than 4 have been dropped. Note that if a computation is done with two terms, one of degree i and one of degree j, and if $i + j - 2 > 4$, then all corrections to the standard exponential identities may be dropped. Consequently, terms of degree 3 and 4 can be interchanged;

$$E \approx e^{[g_2,\circ]}e^{[h_2,\circ]}e^{[\hat{g}_3,\circ]}e^{[h_3,\circ]}e^{[\hat{g}_4,\circ]}e^{[h_4,\circ]}.$$

The second BCH formula of the previous section yields

$$e^{[\hat{g}_3,\circ]}e^{[h_3,\circ]} \approx e^{[\hat{g}_3+h_3+[\hat{g}_3,h_3]/2,\circ]}$$

So

$$E \approx e^{[g_2,\circ]}e^{[h_2,\circ]}e^{[\hat{g}_3+h_3+[\hat{g}_3,h_3]/2+\hat{g}_4+h_4,\circ]}.$$

This gives the correct form for f_3 and f_4. Methods for computing the quadratic terms have been discussed previously.

3.5.6 Iterates of factored maps

This section ends with a result that is important in the analysis of iterates of factored maps.

Theorem. *Let $h_k \in X_k$. Then*

$$\left(e^{[h_2,\circ]}e^{[h_3,\circ]}\dots\right)^k = e^{k[h_2,\circ]}e^{[f_3,\circ]}e^{[f_4,\circ]}\dots, \tag{5.23}$$

where

$$f_3 = \frac{1-e^{-k[h_2,\circ]}}{1-e^{-[h_2,\circ]}}h_3, \tag{5.24}$$

and so forth. It is possible to compute the remaining f_k for any k using elementary means.

Proof. Write the k-th power as a product and then commute all of the h_2 factors to the left to obtain

$$e^{k[h_2,\circ]}e^{[e^{-(k-1)[h_2,\circ]}h_3,\circ]}\dots e^{[e^{-(k-2)[h_2,\circ]}h_3,\circ]}\dots e^{[e^{-[h_2,\circ]}h_3,\circ]}\dots e^{[h_3,\circ]}.$$

The BCH identities then yield:

$$f_3 = e^{-(k-1)[h_2,\circ]}h_3 + \dots + e^{-[h_2,\circ]}h_3 + h_3 = \frac{1-e^{-k[h_2,\circ]}}{1-e^{-[h_2,\circ]}}h_3.$$

Any higher order term can be calculated using the same technique. ∎

3.6 Canonical forms

In this section the problem of transforming a hamiltonian system to a simpler form is examined. Recall that a basic result of hamiltonian mechanics (see Goldstein [140]) says that hamiltonian systems are preserved under canonical transformations. This means that if a system of ordinary differential equations is hamiltonian, and a canonical change of coordinates is made, then the system is still hamiltonian in the new coordinates. In fact, the new Hamiltonian is the transform of the old Hamiltonian. Lie transformations determine canonical transformations, so these can be used to simplify hamiltonian systems.

3.6.1 Canonical transformations

Theorem. *The transformation of $(\mathbf{q},\mathbf{p}) \in (\mathbf{Q}(\epsilon),\mathbf{P}(\epsilon))$ determined by*

$$\mathbf{Q}(\epsilon) = e^{\epsilon[h,\circ]}\mathbf{q}, \qquad \mathbf{P}(\epsilon) = e^{\epsilon[h,\circ]}\mathbf{p} \tag{6.1}$$

is canonical.

Proof. (We have changed from t to ϵ in the Lie transformation to distinguish this work from that in the previous sections.) The definition of canonical (see Goldstein [140]) requires that

$$[Q_i,Q_j] = [q_i,q_j], \qquad [Q_i,P_j] = [q_i,p_j], \qquad [P_i,P_j] = [p_i,p_j],$$

for all i and j satisfying $0 \leq i, j \leq n$. However,

$$[Q_i, Q_j] = [e^{\epsilon[h,\circ]}q_i, e^{\epsilon[h,\circ]}q_j],$$

and the bracket preservation property of Lie series yields

$$[Q_i, Q_j] = e^{\epsilon[h,\circ]}[q_i, q_j].$$

The quantity $[q_i, q_j]$ is a constant; $[q_i, q_j] = \delta_{i,j}$. If c is a constant then $[h, \circ]c = [h, c] = 0$ and consequently $\exp([h, \circ])c = c$. This implies the first of the three conditions; the other two are proved in an identical way. ∎

3.6.2 Simplification of the Hamiltonian

The notion of "simple" we will use requires the new Hamiltonian to have an integral. It is standard to say that g is an integral of h provided that $[g, h] = 0$. It is assumed that g and h are nontrivial. This is a symmetric relationship, so h is also an integral of g. It is also standard to say that g leaves h invariant if $\exp(\epsilon[g, \circ])h = h$ for all ϵ. Our differentiation formulas for Lie transformations and the Lie series definition of the Lie transformation imply that g is an integral of h if and only if g leaves h invariant. A Hamiltonian h is said to be completely integrable if there are n nontrivial integrals g_i, $1 \leq i \leq n$ of h such that $[g_i, h] = 0$.

The simplification problem under consideration is a perturbation problem. Assume that a Hamiltonian, $h(\epsilon)$, of the form

$$h(\epsilon) = h(\epsilon, \mathbf{q}, \mathbf{p}) = \sum_{k=2}^{\infty} \epsilon^k h_k(\mathbf{q}, \mathbf{p}) \tag{6.2}$$

is given. The reason for starting the sum at 2 was given in the introduction. The case of interest is one for which h_k is a homogeneous polynomial of degree k; however, that assumption is not necessary initially.[3] What is desired is a sequence of canonical transformations depending on ϵ that will transform the Hamiltonian, $h(\epsilon)$, into a new Hamiltonian, $\hat{h}(\epsilon)$, which is simpler than the original Hamiltonian $h(\epsilon)$. Since this is a perturbation problem, only a finite number of terms in the series for $h(\epsilon)$ will be simplified. A Hamiltonian which is a power series in ϵ is ***simplified through order*** m if there is a nontrivial function f such that

$$[f, \sum_{k=3}^{m} \epsilon^k h_k] = 0. \tag{6.3}$$

For reasons that will be apparent later, the h_2 term has been omitted from the simplification. There are a procedures for simplifying such terms (see Burgoyne [39], Giacaglia [139], or Goldstein [140]).

The procedure for simplifying Hamiltonians will now be described. Note that the procedure does not depend on h_2 having any special form. The simplification is done inductively. Assume that a function f is given so that

$$[f, h_i] = 0, \qquad 3 \leq i \leq m - 1, \tag{6.4}$$

---

[3]We note that the general case (6.2), when homogeneity is *not* assumed, applies very well to the *rank* expansions of Hamiltonians corresponding to misaligned and misplaced optical elements. These shift the origin of phase space and contain *first*-order polynomials, as described in **4.8.3**, next chapter. (Editor's note.)

that is, $h(\epsilon)$ has been simplified through order $m-1$. Here $m \geq 3$. The problem to be addressed is that of finding a function g_m so that

$$\hat{h}(\epsilon) = e^{\epsilon^{m-2}[g_m, \circ]} h(\epsilon) \tag{6.5}$$

is simplified through order m. Write

$$\hat{h}(\epsilon) = \sum_{k=2}^{\infty} \epsilon^k \hat{h}_k, \tag{6.6}$$

and carry out the power series expansions of the Lie series and $h(\epsilon)$ to obtain:

$$\begin{aligned} \hat{h}(\epsilon) &= \epsilon^2 \hat{h}_2 + \epsilon^3 \hat{h}_3 + \epsilon^4 \hat{h}_4 + \cdots \\ &= \epsilon^2 h_2 + \epsilon^3 h_3 + \epsilon^4 h_4 + \cdots + \epsilon^m h_m + \cdots \\ &\quad + \epsilon^m [g_m, h_2] + \epsilon^{m+1} [g_m, h_3] + \epsilon^{m+2} [g_m, h_4] + \cdots \\ &\quad + \epsilon^{2m-2} [g_m, [g_m, h_2]] + \cdots. \end{aligned} \tag{6.7}$$

Compare powers of ϵ to find:

$$\begin{aligned} \hat{h}_k &= h_k, & 2 \leq k \leq m-1, \\ \hat{h}_m &= [g_m, h_2] + h_m, & k = m, \\ \hat{h}_k &= h.o.t., & k \geq m+1, \end{aligned} \tag{6.8}$$

where *h.o.t.* stands for higher order terms. It is desired that f be an integral of $\hat{h}_m$, that is, $[f, \hat{h}_m] = 0$. If the skew-symmetry of the Poisson bracket is used, then g_m is determined using the equation

$$[f, [h_2, g_m]] = [f, h_m]. \tag{6.9}$$

Once g is determined, the equation

$$\hat{h}(\epsilon) = e^{\epsilon^m [g_m, \circ]} h(\epsilon) \tag{6.10}$$

determines $\hat{h}$.

3.6.3 The perturbation equation

It is easy to construct examples where the equation (6.9), *viz.*,

$$[f, [h_2, g_m]] = [f, h_m] \tag{6.11}$$

is ***inconsistent***, and consequently this simple perturbation procedure fails. In this generality nothing more can be said, so attention will be focused on a situation in which the equation for determining g_m can be analyzed. The last equation is called the ***perturbation equation.***

Assume that the g_k and h_k of the previous discussion are homogeneous polynomials of degree k. Recall that in the section on factored products, the space X_k of homogeneous polynomials of degree k was introduced, so this is the same as assuming that g_k and h_k belong to X_k. The finite dimensionality of X_k permits a complete understanding of the solvability of the perturbation equation. The perturbation equation can be written as follows.

Formulation. *Given $h_2 \in X_2$, $f_2 \in X_2$, and $h_k \in X_k$, find $g_k \in X_k$ such that*

$$[f_2, [h_2, g_k]] = [f_2, h_k]. \tag{6.12}$$

It is necessary to understand the solvability of such equations before the perturbation calculations can be done.

Hopefully the following comments will help in understanding the solvability of the perturbation equation. First, recall that this equation must be solved in X_k, which is a finite-dimensional space. Because $[h_2, \circ]$ is a linear operator on X_k (see the section on factored products), the perturbation equation can be viewed as a matrix equation. Before this equation is described, note that if

$$[h_2, g_k] = h_k \tag{6.13}$$

could be solved for g_k, then the perturbation equation could be solved for any f_2. However, some simple examples show that the matrix corresponding to $[h_2, \circ]$ is not always invertible. Thus, the integral f_2 can be viewed as a consistency condition for the previous equation.

A choice of basis for X_k will allow the perturbation equation to be written as a matrix equation. A convenient basis for X_k is given by

$$x_{ij} = \frac{\mathbf{q}^i \mathbf{p}^j}{\sqrt{i!j!}}, \qquad |i| + |j| = k. \tag{6.14}$$

Also, let $< \circ, \circ >$ be the inner product on X_k which makes x_{ij} orthonormal. This choice of basis may not seem natural, but was found to be convenient after some experimentation. Now, let $\mathbf{v}$ be the vector of components of g_k in this basis, $\mathbf{u}$ be the vector of components of h_k in this basis, and A be the matrix of $[h_2, \circ]$ in this basis. Then the equation

$$[h_2, \circ]g_k = [h_2, g_m] = h_k \tag{6.15}$$

becomes

$$A\mathbf{v} = \mathbf{u}. \tag{6.16}$$

As was said before, in many examples A is not invertible, and the last equation is inconsistent. However, it is well known from matrix theory that the system of equations

$$A^* A\mathbf{v} = A^* \mathbf{u}, \tag{6.17}$$

where A^* is the transpose of A (A is real because H is real), is a consistent set of equations. This is the same thing as projecting $\mathbf{u}$ onto the range of the matrix A, and then solving the resulting equation or finding the least-squares solution of (6.16). If A is not invertible, the solution is not unique. The transpose of the matrix should correspond to the adjoint of the Lie derivative determined by h_2. The next problem is to compute the adjoint of the Lie operator, and show that the matrix of the adjoint is the transpose of the matrix A.

In the section of factored products, the space

$$X = \bigcup_{k=0}^{\infty} X_k \tag{6.18}$$

was introduced. Extend the inner product $\langle \circ \rangle \circ$ to this space by requiring that $\langle X_l \rangle X_m = 0$ for $l \neq m$. Recall that if $h \in X$ then the adjoint $[h, \circ]^*$ of the Lie derivative, $[h, \circ]$, determined by h is uniquely determined by requiring that

$$\langle [h, \circ]^* x, y \rangle = \langle x, [h, \circ] y \rangle, \tag{6.19}$$

for all $x, y \in X$.

Lie derivatives are made up of multiplications by functions and partial derivatives, so first note that

$$\begin{aligned}\left\langle \frac{\partial}{\partial q_k} x_{ij}, x_{rs} \right\rangle &= \langle x_{ij}, q_k x_{rs} \rangle,\\ \left\langle \frac{\partial}{\partial p_k} x_{ij}, x_{rs} \right\rangle &= \langle x_{ij}, p_k x_{rs} \rangle,\end{aligned} \tag{6.20}$$

for all k, i, j, r, s. These formulas are found by simple computations. Because any x or y belonging to X can be written as a linear combination of the basis elements,

$$\begin{aligned}\left\langle \frac{\partial}{\partial q_k} x, y \right\rangle &= \langle x, q_k y \rangle,\\ \left\langle \frac{\partial}{\partial p_k} x, y \right\rangle &= \langle x, p_k y \rangle.\end{aligned} \tag{6.21}$$

Proposition. *If h_2 is a quadratic function (that is, $h_2 \in X_2$), and if h_2^* is defined by*

$$h_2^*(\mathbf{q}, \mathbf{p}) = -h_2(\mathbf{p}, -\mathbf{q}) = -h_2(-\mathbf{p}, \mathbf{q}), \tag{6.22}$$

then

$$[h_2, \circ]^* = [h_2^*, \circ]. \tag{6.23}$$

Also note that if $h, g \in X_2$, then $[h, g]^* = [h^*, g^*]$.

Proof. The function h_2 is a linear combination of the terms $q_i q_j$, $q_i p_j$, and $p_i p_j$. Because the operators are linear, it is enough to prove the proposition for the simple operators. The definition of the Lie derivative gives

$$[q_i q_j, \circ] = q_i \frac{\partial}{\partial p_j} + q_j \frac{\partial}{\partial p_i}.$$

If $x, y \in X$, then the above formulas give

$$\begin{aligned}\langle [q_i q_j, \circ] x, y \rangle &= \langle x, [q_i q_j, \circ]^* y \rangle,\\ &= \left\langle \left(q_i \frac{\partial}{\partial p_j} + q_j \frac{\partial}{\partial p_i} \right) x, y \right\rangle\\ &= \left\langle x, \left(p_i \frac{\partial}{\partial q_j} + p_j \frac{\partial}{\partial q_i} \right) y \right\rangle.\end{aligned}$$

However,

$$[p_i p_j, \circ] = -p_i \frac{\partial}{\partial q_j} - p_j \frac{\partial}{\partial q_i},$$

so

$$[q_i q_j, \circ]^* = -[p_i p_j, \circ].$$

The other two cases are similar. The proof of the note at the end of the proposition is an interesting exercise which uses the definition of the adjoint and the Jacobi identity. ∎

The perturbation equation can be solved if h_2^* is required to be an integral.

Theorem. *If $h_2 \in X_2$ and $h_k \in X_k$, then*

$$[h_2^*, [h_2, g_k]] - [h_2^*, h_k] = 0 \tag{6.24}$$

always has at least one solution $g_k \in X_k$. If $[h_2, \circ]$ is invertible, then g_k is the unique solution of

$$[h_2, g_k] - h_k = 0. \tag{6.25}$$

If $[h_2, \circ]$ is not invertible, then there are nontrivial $\bar{g}_k \in X_k$ such that $[h_2, \bar{g}_k] = 0$ and $g_k + \bar{g}_k$ satisfies

$$[h_2^*, [h_2, g_k + \bar{g}_k]] - [h_k^*, h_k] = 0. \tag{6.26}$$

Proof. The proof is just an abstract version of the discussion of matrices given above. If $f \in X_k$ then f can be written as a sum of two vectors, f_1 in the range of $[h_2, \circ]$ and f_2 perpendicular to the range (that is, $f = f_1 + f_2$ where $[h_2^*, f_2] = 0$), and there exists a $g_k \in X_k$ such that $[h_2, g_k] = f_1$. Here f_1 and f_2 are unique, while g_k is not unique unless $[h_2, \circ]$ is invertible. Thus

$$[h_2^*, [h_2, g_k]] = [h_2^*, f_1] = [h_2^*, f],$$

that is, g_k is a solution of the perturbation equation when the integral is taken to be h_2^*. In addition, if $[h_2, \bar{g}_k] = 0$, then

$$[h_2^*, [h_2, g_k + \bar{g}_k]] = [h_2^*, h_k],$$

so that $g_k + \bar{g}_k$ is also a solution of the perturbation equation. The function $\bar{g}_k$ must be zero if and only if $[h_2, \circ]$ is invertible. ∎

Repeated applications of this result to the perturbation procedure described above gives the following result.

Theorem. *If*

$$h = \sum_{k=2}^{\infty} \epsilon^{k-2} h_k, \tag{6.27}$$

then it is possible to choose $g_k \in X_k$, $3 \leq k \leq m$, such that if

$$\tilde{h} = e^{\epsilon^{m-2}[g_m, \circ]} \ldots e^{\epsilon^2 [g_4, \circ]} e^{\epsilon [g_3, \circ]} h, \tag{6.28}$$

then

$$\tilde{h} = \sum_{k=2}^{\infty} \epsilon^{k-2} \tilde{h}_k, \tag{6.29}$$

and

$$[h_2^*, \sum_{k=3}^{m} \epsilon^{k-2} \tilde{h}_k] = 0. \tag{6.30}$$

Note that the last sum starts at $k = 3$ and is finite.

Proof. $\tilde{h}$ is constructed by iterating the perturbation procedure of the previous section. Define

$$H_0 = h, \qquad H_k = e^{\epsilon^k g_k} H_{k-1}, \qquad k \geq 1,$$

and assume that each H_k can be written as a series:

$$H_k = \sum_{i=2}^{\infty} \epsilon^i h_{k,i}.$$

Use the previous proposition to choose g_k satisfying

$$[h_2^*, [h_2, g_k]] - [g_2, h_{k-1,k-1}] = 0$$

for $1 \leq k \leq m$. Thus h_2^* is an integral of the first $m-2$ terms (except the first) of $\tilde{h}(\epsilon)$. ∎

In many applications $h_2^* = \pm h_2$. In these cases and some others $[h_2^*, h_2] = 0$, and then h_2^* is an integral of the first $m-2$ terms of h.

3.7 Quadratic Hamiltonians and integrals

Now that h_2^*, defined in (6.22), can be made an integral of as many of the terms of $\tilde{h}(\epsilon)$ as desired (except the first), it is of interest to explore what can be said about parts of h_2^* being integrals of $\tilde{h}$. Only the individual terms of $\tilde{h}$ need be considered. Thus assume that $f \in X_k$, and $[h_2^*, f] = 0$.

It is important for the analysis to know when operators commute. Recall that if h and g are two functions, then

$$[[h, \circ], [g, \circ]] = [[h, g], \circ]. \tag{7.1}$$

As before, the outer brackets on the left-hand side of the previous equation are commutator brackets, while all other brackets are Poisson brackets.

3.7.1 Quadratic Hamiltonians

For one degree of freedom, there are three illuminating examples for h_2:

$$h_2 = \tfrac{1}{2}(q^2 + p^2), \qquad h_2 = \tfrac{1}{2}p^2, \qquad h_2 = \tfrac{1}{2}(p^2 - q^2). \tag{7.2}$$

The respective adjoint Hamiltonians are then:

$$h_2^* = -h_2, \qquad h_2^* = -\tfrac{1}{2}q^2, \qquad h_2^* = h_2. \tag{7.3}$$

The first Hamiltonian is skew-adjoint while the last Hamiltonian is self-adjoint.

3.7.1.1 Nilpotent operators

The Lie derivative corresponding to the second Hamiltonian can be written

$$[\tfrac{1}{2}q^2, \circ] = q\frac{\partial}{\partial p}, \tag{7.4}$$

as an operator on X_k. Because $[q^2/2, \circ]$ decreases the degree of a polynomial in p by one, we have

$$[q^2, \circ]^{k+1} X_k = 0. \tag{7.5}$$

An operator such as this, for which some power of the operator is zero is called nilpotent. Such operators cause us some special problems so they will be considered first.

Theorem. *Let $h_2 \in X_2$, $f \in X_k$, and*

$$h_2 = u + v, \tag{7.6}$$

where v is nilpotent and u is diagonalizable. If $[u, v] = 0$ and $[h_2^*, f] = 0$, then

$$[u^*, f] = 0, \qquad [v^*, f] = 0. \tag{7.7}$$

Recall that skew-adjoint and self-adjoint matrices are diagonalizable.

Proof. First note that $h_2^* = u^* + v^*$ and $[u^*, v^*] = [u, v]^* = 0$. If $[v, \circ]$ is nilpotent, so is $[v, \circ]^* = [v^*, \circ]$, and if $[u, \circ]$ is diagonalizable, then so is $[u, \circ]^* = [u^*, \circ]$. Because $[v^*, \circ]$ is nilpotent, there is a k so that $[v^*, \circ]^k = 0$. However,

$$[v^*, \circ]^k f = \sum_{l=0}^{k} \binom{k}{l} [-u^*, \circ]^{k-l} [h_2^*, \circ]^l f = [-u^*, \circ]^k f.$$

Because $[u^*, \circ]$ is diagonalizable, $[u^*, \circ]^k f = 0$ implies that $[u^*, \circ] f = 0$. ∎

3.7.2 Eigenvalues of quadratic operators

A situation frequently appearing in applications is one in which

$$h_2 = \sum_{i=1}^{n} r_i, \tag{7.8}$$

where

$$r_i = \tfrac{1}{2}\alpha_i q_i^2 + \beta_i q_i p_i + \tfrac{1}{2}\gamma_i p_i^2. \tag{7.9}$$

In this case

$$h_2^* = \sum_{i=1}^{n} r_i^*, \qquad r_i^* = -\tfrac{1}{2}\gamma_i q_i^2 + \beta_i q_i p_i - \tfrac{1}{2}\alpha_i p_i^2. \tag{7.10}$$

The first thing to note is that $[r_i, r_j] = [r_i^*, r_j^*] = 0$; the set $[r_i, \circ]$, $1 \le i \le n$, and the set $[r_i^*, \circ]$, $1 \le i \le n$, are commuting sets of operators.

Next all of the eigenvalues and eigenfunctions (eigenvectors) of the operators $[r_i, \circ]$ are computed. There are n degrees of freedom, and it is assumed that the r_i have the form given above. Begin with the space X_1. For now assume that the polynomials in X_k have complex coefficients. The basis $\{q_i, p_i; 1 \le i \le n\}$ for X_1 was previously chosen. Now:

$$\begin{aligned}
[r_i, q_j] &= 0, \qquad \text{for } i \ne j,\\
[r_i, p_j] &= 0, \qquad \text{for } i \ne j,\\
[r_i, q_i] &= -\beta_i q_i - \gamma_i p_i,\\
[r_i, p_i] &= \alpha_i q_i + \beta_i p_i.
\end{aligned} \tag{7.11}$$

Thus, the matrix of the operator $[r_i, \circ]$ is all zeros except for one 2×2 block, which is

$$\begin{pmatrix} -\beta_i & -\gamma_i \\ \alpha_i & \beta_i \end{pmatrix}. \tag{7.12}$$

This matrix has eigenvalues $\pm\lambda_i$, where

$$\lambda_i = \sqrt{\beta_i^2 - \alpha_i \gamma_i}. \tag{7.13}$$

If $\alpha_i = 0$, $\gamma_i = 0$ and $\lambda_i = 0$, then a complete set of eigenfunctions of $[r_i, \circ]$ is:

$$\begin{aligned}
Q_i &= \frac{1}{\sqrt{2}}\left(q_i - \frac{\alpha_i}{\lambda_i - \beta_i} p_i\right), \\
P_i &= \frac{1}{\sqrt{2}}\left(-\frac{\gamma_i}{\lambda_i + \beta_i} q_i + p_i\right), \\
Q_j &= q_j \qquad \text{if } j \neq i, \\
P_j &= p_j \qquad \text{if } j \neq i.
\end{aligned} \tag{7.14}$$

In the cases where one or more of the parameters are zero we choose:

$$\begin{aligned}
Q_i &= \frac{1}{\sqrt{2}}\left(q_i + \frac{2\beta_i}{\gamma_i} p_i\right), && \text{if } \alpha_i = 0, \\
P_i &= \frac{1}{\sqrt{2}}\left(-\frac{\gamma_i}{2\beta_i} p_i + q_i\right), && \text{if } \alpha_i = 0, \\
Q_i &= \frac{1}{\sqrt{2}}\left(q_i - \frac{\alpha_i}{2\beta_i} p_i\right), && \text{if } \gamma_i = 0, \\
P_i &= \frac{1}{\sqrt{2}}\left(\frac{2\beta_i}{\alpha_i} q_i + p_i\right), && \text{if } \gamma_i = 0,
\end{aligned} \tag{7.15}$$

In the case of both α_i and β_i zero, make no change. If $\lambda_i = 0$, the eigenfunctions given above are linearly dependent and, in general, there is no complete set of eigenfunctions. Note that if $\lambda_i = 0$, then the operator $[r_i, \circ]$ is nilpotent.

3.7.3 Eigenbases on X_k

Now find an eigenbasis for $[r_i, \circ]$ on the space X_k for $k > 1$. First, recall that if $x_j \in X_j$ and $x_k \in X_k$, then $x_j x_k \in X_{j+k}$. Now suppose that x_j and x_k are eigenfunctions of $[r_i, \circ]$, that is,

$$[r_i, x_j] = \lambda_{ij} x_j, \qquad [r_i, x_k] = \lambda_{ik} x_k. \tag{7.16}$$

The derivative property of the Poisson bracket gives

$$[r_i, x_j x_k] = [r_i, x_j] x_k + x_j [r_i, x_k] = (\lambda_{ij} + \lambda_{ik}) x_j x_k. \tag{7.17}$$

Thus, the product of eigenfunctions is still an eigenfunction with a new eigenvalue which is the sum of the old eigenvalues.

When $\lambda_i = 0$, that is, when r_i is nilpotent on X_1, the integrability formula gives

$$[r_i, \circ]^k x_i x_j = \sum_{l=0}^{k} \binom{k}{l} ([r_i, \circ]^{k-l} x_i)([r_i, \circ]^l x_j). \tag{7.18}$$

This shows that $[r_i, \circ]$ is nilpotent on X_k for all k. Thus, when $\lambda_i = 0$, $[r_i, \circ]$ has only zero eigenvalues on all X_k.

If it is assumed that all of the operators $[r_i, \circ]$ can be diagonalized on X_1 (which is implied by having $\lambda_i \neq 0$ for all i), then each $[r_i, \circ]$ can be diagonalized on X_k for all k. Because the $[r_i, \circ]$ commute, these operators can be diagonalized simultaneously. In fact, the above discussion shows how to construct the basis. In this case each $[r_i, \circ]$ will have eigenvalues, on X_k, of the form $m_i\lambda_i$ where $|m_i| \leq k$. Consequently $[h_2, \circ]$ will have eigenvalues on X_k of the form

$$\lambda = \sum_{i=1}^{n} m_i\lambda_i. \tag{7.19}$$

Solving the Perturbation Equation will involve divisions by nonzero eigenvalues of $[h_2, \circ]$; a quantity that will estimate these divisors is now introduced. Let

$$d_k = \min_{1 \leq |m| \leq k} (\lambda \circ m), \tag{7.20}$$

where

$$\lambda = (\lambda_1, \lambda_2, \ldots, \lambda_n), \qquad m = (m_1, m_2, \ldots, m_n), \tag{7.21}$$

and m_j is an integer. The number d_k is called a ***divisor***. If $d_k \neq 0$ for all $k \geq 1$, then the λ_j are said to be ***rationally independent***.

Theorem. *Let $h_2 \in X_2$, and*

$$h_2 = \sum_{i=1}^{n} r_i, \tag{7.22}$$

with r_i and d_k defined as above. If $d_k \neq 0$, $f \in X_k$ and

$$[h_2, f] = 0, \tag{7.23}$$

then

$$[r_i, f] = 0, \qquad 1 \leq i \leq n. \tag{7.24}$$

Proof. Suppose that $[h_2, \circ]$ kills an eigenfunction with eigenvalues $\lambda \circ m$. The assumption on d_k then implies that $m_i = 0$ for all i. Thus, each $[r_i, \circ]$ kills the eigenfunction. Writing any null function as a sum of eigenfunctions completes the proof. Note that all $d_k \neq 0$ implies that none of the r_i are nilpotent. ∎

The following result explains some features of the examples presented in the next section.

Theorem. *Suppose h_2 is the same as in the previous theorem, and the eigenvalues λ_i are rationally independent. Then the simplified Hamiltonian produced by our procedure has all of the coefficients of odd powers of ϵ equal to zero.*

Proof. The eigenfunctions of $[h_2, \circ]$, as an operator on X_k, are made up of products of the eigenfunction from X_1. When k is odd this product will contain an odd number of terms. The eigenvalues of $[h_2, \circ]$ are of the form $\sum m_i\lambda_i$, $-k \leq m_i \leq k$. If k is odd then the sum of the m_i must be odd and consequently not all of the m_i can be zero. Because of the rational independence of the λ_i, the eigenvalues of $[h_2, \circ]$ cannot be zero and consequently the coefficient of ϵ^k can be made zero. ∎

3.7.4 Perturbation examples

Two examples are considered which illustrate what is to be expected in applications of the perturbation scheme. The computations were done using the symbol manipulation program MACSYMA running on a VAX computer (where it is called VAXIMA). Most of the formulas were formated by VAXIMA as direct input for the UNIX "eqn" equation formatter. Estimated cpu times are given for the computation; interesting examples take considerable resources. The codes written by Dragt and Finn [83] are considerably faster; we do not get exactly the same answer as Dragt and Finn [83].

3.7.4.1 The cubic potential

The first example has one degree of freedom. Thus, there are two variables (p and q). The Hamiltonian is a linear oscillator with a quadratic force as a perturbation:

$$h = \tfrac{1}{3}q^3\epsilon + \tfrac{1}{2}(q^2 + p^2). \tag{7.25}$$

Recall that the previous theorem guarantees that any term with an odd degree of homogeneity must be zero, so they are not listed; the symbol manipulation programs do confirm this fact. The first step is to find g_3 so that the coefficient of ϵ in the Hamiltonian h has $h_2 = (q^2 + p^2)/2$ as an integral. The program produces:

$$g_3 = \tfrac{1}{3}pq^2 + \tfrac{2}{9}p^3. \tag{7.26}$$

The new Hamiltonian is computed by calculating $\exp(\epsilon[g_3, \circ])h$. Here is a list of the nonzero terms in the new Hamiltonian:

$$\begin{gathered}\tfrac{1}{2}(q^2 + p^2),\\ -\tfrac{1}{6}q^2(q^2 + 2p^2)\epsilon^2,\\ \tfrac{4}{27}q(q^4 + p^2q^2 + 2p^4)\epsilon^3,\\ -\tfrac{1}{54}(5q^6 + 9p^2q^4 - 8p^4q^2 + 4p^6)\epsilon^4.\end{gathered}$$

To simplify the coefficient of ϵ^2, the code produces

$$g_4 = (-\tfrac{1}{16}pq^3 - \tfrac{1}{48}5p^3q)\,\epsilon^2. \tag{7.27}$$

Here is a list of the terms in the new Hamiltonian:

$$\begin{gathered}\tfrac{1}{2}(q^2 + p^2),\\ -\tfrac{5}{48}(q^2 + p^2)^2\epsilon^2,\\ \tfrac{4}{27}q(q^4 + p^2q^2 + 2p^4)\epsilon^3,\\ -\tfrac{1}{3456}(437q^6 + 927p^2q^4 - 257p^4q^2 + 181p^6)\epsilon^4.\end{gathered}$$

Now, for the coefficient of ϵ^3,

$$g_5 = (\tfrac{4}{27}pq^4 + \tfrac{20}{81}p^3q^2 + \tfrac{64}{405}p^5)\,\epsilon^3. \tag{7.28}$$

The terms in the new Hamiltonian are now:

$$\tfrac{1}{2}(q^2 + p^2),$$

$$-\tfrac{5}{48}(q^2+p^2)^2\epsilon^2,$$
$$-\tfrac{1}{3456}(437q^6+927p^2q^4-257p^4q^2+181p^6)\epsilon^4,$$

In the ϵ^4 case

$$g_6=(-\tfrac{101}{1728}pq^5-\tfrac{77}{648}p^3q^3-\tfrac{1}{64}p^5q)\,\epsilon^4. \tag{7.29}$$

Here are the terms in new Hamiltonian:

$$\tfrac{1}{2}(q^2+p^2),$$
$$-\tfrac{5}{48}(q^2+p^2)^2\epsilon^2,$$
$$-\tfrac{235}{3456}(q^2+p^2)^3\epsilon^4.$$

The approximate cpu time required for the above computations was 10 minutes.

The Hamiltonian has the form

$$h=f(\tfrac{1}{2}(q^2+p^2)), \tag{7.30}$$

where

$$f(x)=x-\tfrac{5\times 4}{48}\epsilon^2x^2-\tfrac{235\times 16}{3456}\epsilon^4x^4+\cdots. \tag{7.31}$$

Because the new Hamiltonian is positive for small x, the new system undergoes oscillation for small initial values. When $f(x)=0$, the new system is in equilibrium. For large values of x the new system is still in oscillation; however, the curves in the phase plane are turning in a direction opposite to the direction in which they move for small x. A simple sketch of the phase plane for the old Hamiltonian will show that the motion generated by the old Hamiltonian and the new Hamiltonian are similar for small p and q, and different for large p and q.

3.7.4.2 Mirror machine model

The next example is a model for a simple mirror machine; the quadratic part of the Hamiltonian corresponds to a particle in linear motion and a simple oscillator. This example lies outside much of the standard perturbation theory, but was analyzed by Dragt and Finn [83]. The Hamiltonian is

$$h=q_1^4q_2^2\epsilon^4+\tfrac{1}{2}(q_1^2q_2^2\epsilon^2)+\tfrac{1}{2}(q_2^2+p_2^2+p_1^2). \tag{7.32}$$

Because the coefficient ϵ is zero, $g_3=0$. To simplify the coefficient of ϵ^2, the code finds that

$$g_4=-\tfrac{1}{8}(p_1q_1q_2^2)+\tfrac{1}{4}(q_1^2p_2q_2)-\tfrac{1}{8}(p_1^2p_2q_2)+\tfrac{1}{8}(p_1q_1p_2^2)\,. \tag{7.33}$$

Here is a list of the some of the terms in the new Hamiltonian:

$$\tfrac{1}{2}(q_2^2+p_2^2+p_1^2),$$
$$\tfrac{1}{4}q_1^2(q_2^2+p_2^2)\epsilon^2.$$

The fifth degree term is zero while the sixth degree term contains ten terms. In the next two simplification steps $g_5=0$ and g_6 contains fourteen terms. To complete this calculation thru terms of order ϵ^4 required about 7.2 cpu hours. As the theory predicts, the simplified terms in the new Hamiltonian all have $(q_2^2+p_2^2)/2$ as an integral, and all but the first term of the Hamiltonian have $p_1^2/2$ (or p_1) as an integral. Setting $A_2 = (q_2^2 + p_2^2)/2$, the new Hamiltonian can be written as

$$h = A_2 + \tfrac{1}{2}p_1^2 + \tfrac{1}{2}\epsilon^2 A_2 q_1^2 + \tfrac{1}{16}\epsilon^4 A_2 q_1^2 (A_2 + 14q_1^2) + \cdots.$$

Now A_2 is a constant, so the perturbation algorithm can be applied to the new Hamiltonian, and it can be simplified even further. For more details see [80]. It is also possible to view this Hamiltonian as governing the motion of a particle in a potential:

$$h = \tfrac{1}{2}p_1^2 + V(A_2, q_1),$$

where

$$V(A_2, q_1) = \tfrac{1}{2}\epsilon A_2 q_1^2 + \tfrac{1}{16}\epsilon^4 A_2 q_1^2 (A_2 + 14q_1^2) + \cdots.$$

3.7.5 VAXIMA (MACSYMA) programs for Lie series

The following programs can be used to compute the results in the previous two examples. The easiest way to follow these programs is to start reading the function called *example1()*.

Program Listings

```
/* comments:

     functions defined:
          hamilton
          poisson
          expham
          polyham
          adjointham
          compintham
          minpertham
          solvham
          pertexpham
          doitallham
          exampleN , N=1,...,9

     global variables:
          n, number of degrees of freedom
          t, like epsilon in perturbation series, used to keep track of terms
               of various orders
          p,q, variables if n=1
          p[i],q[i], arrays of variables if n>1
          h,g Hamiltonians
          ntop, largest index in expansions
          c, array of coefficients to be determined
          pert, perturbation term
          termlist, basis for terms in perturbation
          result, list of terms to be simplified
          order, order of term being simplified
          method, choice of methods */
```

```
hamilton(index) := block([ind],
     /* make a general set-up */
     n:index,
     if n = 1 then (
          depends([f,g,h],[q,p])
     ) else (
          ind:[],
          for i thru n do ind:append(ind,[q[i],p[i]]),
          depends([f,g,h],ind)
     ),
     display(n),
     disp(dependencies),
     done)$

poisson(f1,f2) := block(
     /* compute the poisson bracket */
     if n = 1 then (
          return(diff(f1,q)*diff(f2,p)-diff(f1,p)*diff(f2,q))
     ) else (
          return(sum(
               diff(f1,q[i])*diff(f2,p[i])-diff(f1,p[i])*diff(f2,q[i])
          ,i,1,n))
     ))$

expham(f1,f2) := block([term,sum],
     /* compute exp([f1,.])f2 */
     term:f2,
     sum:term,
     for i thru ntop do (
          term:ev(poisson(f1,term),expand,diff),
          sum:sum + term/i!
     ),
     sum)$

polyham(index) := block([term,pol,nn],
     /* make a general homogeneous polynomial of degree index */
     if n = 1 then
          pol:(q+p)
     else
          pol:sum(q[i] + p[i],i,1,n),
     pol:expand(pol↑index),
     nn:nterms(pol),
     array(c,nn),
     termlist:[],
     for i thru nn do (
          term:part(pol,i),
          termlist:endcons(term/numfactor(term),termlist)
     ),
     pert:sum(c[i]*termlist[i],i,1,nn),
     done)$

adjointham(index,f1) := block([temp,temp1,temp2,adj],
     /* calculate the expression that determines the perturbation
                    using the method of Dragt. */
```

```
     temp1:ratcoeff(h,t,0),
     temp2:poisson(f1,temp1) + ratcoeff(h,t,index),
     if n = 1 then
          adj:-subst(-p,temp,subst(q,p,subst(temp,q,temp1)))
     else (
          adj:-temp1,
          for i thru n do (
          adj:-subst(-p[i],temp,subst(q[i],p[i],subst(temp,q[i],adj)))
          )
     ),
     result:[poisson(temp2,adj)]
     )$

compintham(index,f1) := block([temp1,temp2,adj],
     /* calculate the expression that determines the perturbation
        using the complete integrabilty method */
     temp1:ratcoeff(h,t,0),
     temp2:poisson(f1,temp1)+ratcoeff(h,t,index),
     if n = 1 then
          adj[1]:-ratcoeff(expand(temp1),q,2)*p↑2
               +ratcoeff(expand(ratcoeff(expand(temp1),q,1)),p,1)*p*q
               -ratcoeff(expand(temp1),p,2)*q↑2
     else for i thru n do (
          adj[i]:-ratcoeff(expand(temp1),q[i],2)*p[i]↑2
               +ratcoeff(expand(ratcoeff(expand(temp1),q[i],1)),p[i],1)*p[i]*q[i]
               -ratcoeff(expand(temp1),p[i],2)*q[i]↑2
     ),
     result:[],
     for i thru n do (
          result:endcons(poisson(adj[i],temp2),result)
     ),
     remarray(adj),
     done)$

minpertham() := block([nn,temp,min,zz],
     /* minimize the sum of the squares of the coefficients of pert */
     nn:length(termlist),
     temp:expand(pert),
     for i thru nn do (
          temp:expand(ratsubst(zz[i],termlist[i],temp))
     ),
     min:sum(ratcoeff(temp,zz[i])↑2,i,1,nn),
     temp:arrayinfo(c),
     list:[],
     for i:3 thru length(temp) do list:append(temp[i],list),
     varilist:[],
     for i thru nn do (
          if not member(i,list) then varilist:endcons(c[i],varilist)
     ),
     eqnlist:[],
     for i thru length(varilist) do (
         eqnlist:endcons(diff(min,varilist[i]),eqnlist)
     ),
     globalsolve:true,
     linsolve+params:false,
     if length(varilist) > 0 then linsolve(eqnlist,varilist),
     done)$
```

```
solvham() := block([varilist,eqnlist,nn,zz,fac],
     /* find the perturbation in the homogeneous polynomial case */
     varilist:[],
     eqnlist:[],
     nn:length(termlist),
     result:expand(result),
     for i thru nn do (
          result:expand(ratsubst(zz[i],termlist[i],result))
     ),
     for i thru length(result) do (
          for j thru nn do (
               eqnlist:endcons(ratcoeff(result[i],zz[j]),eqnlist)
          )
     ),
     for i thru nn do varilist:endcons(c[i],varilist),
     remarray(c),
     globalsolve:true,
     linsolve+params:false,
     linsolve(eqnlist,varilist),
     pert:ev(pert),
     done)$

pertexpham(f1,f2) := block([term,newterm,sum],
     /* calculate exp([f1,.])f2 in the perturbation case */
     term:sum(t↑k*ratcoeff(f2,t,k),k,0,ntop),
     sum:term,
     for k thru ntop do (
          newterm:0,
          for i:0 thru ntop do (
               newterm:newterm+sum(
                    t↑i*poisson(ratcoeff(f1,t,j),ratcoeff(term,t,i-j))
                         ,j,0,ntop)
          ),
          term:expand(newterm),
          sum:sum+term/k!
     ),
     sum:sum(t↑k*expand(ratcoeff(sum,t,k)),k,0,ntop),
     sum)$

doitallham(order,method) := block([temp],
     /* call all of the programs necessary to complete one step
        of the perturbation procedure.
        method = 1 , Dragt or adjoint method.
        method = 2 , attempt complete integrabilty method. */
     remarray(c),
     polyham(order+2),
     if method = 1 then adjointham(order,pert),
     if method = 2 then compintham(order,pert),
     solvham(),
     minpertham(),
     pert:ratsimp(pert),
     pert:t↑order*ev(pert,infeval,expand),
     typeset:true,
     disp(g[order+2] = pert),
     typeset:false,
     h:pertexpham(pert,h),
```

```
    nn:hipow(h,t),
    temp:0,
    for i:0 thru nn do temp:temp+ratsimp(ratcoeff(h,t,i))*t↑i,
    typeset:true,
    for i:0 thru nn do disp(factor(ratcoeff(h,t,i))*t↑i),
    typeset:false,
    done)$

example1() := block([nn],
  /* first example for this paper */
    gcprint:false,
    programmode:true,
    writefile(text1),
    showtime:true,
    w:1,
    n:1,
    ntop:4,
      h:w*(q↑2+p↑2)/2+t*q↑3/3,
      typeset:true,
      display(h),
      typeset:false,
    method:1,
    for order:1 thru ntop do ( doitallham(order,method),
                                        save(results1,values)),
    done)$

example2() := block([nn],
  /* second example for this paper */
    gcprint:false,
    writefile(text5),
    typeset:true,
    showtime:true,
    programmode:true,
    n:2,
    ntop:6,
    h:(p[1]↑2+q[2]↑2+p[2]↑2)/2
         +t↑2*q[1]↑2*q[2]↑2/2 + t↑4*q[1]↑4*q[2]↑2,
      display(h),
    method:1,
    for order:1 thru ntop do ( doitallham(order,method),
                                        display(status(time),status(runtime)),
                                        save(results5,values)),
    done)$
```

References

We have attempted to collect a wide range of references, to indicate the extent of the interest in Lie techniques and give the reader a point of departure for some applications. Some of the references were found by doing computer searches of library data bases. Because there are so many references, they have been broken into categories. By far the largest number of references are in celestial mechanics; we have omitted many of these.

General hamiltonian Lie techniques

[1] J.R. Cary, Time dependent canonical transformations and the symmetry equals-invariant theorem. *J. Math. Phys.* **18**, 2432–2435 (1977).

[2] J.R. Cary, Lie transform perturbation theory for Hamiltonian systems. *Phys. Rep.* **79**, 129–159 (1981).

[3] R.A. Howland, An accelerated elimination technique for the solution of perturbed Hamiltonian systems. *Celestial Mech.* **15**, 327–352 (1977).

[4] R.A. Howland and D.L. Richardson, The Hamiltonian transformation in quadratic Lie transforms. *Celestial Mech.* **32**, 99–107 (1984).

[5] A.N. Kaufman, The Lie transform: A new approach to classical perturbation theory. *AIP Conf. Proc.* **46**, 268–295 (1978).

[6] A.J. Laub and K. Meyer, Canonical forms for symplectic and Hamiltonian matrices. *Celestial Mech.* **9**, 213–238 (1974).

[7] S. Lie, *Sophus Lie's 1884 Differential Invariants Paper* (translated by M. Ackerman, comments by R. Hermann). *Math. Sci. Press*, Brookline (1976).

[8] S. Lie, *Sophus Lie's 1880 Transformation Group Paper* (translation by M. Ackerman, comments by R. Hermann). *Math. Sci. Press*, Brookline (1976).

[9] V.P. Petruk, The use of Lie series to study nonlinear Hamiltonian systems. *Cosmic Res.* **15**, 800–803 (1977).

[10] R. Ramaswamy and R.A. Marcus, On the onset of chaotic motion in deterministic systems. *J. Chem. Phys.* **74**, 1385–1393 (1981).

[11] C.A. Uzes, Mechanical response and the initial value problem. *J. Math. Phys.* **19**, 2232–2238 (1978).

General non-hamiltonian Lie techniques

[12] R.V. Gamkrelidze, Exponential representation of solutions of ordinary differential equations, Equadiff IV. *Proc. Czechoslovak Conf. Diff. Eqns. and their Appl.*, Prague (1977).

[13] D.M. Greig and M.A. Abd–El–Naby, Iterative solutions of nonlinear initial value differential equations in Chebyshev series using Lie series. *Numer. Math.* **34**, 1–13 (1980).

[14] D.M. Greig and M.A. Abd–El–Naby, A series method for solving nonlinear two-point boundary value problems. *Numer. Math.* **34**, 87–98 (1980).

[15] M. Fliess, M. Lamnabhi, and F. Lamnabhi–Lagarrigue, An algebraic approach to nonlinear functional expansions. *IEEE Trans. on Circuits and Systems* **CAS-30**, 554–570 (1983).

[16] W. Gröbner, Die Lie-Reihen und ihre Anwendungen. *VEB Deutscher Verlag der Wissenschaften*, Berlin (1960).

[17] W. Gröbner and H. Knopp, Contributions to the Method of Lie Series. *Bibliographisches Institut*, Mannheim (1967).

[18] L. Hlavatý, S. Steinberg, and K.B. Wolf, Riccati equations and Lie series. *J. Math. Anal.* **104**, 246–263 (1984).

[19] C.R. Hebd, Development of indeterminate non-commutative functions of solutions to differential equations for nonlinear forces. *Seances Acad. Sci.* **A287**, 1133–1135 (1978).

[20] R.A. Howland, A quadratic transformation technique for the solution of nonlinear systems. *SIAM J. Math. Anal.* **12**, 90–103 (1981).

[21] H. Knapp and G. Wanner, Numerical solution of ordinary differential equations by Groebner's method of Lie-series. *Math. Research Center*, Univ. of Wisconsin, Madison (1968).

[22] K. Ludwig, Topologische Gruppen von Lie-Reihen. *Wiss. Z. Hochsch.* Verkehrswesen Friedrich List, Dresden 27, 799–808 (1980).

[23] J.A. Mitropolsky, Sur la decomposition asymptotique des systemes differentiels fondée sur des transformations de Lie. *Proc. Internat. Conf. Nonlinear Diff. Eqns.*, 283–326. Trento (1980). Academic Press, New York (1981).

[24] J.F. McGarvey, Approximating the general solution of a differential equation. Classroom Notes in Applied Mathematics, *SIAM Review* **24**, 333-337 (1982).

[25] J. Murdock, A unified treatment of some expansion procedures in perturbation theory: Lie series FAA Di Bruno operators, and Arbogast's rule. *Celestial Mech.* **30**, 293–295 (1983).

[26] S. Steinberg, Local Propagators. *Rocky Mountain J. Math.* **10**, 767–798 (1980).

[27] S. Steinberg, Lie series and nonlinear ordinary differential equations. *J. Math. Anal. Appl.* **101**, 39–63 (1984).

[28] S. Steinberg, Factored product expansions of solutions of nonlinear differential equations. *SIAM J. Math. Anal.* **15**, 108–115 (1984).

Baker–Campbell–Hausdorff relations

[29] D. Finkelstein, On relations between commutators. *Comm. Pure Appl. Math.* **8**, 245–250 (1955).

[30] E.Q. Gora, On formulas in closed form for van Vleck expansions. *Inter. J. Quant. Chem.* **6**, 681–700 (1972).

[31] W. Magnus, On the exponential solution of differential equations for a linear operator. *Comm. Pure Appl. Math.* **7**, 649–673 (1954).

[32] R.D. Richtmeyer and S. Greenspan, Expansions of the Campbell–Baker–Hausdorff formula by computer. *Comm. Pure Appl. Math.* **18**, 107–108 (1965).

[33] S. Steinberg, Applications of the Lie algebraic formulas of Baker, Campbell, Hausdorff, and Zassenhaus to the calculation of explicit solutions of partial differential equations. *J. Diff, Eqns.* **26**, 404–434 (1977).

[34] J. Wei and E. Norman, On global representations of the solutions of linear differential equations as a product of exponentials. *Proc. Amer. Math. Soc.* **15**, 327–334 (1964).

[35] K.B. Wolf, On time-dependent quadratic quantum Hamiltonians. *SIAM J. Appl. Math.* **40**, 419–431 (1981).

Celestial mechanics

[36] K. Aksnes, On the choice of reference orbit, canonical variables, and perturbation method in satellite theory. *Celestial Mech.* **8**, 259 (1973).

[37] K.T. Alfriend, A nonlinear stability problem in the three dimensional restricted three body problem. *Celestial Mech.* **5**, 502–511 (1972).

[38] Richard B. Barrar, On the non-existence of transformations to normal form in celestial mechanics. G.E.O. Giacaglia (Editor), *Periodic Orbits, Stability and Resonance*, Reidel, Dordrecht, 228–231 (1970).

[39] N. Burgoyne and R. Cushman, Normal forms for real linear Hamiltonian systems with purely imaginary eigenvalues. *Celestial Mech.* **8**, 435–443 (1974).

[40] J.A. Campbell and W.H. Jefferys, Equivalents of the perturbation theories of Hori and Deprit. *Celestial Mech.* **2**, 467–473 (1970).

[41] H. Claes, Analytical theory of earth's artificial satellites. *Celestial Mech.* **21**, 193–198 (1980).

[42] J.E. Cochran, Long-term motion in a restricted problem of rotational motion (artificial satellites). B.D. Tapley, V. Szebehely (Editors), *Recent Advances in Dynamical Astronomy*, Reidel, Dordrecht, 429–453 (1973).

[43] S. Coffey and K.T. Alfriend, Short period elimination for the tesseral harmonics. *AAS/AIAA Astrodynamics Spec. Conf.*, Lake Tahoe, Nevada (1981).

[44] C.J. Cohen and R.H. Lyddane, Radius of convergence of Lie series for some elliptic elements. *Celestial Mech.* **25**, 221–234 (1981).

[45] A. Deprit, Canonical transformations depending on a small parameter. *Celestial Mech.* **1**, 12–30 (1969).

[46] A. Deprit, The main problem of artificial satellite theory for small and moderate eccentricities. *Celestial Mech.* **2**, 166–206 (1970).

[47] A. Deprit, J. Henrard, J.F. Price, and A. Rom, Birkhoff's normalization. *Celestial Mech.* **1**, 222–251 (1969).

[48] G. Dulinski, The rotational motion of the natural and artificial celestial bodies. I. *Postepy Astron.* **29**, 3–15 (1981).

[49] R. Dvorak and A. Hanslmeir, Numerical integration with Lie-series. S. Ferraz-Mello, P.E. Nacozy (Editors), *Motion of Planets and natural and Artificial Satellites, Proc. of a CNPQ-NSF Symp. and Workshop*, 65–72 (1983).

[50] S. Filippi, A new Lie series method for the numerical integration of ordinary differential equations, with an application to the restricted problem of three bodies. *Technical Report* **NASA-TN-D-3857** (1967).

[51] B. Garfinkel, A. Jupp, and C. Williams, A recursive Von Zeipel algorithm for the ideal resonance problem. *Astronomical J.* **76**, 157–166 (1971).

[52] F. Gustavson, On constructing formal integrals of a Hamiltonian system near an equilibrium point. *Astronomical J.* **71**, 670–686 (1966).

[53] J. Henrard and J. Roels, Equivalence for Lie transforms. *Celestial Mech.* **10**, 497–512 (1974).

[54] J. Henrard and M. Moons, Hamiltonian theory of the libration of the Moon. V. Szebehely (Editor), *Dynamics of Planets and Satellites and Theories of Their Motion, Proc. of the 41st Colloq. of the International Astronomical Union*, Cambridge (1976), Dordrecht (1978), 125–135.

[55] K.V. Holshevnikov, Short-period perturbations of the state vector in the method of Lie transforms. Leningrad. *Gos. Univ. Uchen. Zap.* **402**, *Ser. Mat. Nauk Vyp.* **48**, *Trudy Astronom. Obser.* **36**, 124–134 (1981).

[56] G. Hori, Theory of general perturbations with unspecified canonical variables. *Publ. Astron. Soc. Japan* **18**, 287–296 (1966).

[57] G. Hori, Theory of general perturbations. B.B. Tapley, V. Szebehely (Editors), *Recent Advances in Dynamical Astronomy*, Reidel, Dordrecht, 231–249 (1973).

[58] A.H. Jupp, A comparison of the Bohlin-Von Zeipel and Bohlin-Lie series methods in resonant systems. *Celestial Mech.* **26**, 413–422 (1982).

[59] A.A. Kamel, Perturbation methods in the theory of nonlinear oscillations. *Celestial Mech.* **3**, 90–106 (1970).

[60] B. Kaufman, K.T. Alfriend, and R.R. Dasenbrock, Luni-solar perturbations in the extended phase space representation of the Vinti problem. *Acta Astronuatica* **5**, 727–744 (1978).

[61] B. Kaufman, First order semianalytic satellite theory with recovery of the short period terms due to third body and zonal perturbations. *Acta Astronautica* **8**, 611–623 (1981).

[62] B. Kaufman and W.H. Harr, Implementation of a semianalytic satellite theory with recovery of short period terms. *Acta Astronautica* **11**, 279–286 (1984).

[63] U. Kirchgraber, The property of covariance in Hori's noncanonical perturbation theory. B. B. Tapley, V. Szebehely (Editors), *Recent Advances in Dynamical Astronomy*, Riedel, Dordrecht, 260–261 (1973).

[64] D.Z. Koenov, Solution of canonical equations of perturbed translational-rotational motion of a planet and its satellite by the method of Lie transforms. *Izv. Akad. Nouk Tadzhik. SSR, otdel Fiz.-Mat. Khim. i Geol. Nauk* **83**, 47–54 (1982).

[65] D.J. Lelgemann, A linear solution of the equations of motion of an earth-orbiting satellite based on Lie-series. *Celestial Mech.* **30**, 309–321 (1983).

[66] Liu Lin, Several notes on the transformation methods in non-linear systems. *Acta Astron. Sin.* **23**, 255–263 (1982).

[67] B. McNamara, Super-convergent adiabatic invariants with resonant denominators by Lie series transforms. *J. Math. Phys.* **19**, 2154–2164 (1978).

[68] J. Meffroy, On the elimination of short-period terms in second-order general planetary theory investigated by Hori's method. *Astrophys. and Space Sci.* **25**, 271–354 (1973).

[69] W.A. Mersman, Explicit recursive algorithms for the construction of equivalent canonical transforms. *Celestial Mech.* **3**, 384–389 (1971).

[70] P.J. Message, Asymptotic series for planetary motion in periodic terms in three dimensions. *Celestial Mech.* **26**, 25–39 (1982).

[71] K.R. Meyer, Normal forms for Hamiltonian systems. *Celestial Mech.* **9**, 517–522 (1974).

[72] P. Michaelidis, Orbits near a 2/3 resonance. *Astronom. and Astrophys.* **91**, 165–174 (1980).

[73] B.A. Romanowicz, On the tesseral-harmonics resonance problem in artificial-satellite theory. *Report, Smithsonian Inst. Astrophys Obs.*, Cambridge (1975).

[74] D.S. Schmidt, Literal solution for Hill's lunar problem. *Celestial Mech.* **19**, 279–289 (1979).

[75] Harold Shniad, The equivalence of Von Zeipel mappings and the Lie transforms. *Celestial Mech.* **2**, 114–120 (1970).

[76] M. Sidlichovsky, The inclination changes in the problem of two triaxial rigid spheroids. *Celestial Mech.* **22**, 343–355 (1980).

[77] F. Spirig, Algebraic aspects of perturbation theories. *Celestial Mech.* **20**, 343–354 (1979).

[78] D. Standaert, Direct perturbations of the planets on the Moon's motion. *Celestial Mech.* **22**, 357–369 (1980).

[79] K. Stumpff, On the application of Lie-series to the problems of celestial mechanics. *Nasa Technical Note* **NASA TN D-4460** (1968).

Magnetic optics

[80] D.R. Douglas and A.J. Dragt, Lie algebraic methods for particle tracking calculations. *IEEE Trans. Nuc. Sci* **NS-28**, 2522 (1981).

[81] A.J. Dragt and J.M. Finn, Lie series and invariant functions for analytic symplectic maps. *J. Math. Phys.* **17**, 2215–2227 (1976).

[82] A.J. Dragt, A method of transfer maps for linear and nonlinear beam elements. *IEEE Transactions on Nuclear Science* **NS-26**, 3601–3603 (1979).

[83] A.J. Dragt and J.M. Finn, Normal form for mirror machine Hamiltonians. *J. Math. Phys.* **20**, 2649–2660 (1979).

[84] A.J. Dragt, Transfer map approach to the beam-beam interaction. *Proc. of Symp. on Nonlinear Dynamics and the Beam-Beam Interaction*, Brookhaven National Laboratory (1979).

[85] A.J. Dragt and O.G. Jakubowicz, Analysis of the beam-beam interaction using transfer maps. *Preprint, Dept. of Physics and Astronomy*, Univ. of Maryland (1980).

[86] A.J. Dragt, Charged particle beam transport using Lie algebraic methods. To appear in *IEEE transactions on Nuclear Science.*

[87] A.J. Dragt, Lectures on nonlinear orbit dynamics. *AIP Conference Proceedings* **87** (1982).

[88] A.J. Dragt and E. Forest, Computation of nonlinear behavior of Hamiltonian systems using Lie algebraic methods. *J. Math. Phys.* **24**, 2734–2744 (1984).

[89] J.M. Finn, Integral of Canonical Transformations and Normal Forms for Mirror Machine Hamiltonians. *Thesis*, Univ. Maryland, College Park.

Light optics

[90] A.J. Dragt, A Lie algebraic theory of geometrical optics and optical aberrations. *J. Opt. Soc. America* **72**, 372–379 (1982).

[91] J.C. Garrison and J. Wong, Corrections to the adiabatic approximation for variable parameter free-electron laser. *IEEE J. Quantum Electron* **QE-17**, 1469–1475 (1981).

[92] J.C. Garrison and J. Wong, Nonadiabatic corrections and detrapping for variable parameter free-electron lasers. S.F. Jacobs *et al.* (Editors), *Free-Electron Generators of Coherent Radiation, 3rd Workshop of Free-Electron Laser Devices*, Sun Valley, Idaho, 349–365 (1982).

[93] M. Navarro–Saad and K.B. Wolf, The group-theoretical treatment of aberrating systems. I. Aligned lens systems in third aberration order. *Comunicaciones Técnicas IIMAS* N° 363 (1984). To appear in *J. Math. Phys. (March 1986).*

[94] K.B. Wolf, Approximate canonical transformations and the treatment of aberrations. I. One-dimensional simple n-th order aberrations in optical systems. *Comunicaciones Técnicas IIMAS* N° 352 (1983). (Preliminary version.)

[95] K.B. Wolf, The group-theoretical treatment of aberration system. II. Axis-symmetric inhomogeneous systems and fiber optics in third aberration order. *Comunicaciones Técnicas IIMAS* N° 366 (1984). To appear in *J. Math. Phys. (March 1986).*

[96] K.B. Wolf, On the group-theoretical treatment of Gaussian optics and third-order aberrations. *Proc. of the XII Internat. Colloq. on Group-Theoretical Meth. in Phys.*, Trieste (1983). Lecture Notes in Physics, Vol. 201, (Springer Verlag, 1984); pp. 133–136.

Neutron transport

[97] T. Auerbach, W. Halg, and J. Mennig, The treatment of boundary value problems in optically thick media by means of Lie series. *Nucl. Sci. and Eng.* **49**, 384–387 (1972).

[98] T. Auerbach, J.P. Gandillon, W. Halg, and J. Mennig, Analytical solution of s_4-equations in plane geometry. *Comput. Methods Appl. Mech. Eng.* **2**, 133–146 (1973).

[99] R.A. Axford, Integral-transform Lie-series analysis of transient temperatures in reactor coolant channels. *Technical report*, Los Alamos National Laboratory (1967).

[100] G.H. Cristea, Application of Lie series to resolve the time dependent Boltzmann equation in the diffusion approximation. *Technical Report*, Inst. Atomic Phys., Bucharest (1973).

[101] T. Juillearat, A three-dimensional multigroup p_l-theory with axially variable parameters. *Nukleonik* **12**, 117–124 (1969).

[102] V.P. Korennol and O.V. Khatkevich, Use of the Lie-series method for calculating neutron flux anisotropy in the multi-group p_n-approximation for a flat heterogeneous reactor cell. *Vestsi Akad. Navuk BSSR Ser. Fiz. Energ. Navuk* **4**, 10–14 (1984).

[103] M. Lemanska, Y. Ilamed, and S. Yiftah, The theory of a one-dimensional bare reactor, treated by the Lie series method. *J. Nuclear Energy* **23**, 361–367 (1969).

[104] M. Lemanska, The solution of the multi-dimensiona, multi-group diffusion equation using the Lie series method for a bare reactor. *J. Nucl. Energy* **25**, 397–403 (1971).

[105] M. Lemanska, Exact solution of p_n time-dependent equations with time-dependent cross-sections for slab geometry. *Z. Angw. Math. and Phys.* **26**, 701–711 (1975).

[106] C. Lepori, T. Auerbach, W. Halg, and J. Mennig, Analytical solution of the position-dependent planar neutron transport problem, taking into account anisotropic scatter and several energy groups. *Z. Angew. Math. and Phys.* **25**, 132–135 (1974).

[107] A. Zurkinden, Time dependent neutron transport theory in multigroup p_l-approximation. *Z. Angew. Math. and Phys.* **28**, 393–408 (1977).

Plasma physics

[108] J.R. Cary and A.N. Kaufman, Pondermotive effects in collisionless plasma: A Lie transform approach. *Phys. Fluids* **24**, 1238–1250 (1981).

[109] J.R. Cary and R.G. Littlejohn, Noncanonical Hamiltonian mechanics and its application to magnetic field line flow. *Ann. Phys.* **151**, 1–34 (1983).

[110] S. De, Explosive instabilities in beam plasma system. *Plasma Phys.* **24**, 1043–1050 (1982).

[111] D.H.E. Dubin, J.A. Krommes, C. Oberman, and W.W. Lee, Nonlinear gyrokinetic equations. *Phys. Fluids* **26**, 3524–3535 (1983).

[112] T. Hatori and H. Washimi, Covariant form of the pondermotive potentials in a magnetized plasma. *Phys. Rev. Lett.* **46**, 240–242 (1981).

[113] S. Johnston and A.N. Kaufman, Oscillation centres and mode coupling in non-uniform Vlasov plasma. *J. Plasma Phys.* **22**, 105–119 (1979).

[114] A.N. Kaufman, J. R. Cary, and N. R. Pereira, Universal formula for quasi-static density perturbation by a magnetoplasma wave. *Phys. Fluids* **22**, 790–791 (1979).

[115] R.G. Littlejohn, A guiding center Hamiltonian: A new approach. *J. Math. Phys.* **20**, 2445–2458 (1979).

Other applications

[116] R.K. Bansal and R. Subramanian, Stability analysis of power systems using Lie series and pattern-recognition techniques. *Proc. Inst. Electr. Eng.* **121**, 623–629 (1974).

[117] R.K. Bansal, R. Subramanian, H. Kormanik, and C.C. Li, Comments on 'Decision-surface estimate of nonlinear system stability domain by Lie series method'. *IEEE Trans. Autom. Control* **AC-19**, 629–630 (1974).

[118] D.J. Bell and Q. Ye, A perturbation method for suboptimal feedback control of bilinear systems. *Int. J. Syst. Sci.* **12**, 1157–1168 (1981).

[119] G.H. Cristea, The study of the operating mechanism of Lie series to solve the forward Kolmogorov equation. *Technical Report*, Inst. Atomic Phus., Bucharest (1973).

[120] J.D. Curtis, A Modification of the Brown-Shook Method Based on Lie Series. *Thesis,* Univ. Maryland, Catensville.

[121] R.L. Dewar, Renormalized canonical perturbation theory for stochastic propagators. *J. Phys.* **A9**, 2043–2057 (1976).

[122] R.L. Dewar, Exact oscillation-centre transformations. *J. Phys.* **A11**, 9–26 (1978).

[123] T.K. Hu and J.C. Zink, Lie series perturbation solution of the invariant imbedding reflection equations. *Trans. Am. Nucl. Soc.* **18**, 146–147 (1974).

[124] J. Kormanik and C.C. Li, Decision surface estimate of nonlinear system stability domain by Lie series method. *IEEE Trans. Autom. Contr.* **AC-17**, 666–669 (1972).

[125] J. Kormanik and C.C. Li, On an algorithm for estimating uniform asymptotic stability boundary of nonautonomous system. *Proc. of the 1976 IEEE Conf. on Decision and Control Incl. the 15th Symp. on Adaptive Proc.*, IEEE (1976).

[126] K. Normuratov, Averaging of integro-differential equations using Lie series. *Izv. Akad. Nauk UZSSR Ser. Fiz.-Mat. Nauk*, 6–9 (1983).

[127] R. Rączka, Integration of classical nonlinear relativistic equations by the method of Lie series. *Acta Phys. Pol.* **B4**, 501–520 (1973).

[128] J. Rae, Subdynamics in classical mechanics. B.B. Tapley, V. Szebehely (Editors), *Recent Advances in Celestial Astronomy*, Reidel, Dordrecht, 262–269 (1973).

[129] H.J. Sussmann, Lie brackets and local controllability: A sufficient condition for scalar-input systems. *SIAM J. Control and Optimiz.* **21**, 686–713 (1983).

[130] V.F. Zhuravlev, Method of Lie series in the motion-separation problem in nonlinear mechanics. *Appl. Math. Mech.* **47**, 461–466 (1983).

Symbol manipulation

[131] R.A. Broucke, A FORTRAN-4 system for the manipulation of symbolic Poisson series with applications to celestial mechanics. *Preprint, Institute for Advanced Study in Orbital Mechanics*, Univ. of Texas at Austin (1980).

[132] B. Char, LIEPROC: A MACSYMA program for finding adiabatic invariants of simple Hamiltonian systems via the Lie transform. E. Lewis (Editor), *Proc. of the 1979 MACSYMA Users Conference.*

[133] Bruce Char and Brendon McNamara, LCPT: A program for finding linear canonical transformations. E. Lewis (Editor), *Proc. of the 1979 MACSYMA Users Conference.*

[134] A. Giorgilli, A computer program for integrals of motion. *Computer Phys. Comm.* **16**, 331–343 (1979).

[135] MACSYMA Reference Manual, *Project MAC Mathlab Group*, M.I.T. (July 1977).

[136] R. Pavelle, M. Rothstein, and J. Fitch, Computer algebra, *Scientific American 245* **6** (Dec. 1981).

[137] D.R. Stoutemyer, Symbolic computation comes of age. *SIAM News* **12** (1979).

Texts

[138] G.D. Birkhoff, Dynamical Systems, Colloquium Publications, Vol. 9. *American Math. Soc.*, New York (1927).

[139] G.E.O Giacaglia, Perturbation Methods in Non-Linear Systems. *Springer-Verlag*, New York (1972).

[140] H. Goldstein, Classical Mechanics 2nd Ed. *Addison-Wesley*, Reading (1950).

[141] E. Leimanis, The General Problem of the Motion of Coupled Rigid Bodies about a Fixed Point. *Springer-Verlag*, New York (1965).

[142] A.H. Lichtenberg and M.A. Lieberman, Regular and Stochastic Motion. *Springer-Verlag*, New York (1983).

[143] A.H. Nayfeh, Perturbation Methods. *Wiley*, New York (1973).

[144] B.G. Wybourne, Classical Groups for Physicists. *Wiley*, New York (1974).

Chapter 4

Foundations of a Lie algebraic theory of geometrical optics

by ALEX J. DRAGT, ETIENNE FOREST, and KURT BERNARDO WOLF

ABSTRACT: We present the foundations of a new Lie algebraic method of characterizing optical systems and computing their aberrations. This method represents the action of each separate element of a compound optical system —including all departures from paraxial optics— by a certain operator. The operators can then be concatenated in the same order as the optical elements and, following well-defined rules, we obtain a resultant operator that characterizes the entire system. These include standard aligned optical systems with spherical or aspherical lenses, models of fibers with polynomial z–dependent index profile, and also sharp interfaces between such elements. They are given explicitly to third aberration order.

We generalize a previous result on the factorization of the optical phase-space transformation due to a refraction interface. We also present a group-theoretical classification for aberrations of any order of systems with axial symmetry, applying it to the problem of combining aberrations; new insights are thus provided on the origin and possible correction of these aberrations. We give a fairly complete catalog of the Lie operators corresponding to various simple optical systems. Finally, there is a brief discussion of the possible merits of constructing a computer code, RAYLIE, for the Lie algebraic treatment of geometric ray optics.

4.1 Introduction

Let us consider the optical system illustrated schematically in Figure 1: a ray originates at the general intial point P^{i} with spatial coordinates $\mathbf{r}^{\mathrm{i}}$ and moves in an initial direction specified by the unit vector $\hat{\mathbf{s}}^{\mathrm{i}}$. After passing through an optical device, the ray arrives at a final point P^{f} with coordinates $\mathbf{r}^{\mathrm{f}}$ and in a direction specified by the unit vector $\hat{\mathbf{s}}^{\mathrm{f}}$. The fundamental problem of geometrical optics is: given the initial quantities $(\mathbf{r}^{\mathrm{i}}, \hat{\mathbf{s}}^{\mathrm{i}})$ and a specification of the optical device, to determine the final quantities $(\mathbf{r}^{\mathrm{f}}, \hat{\mathbf{s}}^{\mathrm{f}})$. We may search for the design of an optical device such that the relation between families of initial and final rays have various desired properties, such a focusing, Fourier transformation, or other operations.

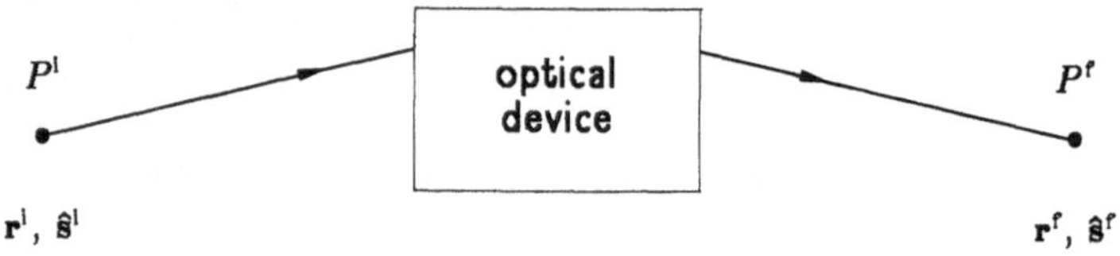

Figure 1. An optical system consisting of an optical device preceded and followed by free propagation in a homogeneous medium. A ray originates at P^i with location $\mathbf{r}^i$ and direction $\hat{\mathbf{s}}^i$, and terminates at P^f with location $\mathbf{r}^f$ and direction $\hat{\mathbf{s}}^f$.

The purpose of this chapter is to provide the foundations for a Lie algebraic approach to the problem of characterizing optical elements and optical systems [1].[1] In this approach, we represent the action of each separate element of a compound optical system, *including all departures from paraxial gaussian optics*, by a corresponding ***operator***. These operators can be concatenated to obtain the resultant operator that characterizes the entire system. Lie algebraic methods provide an operator extension of the matrix methods of gaussian (paraxial) optics to the general case [2].

We believe that the Lie algebraic approach simplifies the calculation of aberrations, and may facilitate their correction. Moreover, this approach is ideally suited to machine computation. Finally, the operator methods developed for geometrical optics may also have extensions to wave optics. This is the subject of ongoing research.

The organization of this chapter is as follows. In Section 2 we show that every optical system gives rise to —and is characterized by— a ***symplectic map***. In Section 3 we present the necessary Lie algebraic tools, and show the way in which symplectic maps can be written as products of Lie transformations. Section 4 describes the computation of Lie transformations for continuous systems, and Section 5 describes the treatment of discontinuous interfaces between two such media. In Section 6 we relate these Lie transformations to paraxial optics and describe aberrations as classified into symplectic group multiplets. We apply this classification in Section 7 to the problem of combining aberrations. In a final section we describe briefly the possible merits of constructing a computer code, **RAYLIE**, for the Lie algebraic treatment of geometric ray optics.

4.2 Optical symplectic maps

In this section we summarize the Fermat–Hamilton formulation of geometrical optics. This will provide the basis for other chapters in the present volume, and is particularly well suited for the introduction of the Lie algebraic methods in the next section. The "black-box" optical device of Figure 1 will be now specified as a three-dimensional medium where an ***index of refraction*** function $n(\mathbf{r}) = n(x, y, z)$ is defined. A ***homogeneous*** medium corresponds to $n(\mathbf{r}) =$ constant. In vacuum $n = 1$, while for any other medium, $n > 1$.

[1]This article gives a preliminary description of the use of Lie algebraic methods in geometrical optics.

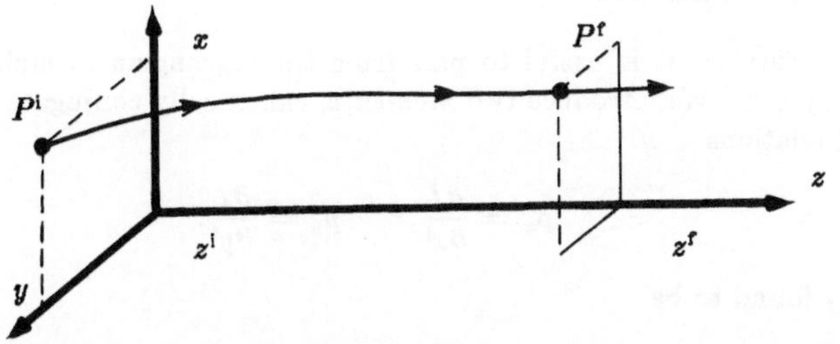

Figure 2. The coordinate system used to describe a light ray: the optical z–axis extends to the right and, normal to it, at $z^{\rm i}$ and $z^{\rm f}$, are the object and image planes parametrized by cartesian coordinates.

We fix the z–coordinate of the initial and final points, $P^{\rm i}$ and $P^{\rm f}$, as shown in Figure 2, and refer to the planes $z = z^{\rm i}$ and $z = z^{\rm f}$ as *object* and *image* planes, respectively. Further, we parametrize the light ray from $P^{\rm i}$ to $P^{\rm f}$ using z as an independent variable, that is, we describe the path of the ray as a line in space specified by the two functions $x(z)$ and $y(z)$. The elements of path length ds along a ray is then given by the expression

$$\begin{aligned} ds &= \sqrt{(dz)^2 + (dx)^2 + (dy)^2} \\ &= \sqrt{1 + (x')^2 + (y')^2}\, dz, \end{aligned} \tag{2.1}$$

where a prime will denote the differentiation with respect to z. As a technical restriction we must thus demand here that the ray coordinates $x(z)$, $y(z)$, be differentiable functions of z.

4.2.1 Fermat's principle

The ***optical path length*** along a ray from $P^{\rm i}$ to $P^{\rm f}$ is defined by the integral

$$A := \int_{z^{\rm i}}^{z^{\rm f}} dz\, n(x, y, z) \sqrt{1 + (x')^2 + (y')^2}. \tag{2.2}$$

Now, ***Fermat's principle*** states that Nature behaves such that the optical path length on the ray be an *extremum*. As in the mechanics of point particles, this leads to the formulation of the Euler–Lagrange equations

$$\begin{aligned} \frac{d}{dz}\frac{\partial L}{\partial x'} - \frac{\partial L}{\partial x} &= 0, \\ \frac{d}{dz}\frac{\partial L}{\partial y'} - \frac{\partial L}{\partial y} &= 0, \end{aligned} \tag{2.3}$$

with an optical Lagrangian L given by the expression

$$L = n(x, y, z) \sqrt{1 + (x')^2 + (y')^2}. \tag{2.4}$$

4.2.2 The hamiltonian formulation

To proceed further, it is useful to pass from the lagrangian formulation to the hamiltonian formulation of the system. We introduce two ***momenta***, canonically conjugate to the coordinates x and y, by the standard relations

$$p_x = \frac{\partial L}{\partial x'}, \qquad p_y = \frac{\partial L}{\partial y'}. \tag{2.5}$$

These are explicitly found to be

$$p_x = n(\mathbf{r})\frac{x'}{\sqrt{1+(x')^2+(y')^2}}, \qquad p_y = n(\mathbf{r})\frac{y'}{\sqrt{1+(x')^2+(y')^2}}. \tag{2.6}$$

and are dimensionless quantities.

We construct a two-component vector $\mathbf{p}$ with entries p_x and p_y, and another one $\mathbf{q}$ with entries $q_x = x$ and $q_y = y$. We should note carefully that whereas $\mathbf{q}$ ranges over the full two-dimensional plane $\Re^2$, $\mathbf{p}$ ranges over a closed circular region[2] of radius $n(q,z)$:

$$p^2 := p_x^2 + p_y^2 = \frac{n^2|\mathbf{q}'|^2}{1+|\mathbf{q}'|^2} = (n\sin\theta)^2 \leq n^2. \tag{2.7}$$

In this expression, θ is the angle between the ray and the optical axis: $|\mathbf{q}'| = |d\mathbf{q}/dz| = \tan\theta$. Rays parallel (or antiparallel) to the optical axis have $\mathbf{p} = \mathbf{0}$; if the ray has a projection advancing in the direction of $+q_x$ or $+q_y$, then the corresponding p_x or p_y is positive; otherwise it is negative. Rays at right angles to the optical axis have their maximal value $|\mathbf{p}| = n$. In general, $|\mathbf{p}| = n\sin\theta$.

According to the standard rule, the ***Hamiltonian*** $H(\mathbf{p},\mathbf{q})$ corresponding to the Lagrangian L is given by the relation $H = p_x x' + p_y y' - L$. In the case of the Lagrangian (2.4) we find the Hamiltonian to be [3]

$$H = -\sqrt{n(\mathbf{q},z)^2 - p^2} = -n\cos\theta. \tag{2.8}$$

A ray leaving the initial point P^{i} is characterized by the initial quantitites $\mathbf{q}^{\mathrm{i}}$ and $\mathbf{p}^{\mathrm{i}}$. The quantity $\mathbf{q}^{\mathrm{i}}$ specifies the point of origin of the ray on the object plane and, according to (2.6–7), $\mathbf{p}^{\mathrm{i}}$ describes the initial *direction* of the ray. Similarly, $\mathbf{q}^{\mathrm{f}}$ and $\mathbf{p}^{\mathrm{f}}$ characterize the ray as it arrives at the final point P^{f} in the image plane. The relation between the initial conditions $(\mathbf{q}^{\mathrm{i}},\mathbf{p}^{\mathrm{i}})$ and the final conditions $(\mathbf{q}^{\mathrm{f}},\mathbf{p}^{\mathrm{f}})$ is found by following, from z^{i} to z^{f}, the trajectory $\big(\mathbf{q}(z),\mathbf{p}(z)\big)$ governed by the Hamilton equations of motion, *viz.*

$$\mathbf{q}' = \nabla_p H, \qquad \mathbf{p}' = -\nabla_q H. \tag{2.9}$$

The constraints on the ranges of $\mathbf{p}$ and H that we noted above, are inherent in the Hamilton formulation of optics. In this chapter we shall not treat them further but will, instead, disregard them judiciously. The locality of the Hamilton equations (2.9) allows us this freedom from global considerations, at the cost of having to watch for outgoing rays with imaginary angles. We are treating metaxial rays which are nevertheless far from the perpendicular. This condition regarding the matter is a *caveat* that naïve Schrödinger-type ***wavization*** of geometrical optics *cannot* be performed consistently since the range of the classical momentum is bounded[3]

[2] A vector along the ray direction in three-dimensional space ranges over a *sphere*. This projects on the x–y plane as *two* disks joined at the circumference.

[3] Further comments on this problem are in Sect. **5.5.3** and Chapter 8, in this volume. (Editor's note.)

4.2.3 The optical transfer map

It is convenient to introduce now a four-component column vector $\mathbf{w}$, with entries $(\mathbf{q}, \mathbf{p})$:

$$\mathbf{w}^\top = (w_1, w_2, w_3, w_4) = (p_x, p_y, q_x, q_y), \tag{2.10}$$

where we indicate by $^\top$ vector or matrix transposition.

Using the terminology of classical mechanics, the set of vectors $\mathbf{w}$ is the ***optical phase space.*** We indicate by $\mathbf{w}^{\mathrm{i}}$ and $\mathbf{w}^{\mathrm{f}}$ the initial and final values of $\mathbf{w}$. The fact that the initial conditions determine the final conditions can be expressed in terms of a functional relationship which we denote formally as the ***optical transfer map*** $\mathfrak{M}$, thus:

$$\mathbf{w}^{\mathrm{f}} = \mathfrak{M}\mathbf{w}^{\mathrm{i}}. \tag{2.11}$$

Fermat's principle, we saw, is equivalent to the statement that the initial conditions $\mathbf{w}^{\mathrm{i}}$ and the final conditions $\mathbf{w}^{\mathrm{f}}$ are on a trajectory governed by the optical Hamiltonian. This statement is equivalent in turn to the statement that $\mathfrak{M}$ is a *symplectic* map. That is, let $\mathbf{M} = ||M_{\alpha\beta}(\mathbf{q}, \mathbf{p})||$ be the Jacobian matrix associated with the mapping $\mathfrak{M}$. The matrix elements are defined by the relation

$$M_{\alpha\beta} := \frac{\partial w^{\mathrm{f}}_\alpha}{\partial w^{\mathrm{i}}_\beta}, \qquad \alpha, \beta = 1, 2, 3, 4. \tag{2.12}$$

Also, let $\mathbf{J}$ be a 4×4 matrix

$$\mathbf{J} = \begin{pmatrix} \mathbf{0} & \mathbf{1} \\ -\mathbf{1} & \mathbf{0} \end{pmatrix}, \tag{2.13}$$

where each entry is a 2×2 matrix, $\mathbf{1}$ denoting the identity and $\mathbf{0}$ the null matrix. Then it can be shown that $\mathbf{M}$ satisfies the matrix equation [4][4]

$$\mathbf{M}^\top \mathbf{J} \mathbf{M} = \mathbf{J}. \tag{2.14}$$

This equation is the condition that $\mathbf{M}$ be a *symplectic matrix*, and in the context of optics it is sometimes called the ***lens equation*** [5].[5] In the same spirit, $\mathfrak{M}$ may be called an ***optical symplectic map.***

4.3 Some Lie algebraic tools

The purpose of this section is to present various Lie algebraic tools which will be required for subsequent discussion. These have been presented ***in extenso*** in references [5] and [6].[6]

4.3.1 Lie operators

Let f and g be any two (differentiable) functions defined on optical phase space, of coordinates $\{w_\alpha\}_{\alpha=1}^4$ as in (2.10). Then, the ***Poisson bracket*** of these two functions will be denoted by the symbol

---

[4] The first article in this reference gives an early discussion on the consequences of what is essentially the *symplectic condition*. The book contains an extensive discussion of optics from a group theoretical perspective.

[5] A symplectic map is in fact a *canonical transformation*, and vice versa. This is shown in Section 4.1 of this reference. A discussion of symplectic matrices is given in chapter 2, and section 4.3 proves that hamiltonian flows produce symplectic maps.

[6] See also the Chapter 3 in this volume, especially sections 3 and 4. (Editor's note.)

$\{f,g\}$, and[7] defined as

$$\{f,g\} := \sum_{i=1}^{2}\left(\frac{\partial f}{\partial q_i}\frac{\partial g}{\partial p_i} - \frac{\partial f}{\partial p_i}\frac{\partial g}{\partial q_i}\right). \tag{3.1}$$

Now let f be a specified function of phase space, and let g be *any* function. Associated with each f there is thus a ***Lie operator*** that acts on the set of functions g. The Lie operator associated with the function f will be denoted by the symbol $:f:$, and[8] it is defined in terms of the Poisson bracket by the rule

$$:f:g = \{f,g\}. \tag{3.2}$$

The Lie operator is *linear*, *i.e.*, $:c_1 f_1 + c_2 f_2: = c_1{:}f_1{:} + c_2{:}f_2{:}$, for c_1 and c_2 constants.

Powers of $:f:$ are defined by taking repeated Poisson brackets. For example, $:f:^2$ is defined by the relation

$$:f:^2 g = :f::f:g = \{f,\{f,g\}\}. \tag{3.3}$$

Finally, $:f:$ to the power zero is defined to be the identity operator

$$:f:^0 g = g, \tag{3.4}$$

and if c is a constant, $:c:$ is the null operator.

Since sums and powers of $:f:$ have been defined, it is also possible to work with power series of $:f:$. Of particular importance is the *exponential* power series, $\exp(:f:)$. This particular object is called the ***Lie transformation*** associated with $:f:$ or f. The Lie transformation is also a linear operator acting on functions of phase space, and is formally defined by the exponential series

$$\exp(:f:) = \sum_{n=0}^{\infty} \frac{1}{n!}:f:^n. \tag{3.5}$$

In particular, the action of $\exp(:f:)$ on any function g is given by

$$\exp(:f:)\,g = g + \{f,g\} + \tfrac{1}{2!}\{f,\{f,g\}\} + \cdots. \tag{3.6}$$

4.3.2 The symplectic map factorization theorem

The stage has now been set for a fundamental result on symplectic maps:[9]

Theorem. (Dragt–Finn) *Suppose that $\mathfrak{M}$ is any symplectic map that sends the origin of phase space into itself,*[10] *that is, if $\mathbf{w}^{\mathrm{i}} = \mathbf{0}$, then $\mathbf{w}^{\mathrm{f}} = \mathbf{0}$. Then $\mathfrak{M}$ can be factored as a product of Lie transformations*

[7] In this volume, $\{f,g\}$ is used for Poisson brackets and $[:f:,:g:]$ or $[A,B]$ for commutators. The *opposite* convention has been used in Refs. [1,4,6]. (Editor's note.)

[8] Alternative notations are plenty. For example, A. Katz, in *Classical Mechanics, Quantum Mechanics, Field Theory*, (Academic Press, 1965), uses f_{op}; Stanly Steinberg uses $\{f,\circ\}$ in this volume; Bernardo Wolf prefers used the accented forms $\hat{f}$, which are shorter for single characters. Much discussion has been spent on this. (Editor's note.)

[9] See section 5.2 of reference [5] and section 2 of reference [6].

[10] Maps that do *not* send the origin into itself may be treated by including an $\exp(:f_1:)$ factor in the product (3.7). See Sect. **4.8.3**. Also, other orders of factorization are possible, generally with different f's. See reference [6].

in the form

$$\mathfrak{M} = \cdots \exp(:f_4:)\exp(:f_3:)\exp(:f_2:), \tag{3.7}$$

*where each function f_n is a **homogeneous polynomial** of degree n in the components of* $\mathbf{w}$. *Moreover, the map is **symplectic** for any set of polynomials. Finally, if the product is truncated at any stage, the result is still a symplectic map.*

When applied to optics, the factorization theorem indicates that the effect of any collection of lenses, prisms, and mirrors, can be characterized by a set of homogeneous polynomials. In particular, the polynomials f_2 reproduce paraxial optics. The higher-order polynomials f_3, f_4, ..., describe departures from paraxial optics and are related to *aberrations* in the case of an imaging system. Thus, from a Lie algebraic perspective, the fundamental problems of geometrical optics are:

- ▸ To study what polynomials *correspond* to various optical elements,
- ▸ to study what polynomials result from *concatenating* various optical elements, and
- ▸ to study what polynomials correspond to various *desired* optical *properties*.[11]

4.3.3 Three operator properties

There are three properties of Lie operators that will be important for subsequent discussion.

First, suppose that $:f:$ and $:g:$ are any two Lie operators. Consider their ***commutator*** $[:f:, :g:]$, defined by the relation

$$[:f:, :g:] = :f::g: - :g::f:. \tag{3.8}$$

Then it can be shown that the commutator of any two Lie operators is again a Lie operator, and in particular one has the result [5,6]

$$[:f:, :g:] = :\{f, g\}:. \tag{3.9}$$

Thus, to compute the commutator of any two Lie operators, it is only necessary to find the Poisson bracket of their associated functions.

Second, suppose that f_m and f_n are two homogeneous polynomials of degrees m and n respectively. Then it is easily checked that the degree of the Poisson bracket $\{f_m, f_n\}$ is given by the relation

$$\deg\{f_m, f_n\} = m + n - 2. \tag{3.10}$$

Third, suppose that $\mathfrak{M}$ is a symplectic map represented as a Lie transformation or a product of Lie transformations. Suppose further that g is any function on phase space. Then it can be shown that

$$\mathfrak{M}\, g(\mathbf{w}^{\mathrm{i}}) = g(\mathfrak{M}\mathbf{w}^{\mathrm{i}}). \tag{3.11}$$

Moreover, as a corollary to this result, suppose that g and h are any two functions on phase space. Then the following relation holds:

$$\mathfrak{M}\big(g(\mathbf{w}^{\mathrm{i}})\, h(\mathbf{w}^{\mathrm{i}})\big) = \big(\mathfrak{M}\, g(\mathbf{w}^{\mathrm{i}})\big)\big(\mathfrak{M}\, h(\mathbf{w}^{\mathrm{i}})\big). \tag{3.12}$$

[11] It is not necessarily the case that every symplectic map corresponds to a physically realizable optical system.

That is, $\mathfrak{M}$ is an *isomorphism* with respect to ordinary function multiplication. Finally, under the same assumption as above, we have the relation

$$\mathfrak{M}\{g,h\} = \{\mathfrak{M}g, \mathfrak{M}h\}. \tag{3.13}$$

That is, $\mathfrak{M}$ is also an ismorphism with respect to "Poisson bracket" multiplication.

4.3.4 Axially symmetric systems and *sp(2,ℜ)*

Suppose, as is often the case, that the optical system under consideration is *axially symmetric* under rotations around some axis, and that it is also *symmetric* with respect to *reflections* through some plane containing the optical axis. Then axial symmetry requires that the f_n be functions only of linear combinations and powers of the variables $\mathbf{p}^2$, $\mathbf{q}^2$, $\mathbf{p}\cdot\mathbf{q}$, and $\mathbf{p}\times\mathbf{q}$; reflection symmetry rules out the variable $\mathbf{p}\times\mathbf{q}$. It follows that all homogeneous polynomials $f_n(\mathbf{p},\mathbf{q})$ with odd n must vanish, since it is impossible to construct an odd-order function by using only $\mathbf{p}^2$, $\mathbf{q}^2$, and $\mathbf{p}\cdot\mathbf{q}$.

Consequently, in any system having the assumed orthogonal-group *o(2)* symmetries, the optical symplectic map $\mathfrak{M}$ must be of the general form [7][12]

$$\mathfrak{M} = \cdots \exp(:f_6:)\exp(:f_4:)\exp(:f_2:), \tag{3.14}$$

The Lie transformations associated with the three quadratic polynomials, $\mathbf{p}^2$, $\mathbf{q}^2$, and $\mathbf{p}\cdot\mathbf{q}$, have simple properties. Elementary calculation gives the following correspondences [1]:

$\mathfrak{M}(\ell) = \exp\left(\frac{-\ell}{2n}:\mathbf{p}^{2}:\right)$	$\leftrightarrow$	Transit by distance ℓ through a medium of refractive index n, in the paraxial gaussian approximation.	(3.15*a*)
$\mathfrak{M}(n_1,n_2,r) = \exp\left(\frac{n_2-n_1}{r}:\mathbf{q}^{2}:\right)$	$\leftrightarrow$	Refraction by a spherical surface of radius r separating media with refracting indices n_1 and n_2, in the paraxial gaussian approximation.	(3.15*b*)
$\mathfrak{M}(\sigma) = \exp(-\sigma:\mathbf{p}\cdot\mathbf{q}:)$	$\leftrightarrow$	A system that is both imaging and telescopic, with magnification $\exp\sigma$.	(3.15*c*)

The action of these operators on phase space $(\mathbf{p},\mathbf{q})$ is *linear*.

[12] Of course, in the general case where no particular symmetries are present, the odd-degree polynomials $f_3, f_5, \ldots$ may also occur, and all the polynomials f_n can in principle, depend on the components of the vectors $\mathbf{p}$ and $\mathbf{q}$ in an unrestricted fashion. These are the problems faced by magnetic or misaligned-lens optics. In the latter, first order polynomials f_1 can occur as well. Lie algebraic tools for the treatment of unaligned systems are currently under development and will appear in the reference.

The Lie operators associated with $\mathbf{p}^2$, $\mathbf{q}^2$, and $\mathbf{p}\cdot\mathbf{q}$, form a very simple Lie algebra under commutation. Equivalently, the polynomials themselves form a Lie algebra under Poisson bracketing, in view of (3.9). We define the operators $K_\pm$ and K_0 by the relations

$$K_+ = \tfrac{1}{2}{:}\mathbf{p}^2{:}, \qquad (3.16a)$$

$$K_0 = \tfrac{1}{2}{:}\mathbf{p}\cdot\mathbf{q}{:}, \qquad (3.16b)$$

$$K_- = \tfrac{1}{2}{:}\mathbf{q}^2{:}. \qquad (3.16c)$$

Then, we find they obey the commutation rules

$$[K_0 K_+] = K_+, \qquad (3.17a)$$

$$[K_0 K_-] = -K_-, \qquad (3.17b)$$

$$[K_+ K_-] = -2K_0. \qquad (3.17c)$$

These are the commutation relations for the two-dimensional real Lie algebra *sp(2,ℜ)*.[13] The Lie transformations generated by (3.16) constitute the group *Sp(2,ℜ)* of linear canonical transformations of phase space. If the sign in (3.17c) were *plus*, we would have the *rotation* algebra *so(3)*. See reference [10, Eq.(3.12)]. This change of sign can be brought about redefining the right-hand sides of (3.16) to a set of $\{\mathcal{L}_\sigma\}_{\sigma=+,0,-}$, where the sign of $\mathcal{L}_+$ is *changed* with respect to K_+. In cartesian components, this is a complex linear combination which essentially performs the Weyl trick on *so(2,1)* ↔ *so(3)*. This will be pursued in Section 7.

4.3.5 The Petzval invariant

We point out that the scalar quantity

$$\mathbf{p}\times\mathbf{q} = p_1 q_2 - p_2 q_1, \qquad (3.18)$$

has vanishing Poisson bracket with $\mathbf{p}^2$, $\mathbf{q}^2$, and $\mathbf{p}\cdot\mathbf{q}$, and is invariant under *sp(2,ℜ)* symplectic maps (3.19). We thus introduce its *square*, the ***Petzval invariant***

$$\Pi := (\mathbf{p}\times\mathbf{q})^2 = \mathbf{p}^2\mathbf{q}^2 - (\mathbf{p}\cdot\mathbf{q})^2 = 4(K_+K_- - K_0^2), \qquad (3.19)$$

where K_σ, $\sigma = \pm, 0$ are the classical functions corresponding to the K_σ in (3.16). This serves, thus, as the *Casimir* operator of the *sp(2,ℜ)* algebra. Classical treatments of the Petzval invariant may be found in references [8, Sect.5.5.3] and [9, Sect.42.6].

It follows that various features of gaussian paraxial optics are intimately related to the properties of *sp(2,ℜ)* and its associated Lie group *Sp(2,ℜ)*. We shall also show below that many tools familiar from the three-dimensional rotation algebra and group [10] may be applied immediately to *sp(2,ℜ)* and *Sp(2,ℜ)*. In addition to its application to gaussian paraxial optics, *sp(2,ℜ)* can be used to describe and classify *departures* from it. As we indicated earlier, departures are described by the polynomials f_4, f_6, etc., in (3.7), corresponding to aberrations of order three, five, etc.

4.3.6 Change of factorization order

Suppose $\mathfrak{M}$ as given by (3.14) is refactored in the form

$$\mathfrak{M} = \exp({:}g_2{:})\exp({:}g_4{:})\exp({:}g_6{:})\cdots. \qquad (3.20)$$

---

[13] See Appendix B. (Editor's note.)

Then, the relation between the first two f_m's in (3.14) and the g_n's here is found through writing

$$\begin{aligned}\mathfrak{M} &= \cdots e^{:f_6:}\, e^{:f_4:}\, e^{:f_2:} \\ &= [e^{:f_2:}\, e^{:-f_2:}][\cdots e^{:f_2:}][e^{:-f_2:}\, e^{:f_6:}\, e^{:f_2:}][e^{:-f_2:}\, e^{:f_4:}\, e^{:f_2:}] \\ &= e^{:f_2:}\,[\cdots e^{:f_6^{\rm tr}:}\, e^{:f_4^{\rm tr}:}] \\ &= e^{:f_2:}\, e^{:f_4^{\rm tr}:}\, e^{:f_6^{\rm tr}:} \cdots .\end{aligned} \tag{3.21}$$

Here use has been made of the Baker-Campbell-Hausdorff formula and the general Lie algebraic relation [6] defining $e^{:f_n^{\rm tr}:}$,

$$e^{:f_n^{\rm tr}:} = e^{:-f_2:}\, e^{:f_n:}\, e^{:f_2:}, \qquad i.e. \;\; f_n^{\rm tr} = e^{:-f_2:} f_n. \tag{3.22}$$

Comparison of (3.20) and (3.7) gives then the relations

$$g_2 = f_2, \tag{3.23a}$$

$$g_4 = f_4^{\rm tr} = e^{:-f_2:} f_4 = \mathfrak{M}_2^{-1} f_4. \tag{3.23b}$$

With the aid of these equations we may pass between the results for third aberration order, factored as (3.14) and as (3.20). The latter have been widely used by Dragt and collaborators in references [1,5,6] and in computer codes. Higher-order f's and g's are subject to relations more difficult to obtain analytically.

4.3.7 Products of symplectic maps

Suppose an optical system is composed of S elements, and let $\mathfrak{M}_{(i)}$ be the optical transfer map for the $i^{\rm th}$ element. Then the optical transfer map $\mathfrak{M}$ for the entire system can be written as a product

$$\mathfrak{M} = \mathfrak{M}_{(1)}\, \mathfrak{M}_{(2)} \cdots \mathfrak{M}_{(S)}. \tag{3.24}$$

The order of the factors follows the order of the elements when we picture as usual the optical axis coordinate z growing to the right. Next observe that each of the $\mathfrak{M}_{(i)}$ has a factorization of the form (3.14). Suppose further that the various $f_n^{(i)}$ for the various $\mathfrak{M}_{(i)}$ are all known. Then the only problem involved in computing $\mathfrak{M}$ is that of combining a collection of known maps and writing the result in factorized form.

This last problem is standard in the theory of Lie algebras, and is solved by means of the Baker-Campbell-Hausdorff formula [6]. Suppose, for example, that $\mathfrak{M}_{(f)}$ and $\mathfrak{M}_{(g)}$ are two maps written in the factorized form (3.7), with polynomials f_n and g_n in the exponents, respectively. Define now the map $\mathfrak{M}_{(h)}$ to be the product of $\mathfrak{M}_{(f)}$ and $\mathfrak{M}_{(g)}$,

$$\mathfrak{M}_{(h)} = \mathfrak{M}_{(f)}\, \mathfrak{M}_{(g)}, \tag{3.25}$$

with polynomials h_n in the exponents, in the factorized form (3.14).[14] Evidently, the h_n's must be computable in terms of the f_m's and g_ℓ's.

Specifically, using the Baker-Campbell-Hausdorff formula, we find the following results [6]. For the paraxial part of the transformations,

$$\exp(:h_2:) = \exp(:f_2:)\, \exp(:g_2:). \tag{3.26}$$

[14] The work of Dragt and collaborators, as was said, uses the factorization order (3.20). The results in the latter may be easily regained from the following formulae through taking the inverse of (3.25) and replacing $g_n \mapsto -\bar{f}_n$, $f_n \mapsto -\bar{g}_n$, $h_n \mapsto -\bar{h}_n$. In this way we reproduce (3.25) with the maps factored in the order (3.20).

This product is best done explicitly in terms of multiplication in the 2×2 matrix representation of $Sp(2,\Re)$ [given, ahead, in (4.26)]. The higher polynomials combine according to order in the following way [6]:

$$h_4 = f_4 + g_4^{\rm tr}, \tag{3.27a}$$
$$h_6 = f_6 + g_6^{\rm tr} + \tfrac{1}{2}\{f_4, g_4^{\rm tr}\}, \tag{3.27b}$$
$$h_8 = f_8 + g_8^{\rm tr} + \{f_4, g_6^{\rm tr}\} + \tfrac{1}{3}\{f_4,\{f_4, g_4^{\rm tr}\}\} - \tfrac{1}{6}\{\{f_4, g_4^{\rm tr}\}, g_4^{\rm tr}\}, \tag{3.27c}$$
$$\ldots \qquad \ldots$$

where we define $g_n^{\rm tr}$ as in (3.22). Inspection of these equations shows that the determination of the remaining h_n's involves carrying out the transformations (3.22) and the evaluation of certain Poisson brackets. These two tasks may be simplified in some cases by the symplectic multiplet decomposition we shall present in Section 7.

4.4 Symplectic maps in continuous systems

The purpose of this section is to compute the symplectic map of a *continuous* system, *i.e.*, an optical medium where the refraction index $n(\mathbf{q}, z)$ is a continuous function of its arguments. This will be done through calculating the first few polynomials f_n of the map in (3.7); then it will applied to the optical Hamiltonian H in (2.8), using the tools developed for the general case in reference [11].

We let $\mathfrak{M}$ be the optical transfer map relating final conditions at $z^{\rm f}$ to initial conditions at $z^{\rm i}$. Then it can be shown that $\mathfrak{M}$ is given in terms of H by the formal expression

$$\mathfrak{M} = \tau \exp\Big(-\int_{z^{\rm i}}^{z^{\rm f}} dz :H(\mathbf{w}^{\rm i}, z):\Big). \tag{4.1}$$

Here τ is a certain ordering operator whose exact definition is not necessary for the present discussion. Moreover, in the case that the Lie operators $:H(\mathbf{w}^{\rm i}, z):$ and $:H(\mathbf{w}^{\rm i}, z'):$ commute for arbitrary z and z' in the interval of interest, it can be shown that the ordering operator τ is unnecessary. One has then the simpler relation

$$\mathfrak{M} = \exp\Big(-\int_{z^{\rm i}}^{z^{\rm f}} dz :H(\mathbf{w}^{\rm i}, z):\Big). \tag{4.2}$$

Both (4.1) and (4.2) are of limited computational utility, and more powerful tools will be presented shortly. However, as an application of (4.2) consider the case of simple transit over a distance ℓ through a homogeneous medium of *constant* refractive index n.

4.4.1 Propagation in a homogeneous medium

We shall treat first the case of free transit over a distance ℓ along the optical axis. In this case the Hamiltonian H is given by (2.8). It is independent of z and (4.2) can be evaluated immediately to give the result

$$\mathfrak{M} = \exp(\ell :\sqrt{n^2 - (\mathbf{p}^{\rm i})^2}:). \tag{4.3}$$

To verify that this $\mathfrak{M}$ does indeed correspond to simple transit, we apply it to $\mathbf{q}^{\rm i}$ and $\mathbf{p}^{\rm i}$, so as to see the effect of the Lie operator $:\sqrt{n^2 - (\mathbf{p}^{\rm i})^2}:$ on phase space. We easily find, using (3.1) and (3.2), that

$$(:\sqrt{n^2-(\mathbf{p}^{\rm i})^2}:)^n\,\mathbf{p}^{\rm i}=0, \qquad n=1,2,3,\ldots, \tag{4.4}$$

$$:\sqrt{n^2-(\mathbf{p}^{\rm i})^2}:\mathbf{q}^{\rm i}=\frac{\mathbf{p}^{\rm i}}{\sqrt{n^2-(\mathbf{p}^{\rm i})^2}}, \tag{4.5a}$$

$$(:\sqrt{n^2-(\mathbf{p}^{\rm i})^2}:)^n\,\mathbf{q}^{\rm i}=0, \qquad n=2,3,4,\ldots. \tag{4.5b}$$

Consequently, in view of (2.10) and (3.6), we have the result

$$\mathbf{p}^{\rm f}=\mathfrak{M}\,\mathbf{p}^{\rm i}=\mathbf{p}^{\rm i}, \tag{4.6a}$$

$$\mathbf{q}^{\rm f}=\mathfrak{M}\,\mathbf{q}^{\rm i}=\mathbf{q}^{\rm i}+\ell\frac{\mathbf{p}^{\rm i}}{\sqrt{n^2-(\mathbf{p}^{\rm i})^2}}. \tag{4.6b}$$

Moreover, by using (2.6) we find the relation

$$\frac{\mathbf{p}^{\rm i}}{\sqrt{n^2-(\mathbf{p}^{\rm i})^2}}=(\mathbf{q}')^{\rm i}.$$

Thus, (4.6) can be written in the final form

$$\mathbf{p}^{\rm f}=\mathfrak{M}\,\mathbf{p}^{\rm i}=\mathbf{p}^{\rm i}, \tag{4.7a}$$

$$\mathbf{q}^{\rm f}=\mathfrak{M}\,\mathbf{q}^{\rm i}=\mathbf{q}^{\rm i}+\ell(\mathbf{q}')^{\rm i}. \tag{4.7b}$$

This is just as expected for simple transit over a distance ℓ, since $\mathbf{q}'$ is a vector in the direction of $\mathbf{p}$ and of magnitude $\tan\theta$, where θ is the angle between the ray and the optical axis.

As we discussed earlier, what we need is a factorization of $\mathfrak{M}$ in the form (3.14). This is readily accomplished for simple transit. Taylor expansion gives the result

$$\begin{aligned}\sqrt{n^2-(\mathbf{p}^{\rm i})^2}&=n^2-\frac{1}{2n}(\mathbf{p}^{\rm i})^2\\&-\frac{1}{8n^3}[(\mathbf{p}^{\rm i})^2]^2-\frac{1}{16n^5}[(\mathbf{p}^{\rm i})^2]^3-\frac{5}{128n^7}[(\mathbf{p}^{\rm i})^2]^4--\frac{7}{256n^9}[(\mathbf{p}^{\rm i})^2]^5-\cdots.\end{aligned} \tag{4.8}$$

Since the Lie operator associated with a constant is zero, it follows that we have the corresponding result for the Lie operators,

$$\begin{aligned}:\sqrt{n^2-(\mathbf{p}^{\rm i})^2}:&=-:\frac{1}{2n}(\mathbf{p}^{\rm i})^2:\\&-:\frac{1}{8n^3}[(\mathbf{p}^{\rm i})^2]^2:-:\frac{1}{16n^5}[(\mathbf{p}^{\rm i})^2]^3:-:\frac{5}{128n^7}[(\mathbf{p}^{\rm i})^2]^4:-:\frac{7}{256n^9}[(\mathbf{p}^{\rm i})^2]^5:-\cdots.\end{aligned} \tag{4.9}$$

Finally, the Lie operators of the form $:(\mathbf{p}^{\rm i})^2:^m$, for various m, all commute. The symplectic map $\mathfrak{M}$ for simple transit can thus be written in the factored product form

$$\mathfrak{M}=\cdots\exp\left(\ell\tfrac{7}{256n^9}[:(\mathbf{p}^{\rm i})^2:]^5\right)\exp\left(\ell\tfrac{5}{128n^7}[:(\mathbf{p}^{\rm i})^2:]^4\right)\exp\left(\ell\tfrac{1}{16n^5}[:(\mathbf{p}^{\rm i})^2:]^3\right)\exp\left(\ell\tfrac{1}{8n^3}[:(\mathbf{p}^{\rm i})^2:]^2\right)\exp\left(\ell\tfrac{1}{2n}:(\mathbf{p}^{\rm i})^2:\right). \tag{4.10}$$

We note that, as expected, the first (*rightmost*) factor above is just the one in (3.15a) for transit in the paraxial approximation. Thus, for example, we verify that simple transit does not change the value of the Petzval invariant (3.19).

4.4.2 Propagation in optical fibers

As a second, and more complicated example, we consider the case of propagation in an axially symmetric graded-index medium. This models, for example, an optical fiber with a z-dependent index profile. Specifically, we assume that the index of refraction n can be expanded as

$$n(\mathbf{q}, z) = \alpha_0(z) + \alpha_2(z)\mathbf{q}^2 + \alpha_4(z)(\mathbf{q}^2)^2 + \alpha_6(z)(\mathbf{q}^2)^3 + \cdots. \tag{4.11}$$

If the medium is to act as a fiber does, then $n(\mathbf{q}, z)$ should have its largest value around $\mathbf{q} = \mathbf{0}$, *i.e.*, we expect $\alpha_0 \geq 1$, $\alpha_2 < 0$. Then, from (2.8) and (4.11), we find that the Hamiltonian H has an expansion in homogeneous polynomials of the form

$$H = H_0 + H_2 + H_4 + H_6 + \cdots, \tag{4.12}$$

with

$$H_0 = -\alpha_0, \tag{4.13a}$$

$$H_2 = \frac{1}{2\alpha_0}\mathbf{p}^2 - \alpha_2\mathbf{q}^2, \tag{4.13b}$$

$$H_4 = \frac{1}{8\alpha_0^3}(\mathbf{p}^2)^2 - \frac{\alpha_2}{2\alpha_0^2}\mathbf{p}^2\mathbf{q}^2 - \alpha_4(\mathbf{q}^2)^2, \tag{4.13c}$$

$$H_6 = \frac{1}{16\alpha_0^5}(\mathbf{p}^2)^3 - \frac{3\alpha_2}{8\alpha_0^4}(\mathbf{p}^2)^2\mathbf{q}^2 + \left(\frac{\alpha_2^2}{2\alpha_0^3} - \frac{\alpha_4}{2\alpha_0^2}\right)\mathbf{p}^2(\mathbf{q}^2)^2 - \alpha_6(\mathbf{q}^2)^3, \tag{4.13d}$$

$$\cdots \qquad \cdots.$$

Let $\mathfrak{M}(\mathbf{w}^{\mathrm{i}}, z)$ be the optical transfer map relating initial conditions at z^{i} to general conditions at the point z. Then this map can be shown to obey the differential equation

$$\mathfrak{M}'(\mathbf{w}^{\mathrm{i}}, z) = \mathfrak{M}(\mathbf{w}^{\mathrm{i}}, z) :-H(\mathbf{w}^{\mathrm{i}}, z): \tag{4.14}$$

with initial conditions

$$\mathfrak{M}(\mathbf{w}^{\mathrm{i}}, z^{\mathrm{i}}) = \mathfrak{I}, \tag{4.15}$$

where $\mathfrak{I}$ is the identity map. In the case of a fiber, when H is z-independent [that is, when $n(\mathbf{q})$ does not depend on z], then (4.14) may be solved for $\mathfrak{M}$ to give

$$\begin{aligned}\mathfrak{M}(\mathbf{w}^{\mathrm{i}}, z) &= \exp(-(z - z^{\mathrm{i}}):H(\mathbf{w}^{\mathrm{i}}):) \\ &= \exp(-(z - z^{\mathrm{i}}):\sqrt{n(\mathbf{q}^{\mathrm{i}})^2 - (\mathbf{p}^{\mathrm{i}})^2}:).\end{aligned} \tag{4.16}$$

This form of writing the result, however, is not particularly useful, because it is not in factored product form. Although there are methods for bringing expressions as (4.16) to such form, it is easier for us to treat the general z-dependent case, and later return the z-independent one as a special instance of the general problem.[15]

4.4.3 The general $(\mathbf{q}, z)$–dependent case

To proceed further, it is convenient to factor $\mathfrak{M}$ in the order given by (3.7), as

$$\mathfrak{M} = \cdots \mathfrak{M}_8\, \mathfrak{M}_6\, \mathfrak{M}_4\, \mathfrak{M}_2, \tag{4.17}$$

---

[15] For methods of factoring (4.16), see reference [5], section 5.4.

where each factor $\mathfrak{M}_m$ is given by an expression of the form

$$\mathfrak{M}_m = \exp(:f_m:), \qquad m = 2, 4, 6, \ldots. \tag{4.18}$$

Then it can be shown that $\mathfrak{M}_2$ obeys the equation of motion

$$\mathfrak{M}'_2(\mathbf{w}^{\mathrm{i}}, z) = \mathfrak{M}_2(\mathbf{w}^{\mathrm{i}}, z) :-H_2(\mathbf{w}^{\mathrm{i}}, z):, \tag{4.19}$$

with the initial condition

$$\mathfrak{M}_2(\mathbf{w}^{\mathrm{i}}, z^{\mathrm{i}}) = \mathfrak{I}. \tag{4.20}$$

Moreover, the *aberration polynomials* $f_4, f_6, \ldots$, are given by formulae of the form

$$f_4(\mathbf{w}^{\mathrm{i}}, z) = -\int_{z^{\mathrm{i}}}^{z} dz'\, H_4^{\mathrm{int}}(\mathbf{w}^{\mathrm{i}}, z'), \tag{4.21a}$$

$$f_6(\mathbf{w}^{\mathrm{i}}, z) = -\int_{z^{\mathrm{i}}}^{z} dz'\, H_6^{\mathrm{int}}(\mathbf{w}^{\mathrm{i}}, z') + \frac{1}{2}\int_{z^{\mathrm{i}}}^{z} dz' \int_{z^{\mathrm{i}}}^{z'} dz''\, \{H_4^{\mathrm{int}}(\mathbf{w}^{\mathrm{i}}, z''), H_4^{\mathrm{int}}(\mathbf{w}^{\mathrm{i}}, z')\}, \tag{4.21b}$$

$$\ldots \qquad \ldots$$

Here H_m^{int} denotes the *interaction* Hamiltonian defined by

$$H_m^{\mathrm{int}}(\mathbf{w}^{\mathrm{i}}, z) = \mathfrak{M}_2 H_m = H_m(\mathfrak{M}_2 \mathbf{w}^{\mathrm{i}}, z). \tag{4.22}$$

4.4.3.1 The paraxial part

Since $\mathfrak{M}_2$ describes linear paraxial optics, we may write its effect in matrix form:[16]

$$\begin{pmatrix} \mathbf{p}(z) \\ \mathbf{q}(z) \end{pmatrix} = \mathfrak{M}_2 \begin{pmatrix} \mathbf{p}^{\mathrm{i}} \\ \mathbf{q}^{\mathrm{i}} \end{pmatrix} = \begin{pmatrix} a(z) & b(z) \\ c(z) & d(z) \end{pmatrix} \begin{pmatrix} \mathbf{p}^{\mathrm{i}} \\ \mathbf{q}^{\mathrm{i}} \end{pmatrix}. \tag{4.23}$$

In this representation, the equation of motion (4.19) becomes the set of ordinary differential equations

$$a' = 2\alpha_2 c, \quad b' = 2\alpha_2 d, \quad c' = a/\alpha_0, \quad d' = b/\alpha_0. \tag{4.24}$$

where α_0 and α_2 depend in general on z. Also, the requirement (4.20) gives the initial conditions

$$a(z^{\mathrm{i}}) = 1, \quad b(z^{\mathrm{i}}) = 0, \quad c(z^{\mathrm{i}}) = 0, \quad d(z^{\mathrm{i}}) = 1. \tag{4.25}$$

Regarding the general z-dependent case, except in very special instances, the solution of (4.24) with the initial conditons (6.25) requires numerical integration. If desired, $\mathfrak{M}_2$ can be written in exponential form once this solution is found![17] This may be done using the correspondence

$$\exp(:\alpha\mathbf{p}^2 + \beta\mathbf{p}\cdot\mathbf{q} + \gamma\mathbf{q}^2:)\begin{pmatrix} \mathbf{p} \\ \mathbf{q} \end{pmatrix} = \begin{pmatrix} \cos\omega + \beta\,\mathrm{sinc}\,\omega & 2\gamma\,\mathrm{sinc}\,\omega \\ -2\alpha\,\mathrm{sinc}\,\omega & \cos\omega - \beta\,\mathrm{sinc}\,\omega \end{pmatrix}\begin{pmatrix} \mathbf{p} \\ \mathbf{q} \end{pmatrix}, \tag{4.26}$$

$$\omega = \pm\sqrt{4\alpha\gamma - \beta^2}, \qquad \mathrm{sinc}\,\omega := \frac{\sin\omega}{\omega}, \tag{4.27}$$

where α, β, and γ are constants. This formula encompasses the three subgroups in (3.15) [*cf.* (*B*.17)].

---

[16]Note that we have here the two-vector $\begin{pmatrix} \mathbf{p} \\ \mathbf{q} \end{pmatrix}$ (for reasons to be seen in Section 7), instead of the two-vector $\begin{pmatrix} \mathbf{q} \\ \mathbf{p} \end{pmatrix}$, as it has usually appeared in several previous references.

[17]However, there is no particular need to do so since there is already the explicit representation (4.23).

4.4.3.2 The aberration part

The computation of the various $H_m^{\rm int}$, $m = 2, 6, 8, \ldots$ is now straightforward. According to (3.12) and (4.22), we only need to apply $\mathfrak{M}_2$ to the constituents of the H_m's. From the linear transformation between $\mathbf{p}$ and $\mathbf{q}$ given in (4.23), we obtain

$$\mathfrak{M}_2 \begin{pmatrix} \mathbf{p}^2 \\ \mathbf{p}\cdot\mathbf{q} \\ \mathbf{q}^2 \end{pmatrix} = \begin{pmatrix} a^2 & 2ab & b^2 \\ ac & ad+bc & bd \\ c^2 & 2cd & d^2 \end{pmatrix} \begin{pmatrix} \mathbf{p}^2 \\ \mathbf{p}\cdot\mathbf{q} \\ \mathbf{q}^2 \end{pmatrix}. \tag{4.28}$$

The matrix carries a 3×3 representation of $Sp(2,\Re)$, such that both the 2×2 identity $\mathbf{1}$ and the $-\mathbf{1}$ matrices correspond to the identity 3×3 matrix; it is the fundamental representation of $SO(2,1)$.[18]

Consequently, we may find $H_4^{\rm int}$ for the z-dependent fiber Hamiltonian (4.11), given by

$$\begin{aligned} H_4^{\rm int}(\mathbf{w}^{\rm i}, z) = {} & u_A(z)\left((\mathbf{p}^{\rm i})^2\right)^2 + u_B(z)\,(\mathbf{p}^{\rm i})^2(\mathbf{p}^{\rm i}\cdot\mathbf{q}^{\rm i}) + u_C(z)\,(\mathbf{p}^{\rm i}\cdot\mathbf{q}^{\rm i})^2 \\ & + u_D(z)\,(\mathbf{p}^{\rm i})^2(\mathbf{q}^{\rm i})^2 + u_E(z)\,(\mathbf{p}^{\rm i}\cdot\mathbf{q}^{\rm i})(\mathbf{q}^{\rm i})^2 + u_F(z)\left((\mathbf{q}^{\rm i})^2\right)^2, \end{aligned} \tag{4.29}$$

where

$$u_A(z) = a^4\frac{1}{8\alpha_0^3} - a^2c^2\frac{\alpha_2}{2\alpha_0^2} - c^4\alpha_4, \tag{4.30a}$$

$$u_B(z) = a^3b\frac{1}{2\alpha_0^3} - (abc^2 + a^2cd)\frac{\alpha_2}{\alpha_0^2} - 4c^3d\,\alpha_4, \tag{4.30b}$$

$$u_C(z) = a^2b^2\frac{1}{2\alpha_0^3} - 2abcd\frac{\alpha_2}{\alpha_0^2} - 4c^2d^2\alpha_4, \tag{4.30c}$$

$$u_D(z) = a^2b^2\frac{1}{4\alpha_0^3} - (b^2c^2 + a^2d^2)\frac{\alpha_2}{2\alpha_0^2} - 2c^2d^2\alpha_4, \tag{4.30d}$$

$$u_E(z) = ab^3\frac{1}{2\alpha_0^3} - (b^2cd + abd^2)\frac{\alpha_2}{\alpha_0^2} - 4cd^3\alpha_4, \tag{4.30e}$$

$$u_F(z) = b^4\frac{1}{8\alpha_0^3} - b^2d^2\frac{\alpha_2}{2\alpha_0^2} - d^4\alpha_4. \tag{4.30f}$$

where α_0, α_2, and α_4 appear in (4.11) with z–dependence in the solutions a, b, c, d of (4.24–25).

The computation of the aberration polynomials f_4, f_6, …, in (4.17–18) can now be also carried out. Suppose, for example, that f_4 is written in the form

$$\begin{aligned} f_4 = {} & A(z)\left((\mathbf{p}^{\rm i})^2\right)^2 + B(z)\,(\mathbf{p}^{\rm i})^2(\mathbf{p}^{\rm i}\cdot\mathbf{q}^{\rm i}) + C(z)\,(\mathbf{p}^{\rm i}\cdot\mathbf{q}^{\rm i})^2 \\ & + D(z)\,(\mathbf{p}^{\rm i})^2(\mathbf{q}^{\rm i})^2 + E(z)\,(\mathbf{p}^{\rm i}\cdot\mathbf{q}^{\rm i})(\mathbf{q}^{\rm i})^2 + F(z)\left((\mathbf{q}^{\rm i})^2\right)^2. \end{aligned} \tag{4.31}$$

Then one has the results

$$X(z) = -\int_{z^{\rm i}}^{z} dz'\, u_X(z'), \qquad X = A, B, \ldots, F. \tag{4.32}$$

At the end of Section 3 we showed that the symplectic algebra $sp(2,\Re)$ appears in the description of the paraxial part. We intend to develop in Section 6 the symplectic classification of aberrations. Then we shall be able to write the six fourth-order summands in (4.31) as a sum of two parts, irreducible under the action of $sp(2,\Re)$, a quintuplet and a singlet. The fourth-order singlet is essentially the Petzval invariant given in (3.19). We shall examine there the separate evolution of these two multiplets, with considerable simplification.

---

[18] See Appendix A. (Editor's note.)

4.4.4 Aberration in graded-index fibers

As a final example of the use of the machinery developed in this section, we consider the case of propagation through a slab of thickness t composed of a z-independent graded-index medium, *i.e.*, $n = n(\mathbf{q})$ only. In this case the quantities α_m in (4.11) are all constants, and equations (4.24) with the initial conditions (4.25) may be integrated immediately through (4.26–7) to give the results

$$a(z) = \cos\big(\Omega(z - z^{\text{i}})\big), \tag{4.33a}$$

$$b(z) = -\Omega\alpha_0 \sin\big(\Omega(z - z^{\text{i}})\big), \tag{4.33b}$$

$$c(z) = \frac{1}{\Omega\alpha_0} \sin\big(\Omega(z - z^{\text{i}})\big), \tag{4.33c}$$

$$d(z) = \cos\big(\Omega(z - z^{\text{i}})\big), \tag{4.33d}$$

where Ω is given by the expression[19]

$$\Omega = \sqrt{-\frac{2\alpha_2}{\alpha_0}}. \tag{4.34}$$

Moreover, the integrals (4.32) can be all evaluated over the interval $z^{\text{i}} \le z' \le z^{\text{i}} + t$ to give the results

$$A = \tfrac{1}{8}(-2\alpha_0\alpha_2)^{-1}[\eta(s_4 - 4s_2) - \theta z], \tag{4.35a}$$

$$B = \tfrac{1}{2}(-2\alpha_0\alpha_2)^{-1/2}\eta[-c_4 + 2c_2], \tag{4.35b}$$

$$C = -\tfrac{1}{2}[\eta s_4 + \tfrac{1}{3}\theta z] - \tfrac{1}{3}\alpha_2 z/\alpha_0^2, \tag{4.35c}$$

$$D = -\tfrac{1}{4}[\eta s_4 + \tfrac{1}{3}\theta z] + \tfrac{1}{3}\alpha_2 z/\alpha_0^2, \tag{4.35d}$$

$$E = \tfrac{1}{2}(-2\alpha_0\alpha_2)^{1/2}\eta[c_4 + 2c_2], \tag{4.35e}$$

$$F = \tfrac{1}{8}(-2\alpha_0\alpha_2)[\eta(s_4 + 4s_2) - \theta z], \tag{4.35f}$$

where

$$s_m := \frac{\sin(m\Omega z)}{m\Omega}, \qquad c_m := \frac{1 - \cos(m\Omega z)}{m\Omega}, \tag{4.36a}$$

$$\eta := -\frac{\alpha_2}{4\alpha_0^2}\left(1 + \frac{2\alpha_0\alpha_4}{\alpha_2^2}\right), \qquad \theta := -\frac{\alpha_2}{4\alpha_0^2}\left(5 - \frac{6\alpha_0\alpha_4}{\alpha_2^2}\right). \tag{4.36b}$$

4.5 Symplectic maps in discontinuous systems

In principle, discontinuous systems can be treated as limiting cases of continuous systems [12]. This seems to have been the ***modus operandi*** of all treatments of symplectic maps in optics, as this reference shows in its introductory section. It is perhaps surprising in view of the basic importance of lens surfaces in applications. However, that approach involves working with highly singular functions (Dirac δ's and its derivatives). Consequently, and in the case of higher-order aberrations in particular, the limiting process can be quite delicate. The purpose of this section is to present a direct approach to discontinuous interfaces with a basic new result.

---

[19] Recall that we assume $\alpha_0 > 1$, $\alpha_2 < 0$.

4.5.1 The surface map factorization theorem

The transformation of optical phase space due to an interface between two media, *i.e.*, a *discontinuity* in the refraction index such as a lens surface or a crack in an optical fiber, will be shown here to *factorize* into a product of two ***root*** transformations. One factor will depend only on the properties of the first medium and the other only on the properties of the second medium. These root transformations are, each, a symplectic map; the refracting-surface transformation is thus also a symplectic map. Furthermore, when a medium is homogeneous, its root transformation is given by a pair of *implicit algebraic* equations which may be solved iteratively to any aberration order.

The factorization of a map produced by a refracting surface was introduced by Wolf in reference [13] to shorten the derivation of the third-order lens aberration coefficients reported before by Dragt [1]. In fact, this first result turned out to be wrong by one coefficient (that of *pocus*, see below), but it was sufficiently close to the actual result to further inquiry into the properties of refracting-surface transformations in general.

The factorization property for differentiable interfaces between two homogeneous media (such as lenses, or of arbitrary "rice-grain" glass surfaces) was stated as a theorem by Navarro-Saad and Wolf in reference [14], based on a simple geometric construction and a short proof by differential forms. This was applied in reference [15] for axially symmetric lens surfaces, and in reference [16] for interfaces between two aligned graded-index fibers; the calculation was done explicitly to third aberration order.[20] The results we present here were derived by Forest and given by Dragt at the León workshop. They generalize the previous results in that the two media need not be z–homogeneous.

4.5.2 The optical path length

The first tool that will be needed to prove the general result is the optical path length. Let $\mathbf{r}^{\mathrm{i}}$ and $\mathbf{r}^{\mathrm{f}}$ be two points in the planes $z = z^{\mathrm{i}}$ and $z = z^{\mathrm{f}}$. They are to be chosen sufficiently close to each other so that there be a *unique* ray joining them. Then, according to (2.2) the ***optical path length*** associated with this ray is

$$A(\mathbf{r}^{\mathrm{i}},\mathbf{r}^{\mathrm{f}}) := \int_{\mathbf{r}^{\mathrm{f}}}^{\mathbf{r}^{\mathrm{i}}} ds\, n(\mathbf{r}) = \int_{z^{\mathrm{i}}}^{z^{\mathrm{f}}} dz\, n(\mathbf{q},z)\sqrt{1+(\mathbf{q}')^2}. \tag{5.1}$$

In view of (2.4), it is also possible to write this in the form

$$A(\mathbf{q}^{\mathrm{i}},z^{\mathrm{i}};\mathbf{q}^{\mathrm{f}},z^{\mathrm{f}}) = \int_{z^{\mathrm{i}}}^{z^{\mathrm{f}}} dz\, L(\mathbf{p},\mathbf{q},z), \tag{5.2}$$

where the Lagrangian may be written in terms of the Hamiltonian and the momentum as

$$\begin{aligned} L(\mathbf{p},\mathbf{q},z) &= \mathbf{p}\cdot\mathbf{q}' - H = \mathbf{p}\cdot\frac{\partial H}{\partial \mathbf{p}} - H \\ &= \frac{n(\mathbf{q},z)^2}{\sqrt{n(\mathbf{q},z)^2-\mathbf{p}^2}}. \end{aligned} \tag{5.3}$$

Finally, let $\mathbf{q}^{\mathrm{i}}$, $\mathbf{p}^{\mathrm{i}}$, and $\mathbf{q}^{\mathrm{f}}$, $\mathbf{p}^{\mathrm{f}}$ be the phase-space coordinates of the end points of the path, and define H^{i} and H^{f} by the rules

$$H^{\mathrm{i}} = H(\mathbf{p}^{\mathrm{i}},\mathbf{q}^{\mathrm{i}},z^{\mathrm{i}}), \qquad H^{\mathrm{f}} = H(\mathbf{p}^{\mathrm{f}},\mathbf{q}^{\mathrm{f}},z^{\mathrm{f}}). \tag{5.4a, b}$$

---

[20] The first draft [15, Subsect. 4.5] contained an algebraic mistake which was spotted by Dragt and Forest.

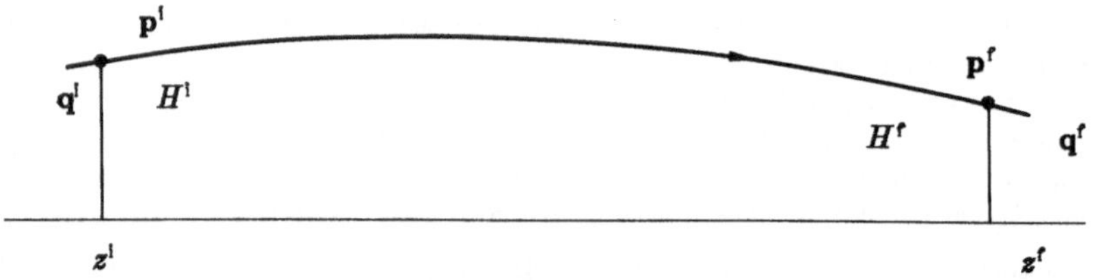

Figure 3. A ray with initial conditions $\mathbf{q}^{\mathrm{i}}$, $\mathbf{p}^{\mathrm{i}}$ at z^{i} and final conditions $\mathbf{q}^{\mathrm{f}}$, $\mathbf{p}^{\mathrm{f}}$ at z^{f}. The quantities H^{i} and H^{f} are the values of the Hamiltonian at the initial and final points, respectively.

See Figure 3. Then, according to standard hamiltonian theory, we have the relations [9, Sect.19.3]:

$$\mathbf{p}^{\mathrm{i}} = -\frac{\partial A}{\partial \mathbf{q}^{\mathrm{i}}}, \qquad H^{\mathrm{i}} = \frac{\partial A}{\partial z^{\mathrm{i}}}, \tag{5.5a, b}$$

$$\mathbf{p}^{\mathrm{f}} = \frac{\partial A}{\partial \mathbf{q}^{\mathrm{f}}}, \qquad H^{\mathrm{f}} = -\frac{\partial A}{\partial z^{\mathrm{f}}}, \tag{5.6a, b}$$

These considerations will be applied now to treat the presence of a discontinuous interface in $n(\mathbf{q}, z)$. As illustrated in Figure 4, we can consider two media having indices of refraction n^- and n^+ respectively, separated by a surface

$$\bar{z} = \varsigma(\bar{\mathbf{q}}). \tag{5.7}$$

Here, the overbar will be used to indicate the values of the quantities *on the surface.* Now consider the optical path length associated to the path consisting of a ray going from $\mathbf{q}^{\mathrm{i}}$, z^{i} to $\bar{\mathbf{q}}$, $\bar{z}$, followed by a ray going from $\bar{\mathbf{q}}$, $\bar{z}$ to $\mathbf{q}^{\mathrm{f}}$, z^{f}. It is evident that this optical path length is additive, and given by the expression

$$A(\mathbf{q}^{\mathrm{i}}, z^{\mathrm{i}}; \bar{\mathbf{q}}; \mathbf{q}^{\mathrm{f}}, z^{\mathrm{f}}) = A^-(\mathbf{q}^{\mathrm{i}}, z^{\mathrm{i}}; \bar{\mathbf{q}}, \bar{z}) + A^+(\bar{\mathbf{q}}, \bar{z}; \mathbf{q}^{\mathrm{f}}, z^{\mathrm{f}}). \tag{5.8}$$

Here, A^- and A^+ are the optical path lengths of the two rays, in the media having refractive indices n^- and n^+, respectively.

According to the Fermat principle, the path under consideration will be that of a *complete ray* provided we have the relation[21]

$$\frac{\partial A}{\partial \bar{\mathbf{q}}} = 0. \tag{5.9}$$

[21] For a similar application of Fermat's principle, see reference [9], Section 25.1.

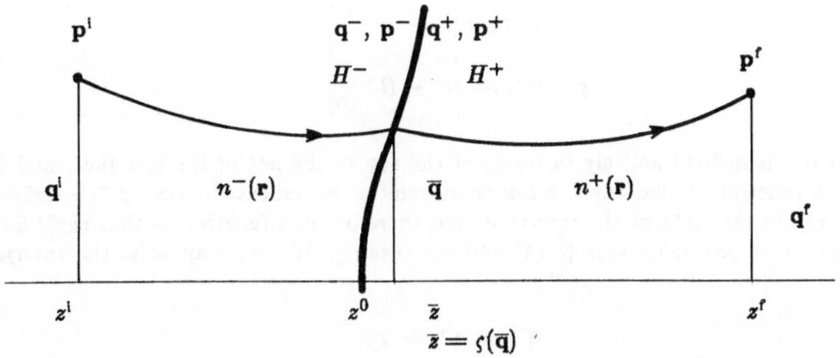

Figure 4. A path consisting of two rays joined at an interface ς. The ray before the interface has initial conditions $\mathbf{q}^i$, $\mathbf{p}^i$ at z^i. It propagates in an inhomogeneous medium with position-dependent index $n^-(\mathbf{r})$, until it intersects the interface ς at $\bar{z}$ with phase-space coordinates $\mathbf{q}^-$, $\mathbf{p}^-$. At this point, the Hamiltonian has the value H^-. The ray after ς leaves the interface with phase-space coordinates $\mathbf{q}^+$, $\mathbf{p}^+$ at $\bar{z}$. There, the Hamiltonian is H^+, with position-dependent index $n^+(\mathbf{r})$. The ray propagates until it arrives at z^f with final conditions $\mathbf{q}^f$, $\mathbf{p}^f$.

Inserting (5.8) into this equation and using the chain rule for partial differentiation, we find the result

$$\frac{\partial A^-}{\partial \bar{\mathbf{q}}} + \frac{\partial A^+}{\partial \bar{\mathbf{q}}} + \frac{\partial A^-}{\partial \bar{z}}\frac{\partial \bar{z}}{\partial \bar{\mathbf{q}}} + \frac{\partial A^+}{\partial \bar{z}}\frac{\partial \bar{z}}{\partial \bar{\mathbf{q}}} = 0, \tag{5.10}$$

where $\dfrac{\partial \bar{z}}{\partial \bar{\mathbf{q}}}$ is obtained from (5.7). It is the projection on the $\mathbf{q}$–plane of the *gradient* of the refracting surface, $\nabla\varsigma(\bar{\mathbf{q}})$.

4.5.3 The interface matching relations

In analogy to equations (5.5) and (5.6), we have the relations

$$\mathbf{p}^- = \frac{\partial A^-}{\partial \bar{\mathbf{q}}}, \qquad H^- = -\frac{\partial A^-}{\partial \bar{z}}, \tag{5.11a, b}$$

$$\mathbf{p}^+ = -\frac{\partial A^+}{\partial \bar{\mathbf{q}}}, \qquad H^+ = \frac{\partial A^+}{\partial \bar{z}}. \tag{5.12a, b}$$

Here the superscripts $^-$ and $^+$ refer to the values of $\mathbf{p}$ and H *just before* and *just after* the interface, respectively. See again Figure 4. It follows that (5.10) can be written in the form

$$\mathbf{p}^- - \mathbf{p}^+ + (H^+ - H^-)\frac{\partial \bar{z}}{\partial \bar{\mathbf{q}}} = \mathbf{0}. \tag{5.13}$$

Now we define the two vectors $\overline{\mathbf{p}}^-$ and $\overline{\mathbf{p}}^+$ by the equations

$$\overline{\mathbf{p}}^-(\mathbf{w}^{\mathrm{i}}) := \mathbf{p}^- - H^- \frac{\partial \overline{z}}{\partial \overline{\mathbf{q}}}, \tag{5.14a}$$

$$\overline{\mathbf{p}}^+(\mathbf{w}^{\mathrm{f}}) := \mathbf{p}^+ - H^+ \frac{\partial \overline{z}}{\partial \overline{\mathbf{q}}}. \tag{5.14b}$$

Observe that $\overline{\mathbf{p}}^-$ is defined entirely in terms of the ray to the left of the interface, and therefore can be viewed as a function of that ray's *initial* conditions $\mathbf{w}^{\mathrm{i}}$ as indicated. Also, $\overline{\mathbf{p}}^+$ is defined entirely in terms of the ray to the right of the interface, and therefore is a function of that ray's *final* conditions $\mathbf{w}^{\mathrm{f}}$. With the aid of the definitions (5.14) and the result (5.13), we may write the ***interface matching relations***

$$\overline{\mathbf{q}}^- = \overline{\mathbf{q}}^+ =: \overline{\mathbf{q}}, \tag{5.15a}$$

$$\overline{\mathbf{p}}^- = \overline{\mathbf{p}}^+ =: \overline{\mathbf{p}}. \tag{5.15b}$$

The first equation simply states the obvious fact that the coordinate associated with a ray does not change in going through the interface, and defines $\overline{\mathbf{q}}$. The *second* equation states the ***not*** so obvious fact that one can find a *second* vector, defined to be $\overline{\mathbf{p}}$, that is *also* conserved across the interface.

4.5.4 The *root* transformation

Even more can be said on the basis of (5.15). We introduce, in analogy to (2.9), a four-component vector $\mathbf{v}^-$, at the surface ς,

$$\mathbf{v}^- := \begin{pmatrix} \overline{\mathbf{p}}^- \\ \overline{\mathbf{q}}^- \end{pmatrix}. \tag{5.16}$$

Next, since $\mathbf{v}^-$ may be defined entirely in terms of the ray to the left of the interface, it must be a function of $\mathbf{w}^{\mathrm{i}}$. Thus, there is a mapping $\mathfrak{R}^-(z^{\mathrm{i}})$ with the property

$$\mathbf{v}^- = \mathfrak{R}^-(z^{\mathrm{i}})\mathbf{w}^{\mathrm{i}}. \tag{5.17}$$

It will now be shown that $\mathfrak{R}^-(z^{\mathrm{i}})$ is a symplectic map. To this end, we consider the first hamiltonian generating function F_1^- defined [17, Sect.9.1] in terms of A^- in (5.8), by the equation

$$F_1^-(\mathbf{q}^{\mathrm{i}}, \overline{\mathbf{q}}^-) := -A^-(\mathbf{q}^{\mathrm{i}}, z^{\mathrm{i}}; \mathbf{q}^-, z^- = \varsigma(\mathbf{q}^-)), \tag{5.18}$$

and find the relations

$$\frac{\partial F_1^-}{\partial \mathbf{q}^{\mathrm{i}}} = -\frac{\partial A^-}{\partial \mathbf{q}^{\mathrm{i}}} = \mathbf{p}^{\mathrm{i}}, \tag{5.19a}$$

$$\begin{aligned} -\frac{\partial F_1^-}{\partial \overline{\mathbf{q}}^-} &= \frac{\partial A^-}{\partial \mathbf{q}^-} + \frac{\partial A^-}{\partial z^-}\frac{\partial z^-}{\partial \mathbf{q}^-} \\ &= \mathbf{p}^- - H^- \frac{\partial z^-}{\partial \mathbf{q}^-} = \overline{\mathbf{p}}^-. \end{aligned} \tag{5.19b}$$

Here we have made use of (5.5*a*), (5.11*a*), (5.12*a*), (5.14*a*), and the fact that $\mathbf{q}^- = \overline{\mathbf{q}}$. Equations (5.19) are the relations for a canonical transformation produced by the function F_1^-, from the ***old*** variables $\mathbf{q}^{\mathrm{i}}$, $\mathbf{p}^{\mathrm{i}}$ to the ***new*** variables $\overline{\mathbf{q}}^-$, $\overline{\mathbf{p}}^-$. Consequently, $\mathfrak{R}^-(z^{\mathrm{i}})$ is a symplectic map [5] that relates $\mathbf{v}^-$ and $\mathbf{w}^{\mathrm{i}}$.

Similarly, let $\mathbf{v}^+$ be a four-component vector at the surface ς, with entries

$$\mathbf{v}^+ := \begin{pmatrix} \mathbf{p}^+ \\ \mathbf{q}^+ \end{pmatrix}. \tag{5.20}$$

Then there is a map $\mathfrak{R}^+(z^{\rm f})$ with the property

$$\mathbf{v}^+ = \mathfrak{R}^+(z^{\rm f})\mathbf{w}^{\rm f}. \tag{5.21}$$

This map $\mathfrak{R}^+(z^{\rm f})$ is also symplectic. For verification, we consider the hamiltonian transformation function F_1^+ defined in terms of A^+ by the equation

$$F_1^+(\mathbf{q}^{\rm f}, \mathbf{q}^+) := A^+(\mathbf{q}^+, z^+ = \varsigma(\mathbf{q}^+); \mathbf{q}^{\rm f}, z^{\rm f}). \tag{5.22}$$

Then we have the relations

$$\frac{\partial F_1^+}{\partial \mathbf{q}^{\rm f}} = \frac{\partial A^+}{\partial \mathbf{q}^{\rm f}} = \mathbf{p}^{\rm f}, \tag{5.23a}$$

$$\begin{aligned} -\frac{\partial F_1^+}{\partial \mathbf{q}^+} &= -\frac{\partial A^+}{\partial \mathbf{q}^+} - \frac{\partial A^+}{\partial z^+}\frac{\partial z^+}{\partial \mathbf{q}^+} \\ &= \mathbf{p}^+ - H^+\frac{\partial z^+}{\partial \mathbf{q}^+} = \overline{\mathbf{p}}^+. \end{aligned} \tag{5.23b}$$

These equations are again the relations for a canonical transformation between the *old* variables $\mathbf{q}^{\rm f}$, $\mathbf{p}^{\rm f}$ and the *new* variables $\mathbf{q}^+$, $\overline{\mathbf{p}}^+$. Consequently, $\mathfrak{R}^+(z^{\rm f})$ is a symplectic map.

4.5.5 The optical transfer map

We write the expressions (2.10), (5.17), and (5.21) as Lie transformations. Then, being explicit, there are the three relations

$$\mathbf{w}^{\rm f}(\mathbf{w}^{\rm i}) = \mathfrak{M}(\mathbf{w}^{\rm i}, z^{\rm i}, z^{\rm f})\,\mathbf{w}^{\rm i}, \tag{5.24a}$$

$$\mathbf{v}^-(\mathbf{w}^{\rm i}) = \mathfrak{R}^-(\mathbf{w}^{\rm i}, z^{\rm i})\,\mathbf{w}^{\rm i}, \tag{5.24b}$$

$$\mathbf{v}^+(\mathbf{w}^{\rm f}) = \mathfrak{R}^+(\mathbf{w}^{\rm f}, z^{\rm f})\,\mathbf{w}^{\rm f}. \tag{5.24c}$$

Here the arguments that occur in the various operators have been written out in full in order to indicate the variables that would occur in the various generating polynomials f_n of a factorized representation (4.17–18). Now insert (5.24a) into (7.24c) to get the result

$$\begin{aligned} \mathbf{v}^+(\mathbf{w}^{\rm f}) = \mathbf{v}^+(\mathfrak{M}\mathbf{w}^{\rm i}) &= \mathfrak{R}^+(\mathfrak{M}\mathbf{w}^{\rm i}, z^{\rm f})\,\mathfrak{M}\mathbf{w}^{\rm i} \\ &= \mathfrak{M}\mathfrak{R}^+(\mathbf{w}^{\rm i}, z^{\rm f})\,\mathfrak{M}^{-1}\mathfrak{M}\mathbf{w}^{\rm i} \\ &= \mathfrak{M}\mathfrak{R}^+(\mathbf{w}^{\rm i}, z^{\rm f})\,\mathbf{w}^{\rm i}. \end{aligned} \tag{5.25}$$

We have made use of Lie algebraic manipulations to obtain the relation

$$\mathfrak{M}\mathfrak{R}^+(\mathbf{w}^{\rm i}, z^{\rm f})\mathfrak{M}^{-1} = \mathfrak{R}^+(\mathfrak{M}\mathbf{w}^{\rm i}, z^{\rm f}). \tag{5.26}$$

Next, we combine (5.15) and (5.24b) to obtain

$$\mathbf{v}^+(\mathbf{w}^{\rm f}) = \mathbf{v}^-(\mathbf{w}^{\rm i}) = \mathfrak{R}^-(\mathbf{w}^{\rm i}, z^{\rm i})\,\mathbf{w}^{\rm i}. \tag{5.27}$$

Now *watch closely!* The left-hand sides of (5.25) and (5.27) are equal; the right-hand sides are thus related through

$$\mathfrak{M}(\mathbf{w}^{\mathrm{i}}; z^{\mathrm{i}}, z^{\mathrm{f}})\,\mathfrak{R}^{+}(\mathbf{w}^{\mathrm{i}}; z^{\mathrm{f}})\,\mathbf{w}^{\mathrm{i}} = \mathfrak{R}^{-}(\mathbf{w}^{\mathrm{i}}; z^{\mathrm{i}})\,\mathbf{w}^{\mathrm{i}}. \tag{5.28}$$

Since the arguments of all the operators are the same and $\mathbf{w}^{\mathrm{i}}$ is arbitrary, this is equivalent to the *operator* relation

$$\mathfrak{M}(z^{\mathrm{i}}, z^{\mathrm{f}})\,\mathfrak{R}^{+}(z^{\mathrm{f}}) = \mathfrak{R}^{-}(z^{\mathrm{i}}) \tag{5.29}$$

We solve this relation for $\mathfrak{M}$ to obtain the optical transfer map of the refracting surface in the final, *factorized* form

$$\mathfrak{M}(z^{\mathrm{i}}, z^{\mathrm{f}}) = \mathfrak{R}^{-}(z^{\mathrm{i}})\left(\mathfrak{R}^{+}(z^{\mathrm{f}})\right)^{-1}. \tag{5.30}$$

What we have accomplished is that the optical transfer map for a system composed of two media separated by an interface ς, has been written in terms of two symplectic ***root*** maps $\mathfrak{R}^{-}$ and $\mathfrak{R}^{+}$. The inverse of a symplectic map is also a symplectic map [5, Sect.4.2]. Thus the product $\mathfrak{R}^{-}(\mathfrak{R}^{+})^{-1}$ is a symplectic map. Moreover, by construction, the first map depends *only* on the properties of the medium *before* the interface, and the second depends *only* on the properties of the medium *after* the interface. Both maps, depend, of course, on the shape of the interface ς.

We may eliminate the free-flight part of the root transformation, moving z^{f} and z^{i} up to the optical center of the surface ς. The map $\mathfrak{M}$ due to the interface *alone* can be found thus by a simple limiting process. The interface ς intersects the optical z axis at the point z^0 given by (5.7), as

$$z^0 = \varsigma(\mathbf{0}). \tag{5.31}$$

Then the effect of the interface itself is given by the ***refracting surface map*** $\mathfrak{S}$, defined as the limit

$$\mathfrak{S} = \lim_{\substack{z^{\mathrm{i}} \to z^0 \\ z^{\mathrm{f}} \to z^0}} \mathfrak{M}(z^{\mathrm{i}}, z^{\mathrm{f}}) = \mathfrak{R}^{-}(z^0)\left(\mathfrak{R}^{+}(z^0)\right)^{-1}. \tag{5.32}$$

Although root maps are generally of interest only insofar as they appear in combination to make up complete map $\mathfrak{S}$, they may be of use by themselves in situations where either the object or image sufaces are *curved* rather than planar.

4.5.6 The *root* map in homogeneous media

As a concrete application of the above factorization theorem, consider the simplest case in which ς is the boundary between two homogeneous media, whose refraction indices, n^- and n^+, are constant [14]. First we look at the ray *before* the interface. From (4.6) we see that the incoming ray satisfies the free-propagation equations

$$\mathbf{p}(z) = \mathbf{p}^{\mathrm{i}}, \tag{5.33a}$$

$$\mathbf{q}(z) = \mathbf{q}^{\mathrm{i}} + (z - z^{\mathrm{i}})\frac{\mathbf{p}^{\mathrm{i}}}{\sqrt{(n^-)^2 - (\mathbf{p}^{\mathrm{i}})^2}}. \tag{5.33b}$$

It follows from this, valuated at $z = \varsigma(\mathbf{q}^-)$, that

$$\mathbf{p}^- = \mathbf{p}^{\mathrm{i}}, \tag{5.34a}$$

$$\mathbf{q}^- = \mathbf{q}^{\mathrm{i}} + \left(\varsigma(\mathbf{q}^-) - z^{\mathrm{i}}\right)\frac{\mathbf{p}^{\mathrm{i}}}{\sqrt{(n^-)^2 - (\mathbf{p}^{\mathrm{i}})^2}}. \tag{5.34b}$$

Now take the limit of the second equation as $z^{\mathrm{i}} \to z^0$, and evaluate (5.14a), to give the pair of equations

$$\overline{\mathbf{p}}^- = \mathbf{p}^{\mathrm{i}} + \sqrt{(n^-)^2 - (\mathbf{p}^{\mathrm{i}})^2}\,\frac{\partial \varsigma}{\partial \mathbf{q}^-}, \tag{5.35a}$$

$$\mathbf{q}^- = \mathbf{q}^{\mathrm{i}} + \big(\varsigma(\mathbf{q}^-) - z^0\big)\frac{\mathbf{p}^{\mathrm{i}}}{\sqrt{(n^-)^2 - (\mathbf{p}^{\mathrm{i}})^2}}, \tag{5.35b}$$

These equations implicitly define $\mathfrak{R}^-(z^0)$.

Next we look at the ray *after* the interface. It satisfies the free-propagation equations

$$\mathbf{p}(z) = \mathbf{p}^{\mathrm{f}}, \tag{5.36a}$$

$$\mathbf{q}(z) = \mathbf{q}^{\mathrm{f}} + (z - z^{\mathrm{f}})\frac{\mathbf{p}^{\mathrm{f}}}{\sqrt{(n^+)^2 - (\mathbf{p}^{\mathrm{f}})^2}}. \tag{5.36b}$$

Consequently, we have the relations

$$\mathbf{p}^+ = \mathbf{p}^{\mathrm{f}}, \tag{5.37a}$$

$$\mathbf{q}^+ = \mathbf{q}^{\mathrm{f}} + \big(\varsigma(\mathbf{q}^+) - z^{\mathrm{f}}\big)\frac{\mathbf{p}^{\mathrm{f}}}{\sqrt{(n^+)^2 - (\mathbf{p}^{\mathrm{f}})^2}}. \tag{5.37b}$$

Finally, taking the limit of (5.37*b*) as $z^{\mathrm{f}} \to z^0$, and evaluating (5.14*b*), we obtain the pair of relations

$$\overline{\mathbf{p}}^+ = \mathbf{p}^{\mathrm{f}} + \sqrt{(n^+)^2 - (\mathbf{p}^{\mathrm{f}})^2}\,\frac{\partial \varsigma}{\partial \mathbf{q}^+}. \tag{5.38a}$$

$$\overline{\mathbf{q}}^+ = \mathbf{q}^{\mathrm{i}} + \big(\varsigma(\mathbf{q}^+) + z^0\big)\frac{\mathbf{p}^{\mathrm{f}}}{\sqrt{(n^+)^2 - (\mathbf{p}^{\mathrm{f}})^2}}, \tag{5.38b}$$

These equations implicitly define $\mathfrak{R}^+(z^0)$.

4.6 Refraction at surfaces

We note that the equations defining $\mathfrak{R}^-$ in (5.35) and those defining $\mathfrak{R}^+$ in (5.38) are identical in form. We set now n to be n^- or n^+; for simplicity, also, we take the optical center of the surface (*i.e.*, the intersection of the surface with the optical axis) to define the $z = 0$ origin, so we set $z^0 = 0$ and we may speak simply of *the* root map $\mathfrak{R}$. Because of (5.15) we may write the result in the form [14]:

$$\mathfrak{R}\mathbf{p} = \overline{\mathbf{p}} = \mathbf{p} + \sqrt{n^2 - \mathbf{p}^2}\,\frac{\partial \varsigma(\overline{\mathbf{q}})}{\partial \overline{\mathbf{q}}}, \tag{6.1a}$$

$$\mathfrak{R}\mathbf{q} = \overline{\mathbf{q}} = \mathbf{q} + \varsigma(\overline{\mathbf{q}})\frac{\mathbf{p}}{\sqrt{n^2 - \mathbf{p}^2}}. \tag{6.1b}$$

We remind the reader that $\varsigma(\mathbf{q})$ may be a quite arbitrary surface. In this section, we shall solve this pair of equations iteratively for axially-symmetric polynomial surfaces $\varsigma(\mathbf{q}^2)$.

It must be emphasized that this pair of (Navarro-Saad–Wolf) equations have not yet been subjected to an exhaustive analysis. They represent a basic and ubiquitous physical process —refraction at a surface— with the very specific property of factorization. For instance: the mappings of phase space which $\mathfrak{R}$ defines, have necessarily (for $\varsigma \neq$ constant) singular lines, for which we may find algebraic expressions. These, most probably, correspond to *caustics* in lens optics: for any curved surface ς, we may draw light beams which are *tangent* to the surface. In crossing these lines, the nature of the solution changes drastically: multivalued functions and/or complex angles may appear. In reference [18] we initiated the study of the *exact* solutions to (6.1) for *spherical* surfaces and the concomitant *symmetries* of the system. These correspond obviously to rotations of the sphere around its center, and allow for abbreviated computation of the aberration coefficients. Similar techniques should apply to other conics of revolution. The results we present here apply for *general* polynomial surfaces of revolution.

4.6.1 The root map for polynomial surfaces

We will show the application of the root transformation for the case when the refracting interface is axially symmetric and described by the polynomial

$$z = \varsigma(\mathbf{q}) = \varsigma_2 \mathbf{q}^2 + \varsigma_4 (\mathbf{q}^2)^2 + \varsigma_6 (\mathbf{q}^2)^3 + \varsigma_8 (\mathbf{q}^2)^4 + \cdots. \tag{6.2}$$

This surface, we note, has its optical center at the origin, $\varsigma(\mathbf{0}) = 0$, and is tangent there to the $z = 0$ plane, *i.e.*, $\partial\varsigma(\mathbf{q})/\partial\mathbf{q}|_{\mathbf{q}=\mathbf{0}} = 0$.

First we *iterate* equation (6.1*b*), replacing the expression for $\overline{\mathbf{q}}$ in the left-hand side into the function $\varsigma(\overline{\mathbf{q}})$ in the right-hand side, in its explicit series form (6.2), and use the series expansion

$$\frac{\mathbf{p}}{\sqrt{n^2 - \mathbf{p}^2}} = \frac{1}{n}\mathbf{p} + \frac{1}{2n^3}\mathbf{p}^2\,\mathbf{p} + \frac{3}{8n^5}(\mathbf{p}^2)^2\,\mathbf{p} + \frac{5}{16n^5}(\mathbf{p}^2)^4\,\mathbf{p} + \cdots + \frac{(2m-1)!!}{(2m)!!\,n^{2m+1}}(\mathbf{p}^2)^m\,\mathbf{p} + \cdots. \tag{6.3}$$

Since this process does not terminate by itself, we may retain in the expression the terms *up to the* N^{th} *order*.

The result, for aberration order *seven*, is

$$\begin{array}{ll}
1^{\text{st}} \text{ order} & \overline{\mathbf{q}} = \mathbf{q} \\
3^{\text{d}} \text{ order} & \quad + \dfrac{\varsigma_2}{n}\mathbf{q}^2\,\mathbf{p} \\
5^{\text{th}} \text{ order} & \quad + \dfrac{\varsigma_2}{2n^3}\mathbf{p}^2\mathbf{q}^2\,\mathbf{p} + \dfrac{2\varsigma_2^2}{n^2}\mathbf{p}\cdot\mathbf{q}\,\mathbf{q}^2\mathbf{p} + \dfrac{\varsigma_4}{n}(\mathbf{q}^2)^2\,\mathbf{p} \\
7^{\text{th}} \text{ order} & \quad + \dfrac{3\varsigma_2}{8n^5}(\mathbf{p}^2)^2\mathbf{q}^2\,\mathbf{p} + \dfrac{2\varsigma_2^2}{n^4}\mathbf{p}^2\mathbf{p}\cdot\mathbf{q}\,\mathbf{q}^2\mathbf{p} + \dfrac{2\varsigma_2^3 + \varsigma_4}{2n^3}\mathbf{p}^2(\mathbf{q}^2)^2\,\mathbf{p} \\
 & \quad + \dfrac{4\varsigma_2^3}{n^3}(\mathbf{p}\cdot\mathbf{q})^2\mathbf{q}^2\mathbf{p} + \dfrac{6\varsigma_2\varsigma_4}{n^2}\mathbf{p}\cdot\mathbf{q}\,(\mathbf{q}^2)^2\mathbf{p} + \dfrac{\varsigma_6}{n}(\mathbf{q}^2)^3\,\mathbf{p} + O_9.
\end{array} \tag{6.4}$$

Next, we *substitute* this result for $\overline{\mathbf{q}}$ into (6.1*a*) to obtain $\overline{\mathbf{p}}$. Here we need the $\mathbf{q}$-derivative of the series (6.2), and the expansion

$$\sqrt{n^2 - \mathbf{p}^2} = n - \frac{1}{2n}\mathbf{p}^2 - \frac{1}{8n^3}(\mathbf{p}^2)^2 - \frac{1}{16n^5}(\mathbf{p}^2)^3 - \cdots - \frac{(2m-3)!!}{(2m)!!\,n^{2m-1}}(\mathbf{p}^2)^m - \cdots. \tag{6.5}$$

Here too, we retain the resulting series for $\overline{\mathbf{p}}$ up to N^{th} aberration order.

We find

$$
\begin{array}{ll}
1^{\rm st}\ \text{order} & \overline{\mathbf{p}} = \mathbf{p} + 2\varsigma_2 n\,\mathbf{q} \\
3^{\rm d}\ \text{order} & \quad -\frac{\varsigma_2}{n}\mathbf{p}^2\,\mathbf{q} + 2\varsigma_2^2\mathbf{q}^2\,\mathbf{p} + 4\varsigma_4 n\mathbf{q}^2\,\mathbf{q} \\
5^{\rm th}\ \text{order} & \quad -\frac{\varsigma_2}{4n^3}(\mathbf{p}^2)^2\,\mathbf{q} - \frac{2\varsigma_4}{n}\mathbf{p}^2\mathbf{q}^2\,\mathbf{q} + \frac{4\varsigma_2^3}{n}\mathbf{p}\cdot\mathbf{q}\,\mathbf{q}^2\mathbf{p} \\
 & \quad + 6\varsigma_2\varsigma_4(\mathbf{q}^2)^2\,\mathbf{p} + 8\varsigma_2\varsigma_4\mathbf{p}\cdot\mathbf{q}\,\mathbf{q}^2\mathbf{q} + 6\varsigma_6 n(\mathbf{q}^2)^2\,\mathbf{q} \\
7^{\rm th}\ \text{order} & \quad -\frac{\varsigma_2}{8n^5}(\mathbf{p}^2)^3\,\mathbf{q} - \frac{\varsigma_4}{2n^3}(\mathbf{p}^2)^2\mathbf{q}^2\,\mathbf{q} + \frac{2\varsigma_2^3}{n^3}\mathbf{p}^2\mathbf{p}\cdot\mathbf{q}\,\mathbf{q}^2\mathbf{p} + \frac{2\varsigma_2^4}{n^2}\mathbf{p}^2(\mathbf{q}^2)^2\,\mathbf{p} \\
 & \quad + \frac{4\varsigma_2^2\varsigma_4 - 3\varsigma_6}{n}\mathbf{p}^2(\mathbf{q}^2)^2\,\mathbf{q} + \frac{8\varsigma_2^4}{n^2}(\mathbf{p}\cdot\mathbf{q})^2\mathbf{q}^2\mathbf{p} + \frac{28\varsigma_2^2\varsigma_4}{n}\mathbf{p}\cdot\mathbf{q}\,(\mathbf{q}^2)^2\mathbf{p} \\
 & \quad + 4(2\varsigma_2\varsigma_6 + \varsigma_4^2)(\mathbf{q}^2)^3\,\mathbf{p} + \frac{16\varsigma_2^2\varsigma_4}{n}(\mathbf{p}\cdot\mathbf{q})^2\mathbf{q}^2\mathbf{q} \\
 & \quad + 8(3\varsigma_2\varsigma_6 + \varsigma_4^2)\mathbf{p}\cdot\mathbf{q}\,(\mathbf{q}^2)^2\mathbf{q} + 8\varsigma_8 n\,(\mathbf{q}^2)^3\,\mathbf{q} + O_9.
\end{array}
\tag{6.6}
$$

Now we must find the *inverse* of these transformations, changing in (6.1) the refraction index $n = n^-$ of the first medium, by $n = n^+$ in the second medium. For this we write the pair of equations as

$$\mathfrak{R}^{-1}\,\overline{\mathbf{p}} = \mathbf{p} = \overline{\mathbf{p}} - \sqrt{n^2 - \mathbf{p}^2}\,\frac{\partial\varsigma(\overline{\mathbf{q}})}{\partial\overline{\mathbf{q}}}, \tag{6.7a}$$

$$\mathfrak{R}^{-1}\,\overline{\mathbf{q}} = \mathbf{q} = \overline{\mathbf{q}} - \varsigma(\overline{\mathbf{q}})\frac{\mathbf{p}}{\sqrt{n^2 - \mathbf{p}^2}}. \tag{6.7b}$$

Now we *iterate* first (6.7*a*) for $\mathbf{p}(\overline{\mathbf{p}}, \overline{\mathbf{q}})$, and then *substitute* the result in (6.7*b*) for $\mathbf{q}(\overline{\mathbf{p}}, \overline{\mathbf{q}})$.

4.6.2 The refracting surface map

We now concatenate $\mathfrak{R}$ with n^- [Eqs. (6.4) and (6.6)] with the similar expression for $\mathfrak{R}^{-1}$ with n^+ that stems from (6.7). To third aberration order this may be done confortably by hand [14], and fifth order may be tackled with patience after a couple of pages of calculations. The results shown below for seventh order have been calculated by computer, at the University of Maryland, by Forest [19] with the aid of **MACSYMA** symbolic comptation. Independently, they were calculated by Navarro-Saad [20] at IIMAS–UNAM to ninth order using **REDUCE**. This has been reported in reference [21].

In the following expressions, the refraction indices n^- and n^+ are called n and m, $\mathbf{p}$, $\mathbf{q}$ and $\mathbf{p}'$, $\mathbf{q}'$ stand for the initial and final phase-space coordinates *at* $z = 0$. This is to keep superindices to a minimum. The results printed here were taken from reference [20]. Navarro-Saad obtained a **REDUCE** symbolic computation output file, and we doctored it by **EMACS** editor to turn it into a well-formatted TEX input file, to be used here for typographic composition. In this way, all copying errata are removed. The results are:

$$\mathbf{p}' = \mathbf{p} \qquad \text{(6.8}a\text{, first order)}$$

$$+\mathbf{q}\left(2n\varsigma_2 - 2m\varsigma_2\right)$$

$$+p^2\mathbf{q}\left(-\tfrac{1}{n}\varsigma_2 + \tfrac{1}{m}\varsigma_2\right) \qquad \text{(6.8}a\text{, third order)}$$

$$+q^2\mathbf{p}\left(-2\tfrac{1}{n}m{\varsigma_2}^2 + 2{\varsigma_2}^2\right)$$

$$+q^2\mathbf{q}\left(4n^2\tfrac{1}{m}{\varsigma_2}^3 - 8n{\varsigma_2}^3 + 4n\varsigma_4 + 4m{\varsigma_2}^3 - 4m\varsigma_4\right)$$

$$\begin{aligned}
&+(\mathbf{p}\cdot\mathbf{q})\mathbf{q}\left(4n\tfrac{1}{m}\varsigma_2{}^2 - 4\varsigma_2{}^2\right)\\
&+p^4\mathbf{q}\left(-\tfrac{1}{4}\tfrac{1}{n^3}\varsigma_2 + \tfrac{1}{4}\tfrac{1}{m^3}\varsigma_2\right)\\
&+p^2q^2\mathbf{p}\left(\tfrac{1}{n}\tfrac{1}{m}\varsigma_2{}^2 - \tfrac{1}{n^3}m\varsigma_2{}^2\right)\\
&+p^2q^2\mathbf{q}\left(2n^2\tfrac{1}{m^3}\varsigma_2{}^3 - 2\tfrac{1}{n}\varsigma_4 - 2\tfrac{1}{m}\varsigma_2{}^3 + 2\tfrac{1}{m}\varsigma_4\right)\\
&+p^2(\mathbf{p}\cdot\mathbf{q})\mathbf{q}\left(2n\tfrac{1}{m^3}\varsigma_2{}^2 - 2\tfrac{1}{n}\tfrac{1}{m}\varsigma_2{}^2\right)\\
&+q^4\mathbf{p}\left(4n\tfrac{1}{m}\varsigma_2{}^4 + 4\tfrac{1}{n}m\varsigma_2{}^4 - 6\tfrac{1}{n}m\varsigma_2\varsigma_4 - 8\varsigma_2{}^4 + 6\varsigma_2\varsigma_4\right)\\
&+q^4\mathbf{q}\left(4n^4\tfrac{1}{m^3}\varsigma_2{}^5 - 24n^2\tfrac{1}{m}\varsigma_2{}^5 + 24n^2\tfrac{1}{m}\varsigma_2{}^2\varsigma_4 + 32n\varsigma_2{}^5\right.\\
&\qquad\left. -48n\varsigma_2{}^2\varsigma_4 + 6n\varsigma_6 - 12m\varsigma_2{}^5 + 24m\varsigma_2{}^2\varsigma_4 - 6m\varsigma_6\right)\\
&+q^2(\mathbf{p}\cdot\mathbf{q})\mathbf{p}\left(-4\tfrac{1}{n^2}m\varsigma_2{}^3 + 4\tfrac{1}{m}\varsigma_2{}^3\right)\\
&+q^2(\mathbf{p}\cdot\mathbf{q})\mathbf{q}\left(8n^3\tfrac{1}{m^3}\varsigma_2{}^4 - 16n\tfrac{1}{m}\varsigma_2{}^4 + 16n\tfrac{1}{m}\varsigma_2\varsigma_4 + 8\tfrac{1}{n}m\varsigma_2{}^4 - 8\tfrac{1}{n}m\varsigma_2\varsigma_4 - 8\varsigma_2\varsigma_4\right)\\
&+(\mathbf{p}\cdot\mathbf{q})^2\mathbf{q}\left(4n^2\tfrac{1}{m^3}\varsigma_2{}^3 - 4\tfrac{1}{m}\varsigma_2{}^3\right)
\end{aligned}$$

(6.8*a*, fifth order)

$$\begin{aligned}
&+p^6\mathbf{q}\left(-\tfrac{1}{8}\tfrac{1}{n^5}\varsigma_2 + \tfrac{1}{8}\tfrac{1}{m^5}\varsigma_2\right)\\
&+p^4q^2\mathbf{p}\left(\tfrac{1}{4}\tfrac{1}{n}\tfrac{1}{m^3}\varsigma_2{}^2 + \tfrac{1}{2}\tfrac{1}{n^3}\tfrac{1}{m}\varsigma_2{}^2 - \tfrac{3}{4}\tfrac{1}{n^5}m\varsigma_2{}^2\right)\\
&+p^4q^2\mathbf{q}\left(\tfrac{3}{2}n^2\tfrac{1}{m^5}\varsigma_2{}^3 - \tfrac{1}{n^3}\varsigma_2{}^3 - \tfrac{1}{2}\tfrac{1}{n^3}\varsigma_4 - \tfrac{1}{2}\tfrac{1}{m^3}\varsigma_2{}^3 + \tfrac{1}{2}\tfrac{1}{m^3}\varsigma_4\right)\\
&+p^4(\mathbf{p}\cdot\mathbf{q})\mathbf{q}\left(\tfrac{3}{2}n\tfrac{1}{m^5}\varsigma_2{}^2 - \tfrac{1}{n}\tfrac{1}{m^3}\varsigma_2{}^2 - \tfrac{1}{2}\tfrac{1}{n^3}\tfrac{1}{m}\varsigma_2{}^2\right)\\
&+p^2q^4\mathbf{p}\left(2n\tfrac{1}{m^3}\varsigma_2{}^4 + 3\tfrac{1}{n}\tfrac{1}{m}\varsigma_2\varsigma_4 - 2\tfrac{1}{n^2}\varsigma_2{}^4 - 3\tfrac{1}{n^3}m\varsigma_2\varsigma_4\right)\\
&+p^2q^4\mathbf{q}\left(6n^4\tfrac{1}{m^5}\varsigma_2{}^5 - 12n^2\tfrac{1}{m^3}\varsigma_2{}^5 + 12n^2\tfrac{1}{m^3}\varsigma_2{}^2\varsigma_4 - 8\tfrac{1}{n}\varsigma_2{}^5 + 8\tfrac{1}{n}\varsigma_2{}^2\varsigma_4\right.\\
&\qquad\left. -3\tfrac{1}{n}\varsigma_6 + 4\tfrac{1}{n^2}m\varsigma_2{}^5 - 4\tfrac{1}{n^2}m\varsigma_2{}^2\varsigma_4 + 10\tfrac{1}{m}\varsigma_2{}^5 - 16\tfrac{1}{m}\varsigma_2{}^2\varsigma_4 + 3\tfrac{1}{m}\varsigma_6\right)\\
&+p^2q^2(\mathbf{p}\cdot\mathbf{q})\mathbf{p}\left(2\tfrac{1}{n^2}\tfrac{1}{m}\varsigma_2{}^3 - 4\tfrac{1}{n^4}m\varsigma_2{}^3 + 2\tfrac{1}{m^3}\varsigma_2{}^3\right)\\
&+p^2q^2(\mathbf{p}\cdot\mathbf{q})\mathbf{q}\left(12n^3\tfrac{1}{m^5}\varsigma_2{}^4 - 12n\tfrac{1}{m^3}\varsigma_2{}^4 + 8n\tfrac{1}{m^3}\varsigma_2\varsigma_4 + 4\tfrac{1}{n}\tfrac{1}{m}\varsigma_2{}^4\right.\\
&\qquad\left. -4\tfrac{1}{n}\tfrac{1}{m}\varsigma_2\varsigma_4 - 8\tfrac{1}{n^2}\varsigma_2{}^4 + 4\tfrac{1}{n^3}m\varsigma_2{}^4 - 4\tfrac{1}{n^3}m\varsigma_2\varsigma_4\right)\\
&+p^2(\mathbf{p}\cdot\mathbf{q})^2\mathbf{q}\left(6n^2\tfrac{1}{m^5}\varsigma_2{}^3 - 6\tfrac{1}{m^3}\varsigma_2{}^3\right)\\
&+q^6\mathbf{p}\left(4n^3\tfrac{1}{m^3}\varsigma_2{}^6 - 24n\tfrac{1}{m}\varsigma_2{}^6 + 28n\tfrac{1}{m}\varsigma_2{}^3\varsigma_4 - 12\tfrac{1}{n}m\varsigma_2{}^6 + 28\tfrac{1}{n}m\varsigma_2{}^3\varsigma_4\right.\\
&\qquad\left. -8\tfrac{1}{n}m\varsigma_2\varsigma_6 - 4\tfrac{1}{n}m\varsigma_4{}^2 + 32\varsigma_2{}^6 - 56\varsigma_2{}^3\varsigma_4 + 8\varsigma_2\varsigma_6 + 4\varsigma_4{}^2\right)\\
&+q^6\mathbf{q}\left(8n^6\tfrac{1}{m^5}\varsigma_2{}^7 - 40n^4\tfrac{1}{m^3}\varsigma_2{}^7 + 40n^4\tfrac{1}{m^3}\varsigma_2{}^4\varsigma_4 + 120n^2\tfrac{1}{m}\varsigma_2{}^7 - 240n^2\tfrac{1}{m}\varsigma_2{}^4\varsigma_4\right.\\
&\qquad +36n^2\tfrac{1}{m}\varsigma_2{}^2\varsigma_6 + 48n^2\tfrac{1}{m}\varsigma_2\varsigma_4{}^2 - 128n\varsigma_2{}^7 + 320n\varsigma_2{}^4\varsigma_4 - 72n\varsigma_2{}^2\varsigma_6\\
&\qquad\left. -96n\varsigma_2\varsigma_4{}^2 + 8n\varsigma_8 + 40m\varsigma_2{}^7 - 120m\varsigma_2{}^4\varsigma_4 + 36m\varsigma_2{}^2\varsigma_6 + 48m\varsigma_2\varsigma_4{}^2 - 8m\varsigma_8\right)\\
&+q^4(\mathbf{p}\cdot\mathbf{q})\mathbf{p}\left(8n^2\tfrac{1}{m^3}\varsigma_2{}^5 - 16\tfrac{1}{n}\varsigma_2{}^5 + 8\tfrac{1}{n}\varsigma_2{}^2\varsigma_4 + 16\tfrac{1}{n^2}m\varsigma_2{}^5\right.\\
&\qquad\left. -28\tfrac{1}{n^2}m\varsigma_2{}^2\varsigma_4 - 8\tfrac{1}{m}\varsigma_2{}^5 + 20\tfrac{1}{m}\varsigma_2{}^2\varsigma_4\right)\\
&+q^4(\mathbf{p}\cdot\mathbf{q})\mathbf{q}\left(24n^5\tfrac{1}{m^5}\varsigma_2{}^6 - 64n^3\tfrac{1}{m^3}\varsigma_2{}^6 + 64n^3\tfrac{1}{m^3}\varsigma_2{}^3\varsigma_4 + 24n\tfrac{1}{m}\varsigma_2{}^6\right.\\
&\qquad -88n\tfrac{1}{m}\varsigma_2{}^3\varsigma_4 + 24n\tfrac{1}{m}\varsigma_2\varsigma_6 + 16n\tfrac{1}{m}\varsigma_4{}^2 - 48\tfrac{1}{n}m\varsigma_2{}^6\\
&\qquad\left. +104\tfrac{1}{n}m\varsigma_2{}^3\varsigma_4 - 24\tfrac{1}{n}m\varsigma_2\varsigma_6 - 8\tfrac{1}{n}m\varsigma_4{}^2 + 64\varsigma_2{}^6 - 80\varsigma_2{}^3\varsigma_4 - 8\varsigma_4{}^2\right)
\end{aligned}$$

(6.8*a*, seventh order)

$$
\begin{aligned}
&+q^2(\mathbf{p}\cdot\mathbf{q})^2\mathbf{p}\left(4n\tfrac{1}{m^3}\varsigma_2{}^4+4\tfrac{1}{n}\tfrac{1}{m}\varsigma_2{}^4-8\tfrac{1}{n^3}m\varsigma_2{}^4\right)\\
&+q^2(\mathbf{p}\cdot\mathbf{q})^2\mathbf{q}\left(24n^4\tfrac{1}{m^5}\varsigma_2{}^5-32n^2\tfrac{1}{m^3}\varsigma_2{}^5+24n^2\tfrac{1}{m^3}\varsigma_2{}^2\varsigma_4-16\tfrac{1}{n}\varsigma_2{}^2\varsigma_4\right.\\
&\qquad\qquad\left.+16\tfrac{1}{n^2}m\varsigma_2{}^5-16\tfrac{1}{n^2}m\varsigma_2{}^2\varsigma_4-8\tfrac{1}{m}\varsigma_2{}^5+8\tfrac{1}{m}\varsigma_2{}^2\varsigma_4\right)\\
&+(\mathbf{p}\cdot\mathbf{q})^3\mathbf{q}\left(8n^3\tfrac{1}{m^5}\varsigma_2{}^4-8n\tfrac{1}{m^3}\varsigma_2{}^4\right)+O_9;
\end{aligned}
$$

$$\mathbf{q}'=\mathbf{q}$$

(6.8b, first order)

$$+q^2\mathbf{p}\left(\tfrac{1}{n}\varsigma_2-\tfrac{1}{m}\varsigma_2\right)$$

(6.8b, third order)

$$
\begin{aligned}
&+q^2\mathbf{q}\left(-2n\tfrac{1}{m}\varsigma_2{}^2+2\varsigma_2{}^2\right)\\
&+p^2q^2\mathbf{p}\left(\tfrac{1}{2}\tfrac{1}{n^3}\varsigma_2-\tfrac{1}{2}\tfrac{1}{m^3}\varsigma_2\right)
\end{aligned}
$$

(6.8b, fifth order)

$$
\begin{aligned}
&+p^2q^2\mathbf{q}\left(-n\tfrac{1}{m^3}\varsigma_2{}^2+\tfrac{1}{n}\tfrac{1}{m}\varsigma_2{}^2\right)\\
&+q^4\mathbf{p}\left(-2n^2\tfrac{1}{m^3}\varsigma_2{}^3+4n\tfrac{1}{m^2}\varsigma_2{}^3+2\tfrac{1}{n}\varsigma_2{}^3+\tfrac{1}{n}\varsigma_4-4\tfrac{1}{m}\varsigma_2{}^3-\tfrac{1}{m}\varsigma_4\right)\\
&+q^4\mathbf{q}\left(-4n^3\tfrac{1}{m^3}\varsigma_2{}^4+8n^2\tfrac{1}{m^2}\varsigma_2{}^4-4n\tfrac{1}{m}\varsigma_2{}^4-6n\tfrac{1}{m}\varsigma_2\varsigma_4+6\varsigma_2\varsigma_4\right)\\
&+q^2(\mathbf{p}\cdot\mathbf{q})\mathbf{p}\left(-2n\tfrac{1}{m^3}\varsigma_2{}^2-2\tfrac{1}{n}\tfrac{1}{m}\varsigma_2{}^2+2\tfrac{1}{n^2}\varsigma_2{}^2+2\tfrac{1}{m^2}\varsigma_2{}^2\right)\\
&+q^2(\mathbf{p}\cdot\mathbf{q})\mathbf{q}\left(-4n^2\tfrac{1}{m^3}\varsigma_2{}^3+4n\tfrac{1}{m^2}\varsigma_2{}^3+4\tfrac{1}{n}\varsigma_2{}^3-4\tfrac{1}{m}\varsigma_2{}^3\right)\\
&+p^4q^2\mathbf{p}\left(\tfrac{3}{8}\tfrac{1}{n^5}\varsigma_2-\tfrac{3}{8}\tfrac{1}{m^5}\varsigma_2\right)
\end{aligned}
$$

(6.8b, seventh order)

$$
\begin{aligned}
&+p^4q^2\mathbf{q}\left(-\tfrac{3}{4}n\tfrac{1}{m^5}\varsigma_2{}^2+\tfrac{1}{2}\tfrac{1}{n}\tfrac{1}{m^3}\varsigma_2{}^2+\tfrac{1}{4}\tfrac{1}{n^3}\tfrac{1}{m}\varsigma_2{}^2\right)\\
&+p^2q^4\mathbf{p}\left(-3n^2\tfrac{1}{m^5}\varsigma_2{}^3+4n\tfrac{1}{m^4}\varsigma_2{}^3-\tfrac{1}{n^2}\tfrac{1}{m}\varsigma_2{}^3+2\tfrac{1}{n^3}\varsigma_2{}^3+\tfrac{1}{2}\tfrac{1}{n^3}\varsigma_4-2\tfrac{1}{m^3}\varsigma_2{}^3-\tfrac{1}{2}\tfrac{1}{m^3}\varsigma_4\right)\\
&+p^2q^4\mathbf{q}\left(-6n^3\tfrac{1}{m^5}\varsigma_2{}^4+8n^2\tfrac{1}{m^4}\varsigma_2{}^4-3n\tfrac{1}{m^3}\varsigma_2\varsigma_4+3\tfrac{1}{n}\tfrac{1}{m}\varsigma_2\varsigma_4+2\tfrac{1}{n^2}\varsigma_2{}^4-4\tfrac{1}{m^2}\varsigma_2{}^4\right)\\
&+p^2q^2(\mathbf{p}\cdot\mathbf{q})\mathbf{p}\left(-3n\tfrac{1}{m^5}\varsigma_2{}^2-\tfrac{1}{n^3}\tfrac{1}{m}\varsigma_2{}^2+2\tfrac{1}{n^4}\varsigma_2{}^2+2\tfrac{1}{m^4}\varsigma_2{}^2\right)\\
&+p^2q^2(\mathbf{p}\cdot\mathbf{q})\mathbf{q}\left(-6n^2\tfrac{1}{m^5}\varsigma_2{}^3+4n\tfrac{1}{m^4}\varsigma_2{}^3-2\tfrac{1}{n}\tfrac{1}{m^2}\varsigma_2{}^3+2\tfrac{1}{n^3}\varsigma_2{}^3+2\tfrac{1}{m^3}\varsigma_2{}^3\right)\\
&+q^6\mathbf{p}\left(-6n^4\tfrac{1}{m^5}\varsigma_2{}^5+16n^3\tfrac{1}{m^4}\varsigma_2{}^5-16n^2\tfrac{1}{m^3}\varsigma_2{}^5-10n^2\tfrac{1}{m^3}\varsigma_2{}^2\varsigma_4\right.\\
&\qquad\left.+8n\tfrac{1}{m^2}\varsigma_2{}^5+20n\tfrac{1}{m^2}\varsigma_2{}^2\varsigma_4+8\tfrac{1}{n}\varsigma_2{}^2\varsigma_4+\tfrac{1}{n}\varsigma_6-2\tfrac{1}{m}\varsigma_2{}^5-18\tfrac{1}{m}\varsigma_2{}^2\varsigma_4-\tfrac{1}{m}\varsigma_6\right)\\
&+q^6\mathbf{q}\left(-12n^5\tfrac{1}{m^5}\varsigma_2{}^6+32n^4\tfrac{1}{m^4}\varsigma_2{}^6-24n^3\tfrac{1}{m^3}\varsigma_2{}^6-28n^3\tfrac{1}{m^3}\varsigma_2{}^3\varsigma_4\right.\\
&\qquad\left.+56n^2\tfrac{1}{m^2}\varsigma_2{}^3\varsigma_4+4n\tfrac{1}{m}\varsigma_2{}^6-28n\tfrac{1}{m}\varsigma_2{}^3\varsigma_4-8n\tfrac{1}{m}\varsigma_2\varsigma_6-4n\tfrac{1}{m}\varsigma_4{}^2+8\varsigma_2\varsigma_6+4\varsigma_4{}^2\right)\\
&+q^4(\mathbf{p}\cdot\mathbf{q})\mathbf{p}\left(-12n^3\tfrac{1}{m^5}\varsigma_2{}^4+24n^2\tfrac{1}{m^4}\varsigma_2{}^4-24n\tfrac{1}{m^3}\varsigma_2{}^4-6n\tfrac{1}{m^3}\varsigma_2\varsigma_4\right.\\
&\qquad\qquad\left.-16\tfrac{1}{n}\tfrac{1}{m}\varsigma_2{}^4-6\tfrac{1}{n}\tfrac{1}{m}\varsigma_2\varsigma_4+8\tfrac{1}{n^2}\varsigma_2{}^4+6\tfrac{1}{n^2}\varsigma_2\varsigma_4+20\tfrac{1}{m^2}\varsigma_2{}^4+6\tfrac{1}{m^2}\varsigma_2\varsigma_4\right)\\
&+q^4(\mathbf{p}\cdot\mathbf{q})\mathbf{q}\left(-24n^4\tfrac{1}{m^5}\varsigma_2{}^5+48n^3\tfrac{1}{m^4}\varsigma_2{}^5-40n^2\tfrac{1}{m^3}\varsigma_2{}^5-20n^2\tfrac{1}{m^3}\varsigma_2{}^2\varsigma_4\right.\\
&\qquad\qquad\qquad\left.+32n\tfrac{1}{m^2}\varsigma_2{}^5+20n\tfrac{1}{m^2}\varsigma_2{}^2\varsigma_4+28\tfrac{1}{n}\varsigma_2{}^2\varsigma_4-16\tfrac{1}{m}\varsigma_2{}^5-28\tfrac{1}{m}\varsigma_2{}^2\varsigma_4\right)\\
&+q^2(\mathbf{p}\cdot\mathbf{q})^2\mathbf{p}\left(-6n^2\tfrac{1}{m^5}\varsigma_2{}^3+8n\tfrac{1}{m^4}\varsigma_2{}^3+4\tfrac{1}{n}\tfrac{1}{m^2}\varsigma_2{}^3-4\tfrac{1}{n^2}\tfrac{1}{m}\varsigma_2{}^3+4\tfrac{1}{n^3}\varsigma_2{}^3-6\tfrac{1}{m^3}\varsigma_2{}^3\right)\\
&+q^2(\mathbf{p}\cdot\mathbf{q})^2\mathbf{q}\left(-12n^3\tfrac{1}{m^5}\varsigma_2{}^4+16n^2\tfrac{1}{m^4}\varsigma_2{}^4-12n\tfrac{1}{m^3}\varsigma_2{}^4-8\tfrac{1}{n}\tfrac{1}{m}\varsigma_2{}^4+8\tfrac{1}{n^2}\varsigma_2{}^4+8\tfrac{1}{m^2}\varsigma_2{}^4\right)+O_9.
\end{aligned}
$$

The ninth order terms run for seven pages [20], so they are omitted here.

4.6.3 Surface tangency conditions

The structure of equations (6.4) [and (6.8)] is

$$\tilde{\mathbf{p}} = \sum_{k=0}^{\infty} \sum_{k_+ + k_0 + k_- = k} M_{k_+ k_0 k_-} (\rho_{k_+ k_0 k_-} \mathbf{p} + \sigma_{k_+ k_0 k_-} \mathbf{q}), \tag{6.9a}$$

$$\tilde{\mathbf{q}} = \sum_{k=0}^{\infty} \sum_{k_+ + k_0 + k_- = k} M_{k_+ k_0 k_-} (\tau_{k_+ k_0 k_-} \mathbf{p} + \upsilon_{k_+ k_0 k_-} \mathbf{q}), \tag{6.9b}$$

where $\rho_{k_+ k_0 k_-}$, $\sigma_{k_+ k_0 k_-}$, $\tau_{k_+ k_0 k_-}$, and $\upsilon_{k_+ k_0 k_-}$ are coefficients depending on the refraction indices and the polynomial constants $\{\varsigma_m\}_{m=2(2)}^{\infty}$ in (6.2). The $M_{k_+ k_0 k_-}$ are the axially-symmetric *monomials* of order $2k$ in the components of $\mathbf{p}$ and $\mathbf{q}$, given by

$$M_{k_+ k_0 k_-} := (\mathbf{p}^2)^{k_+} (\mathbf{p}\cdot\mathbf{q})^{k_0} (\mathbf{q}^2)^{k_-}. \tag{6.10}$$

Comparison of the forms (6.9) with the result of the root transformation (6.4)–(6.6) [or surface transformation, (6.8)], shows that many of the possible monomials allowed to appear on the basis of axial symmetry, do not in fact do so. They are *absent*. We now proceed to explain the absence of some of them. In reference [18], the argument is applied to the aberration coefficients of the full surface transformation. Here we apply it to the phase-space transformation of the root map.

We remarked before that the surface $z = \varsigma(\mathbf{q})$ is such that $\varsigma(\mathbf{0}) = 0$ and $\partial\varsigma(\mathbf{q})/\partial\mathbf{q}|_{\mathbf{q}=\mathbf{0}} = \mathbf{0}$, *i.e.*, the surface is *tangent* to the $z = 0$ plane. This entails the following *first tangency* conditions for (6.1) at the optical center:

$$\overline{\mathbf{p}}|_{\mathbf{q}=\mathbf{0}} = \mathbf{p}, \qquad \overline{\mathbf{q}}|_{\mathbf{q}=\mathbf{0}} = \mathbf{0}. \tag{6.11a, b}$$

The differentials of the cartesian components of (6.1) are

$$d\overline{p}_i = dp_i - \frac{p_j}{\sqrt{n^2 - \mathbf{p}^2}} \frac{\partial\varsigma}{\partial\overline{q}_i} dp_j + \sqrt{n^2 - \mathbf{p}^2} \frac{\partial^2\varsigma}{\partial\overline{q}_i \partial\overline{q}_j} d\overline{q}_j, \tag{6.12a}$$

$$d\overline{q}_i = dq_i + \varsigma(\overline{\mathbf{q}}) \frac{(n^2 - \mathbf{p}^2)\delta_{i,j} + p_i p_j}{(n^2 - p^2)^{3/2}} dp_j + \frac{p_i}{\sqrt{n^2 - \mathbf{p}^2}} \frac{\partial\varsigma}{\partial\overline{q}_j} d\overline{q}_j, \tag{6.12b}$$

where the summation convention applies over repeated indices. We valuate (6.12) at $\mathbf{q} = \mathbf{0}$ to find a *second* tangency conditions on the root map

$$\left.\frac{\partial\overline{p}_i}{\partial p_j}\right|_{\mathbf{q}=\mathbf{0}} = \delta_{i,j}, \qquad \left.\frac{\partial\overline{p}_i}{\partial q_j}\right|_{\mathbf{q}=\mathbf{0}} = 2\varsigma_2\sqrt{n^2 - \mathbf{p}^2}\,\delta_{i,j}, \tag{6.13a, b}$$

$$\left.\frac{\partial\overline{q}_i}{\partial p_j}\right|_{\mathbf{q}=\mathbf{0}} = 0, \qquad \left.\frac{\partial\overline{q}_i}{\partial q_j}\right|_{\mathbf{q}=\mathbf{0}} = \delta_{i,j}. \tag{6.13c, d}$$

We now impose these tangency conditions on the general forms (6.9), aided by the relations

$$M_{k_+ k_0 k_-}\big|_{\mathbf{q}=\mathbf{0}} = (\mathbf{p}^2)^{k_+} \delta_{k_0,0}\, \delta_{K_-,0}, \tag{6.14}$$

$$\frac{\partial}{\partial p_j} M_{k_+ k_0 k_-} = 2m_+ M_{k_+-1,k_0,k_-} p_j + m_0 M_{k_+,k_0-1,k_-} q_j, \tag{6.15a}$$

$$\frac{\partial}{\partial q_j} M_{k_+ k_0 k_-} = m_0 M_{k_+,k_0-1,k_-} p_j + 2m_- M_{k_+,k_0,k_--1} q_j, \tag{6.15b}$$

The first equation is a simple consequence of (6.10) and the two others come from the Leibnitz rule; they will be used also in the next section. We thus get the following *selection rules* on the coefficients in (6.9):

$$(6.11a,b) \quad \Rightarrow \quad \tau_{k,0,0} = 0, \qquad \rho_{k,0,0} = \delta_{k,0}, \tag{6.16a, b}$$

$$(6.13b) \quad \Rightarrow \quad \rho_{k,1,0} = 0, \qquad \sigma_{k,0,0} = -\frac{2\varsigma_2}{n^{2k-1}}\frac{(2k-3)!!}{(2k)!!}, \tag{6.16c, d}$$

$$(6.13d) \quad \Rightarrow \quad \tau_{k,1,0} = 0, \qquad \upsilon_{k,0,0} = \delta_{k,0}, \tag{6.16e, f}$$

for $k = 0, 1, 2, \ldots$. In obtaining (6.16d) we have used the series expansion (6.5), where we recall that $1!! = 1$, $0!! = 1$, $(-1)!! = 1$, and $(-3)!! = -1$.

For rays parallel to the optical axis, $\mathbf{p} = 0$,

$$\overline{\mathbf{q}}|_{\mathbf{p}=0} = \mathbf{q}, \qquad \overline{\mathbf{p}}|_{\mathbf{p}=0} = n\frac{\partial \varsigma}{\partial \mathbf{q}}; \tag{6.17}$$

they imply, similarly, the selection rules

$$\upsilon_{0,0,k} = \delta_{k,0}, \qquad \sigma_{0,0,k} = (2k+1)\, n\varsigma_{2k}. \tag{6.18}$$

These selection rules, (6.16) and (6.18), explain the absence of terms with 'left- and right-concentrated' index sets (k_+, k_0, k_-). They are probably not exhaustive, however. One may yet find closed expressions and/or recurrence relations for the other coefficients.

4.6.4 The aberration polynomials

Once we have the phase-space transformation given by an operator $\mathfrak{M}$ in explicit form, we may find the gaussian and aberration polynomials f_{2k}, $k = 1, 2, 3, \ldots$, in the factorized form (3.7)–(3.14). This we may do iteratively on the order m, once we have found f_2, first for the root transformation $\mathfrak{R}$, and then for the surface transformation $\mathfrak{S}$.

Indeed, from (3.15*b*) we know that the *linear* terms in the root transformation $\mathfrak{R}$ given by (6.4)–(6.6), are produced by[22]

$$f_2^{\varrho} = \varsigma_2 n\, \mathbf{q}^2. \tag{6.19}$$

The series $\mathfrak{R} = \exp(:f_2^{\varrho}:)$ terminates after the first Poisson bracket, and $\mathfrak{R}_2\mathbf{p} = \mathbf{p} + 2\varsigma_2 n\mathbf{q}$, $\mathfrak{R}_2\mathbf{q} = \mathbf{q}$.

4.6.4.1 The linear part of the transformation

Above we found the linear part of the root transformation. The general case is the following. In a nonlinear transformation of optical phase space such as (6.4)–(6.6) or (6.8), $\mathbf{w} \mapsto \mathfrak{M}\mathbf{w} = \mathbf{w}'(\mathbf{w})$, $\mathbf{w}'$ is expanded as a series $\mathbf{w}' = \mathbf{w}_1 + O_2$, where O_2 contains all terms of order two or higher in the components of $\mathbf{w}$, and $\mathbf{w}_1$ isolates the *linear part*

$$\mathbf{w}_1 = \begin{pmatrix} \mathbf{p}_1(\mathbf{p}, \mathbf{q}) \\ \mathbf{q}_1(\mathbf{p}, \mathbf{q}) \end{pmatrix} = \begin{pmatrix} a & b \\ c & d \end{pmatrix} \begin{pmatrix} \mathbf{p} \\ \mathbf{q} \end{pmatrix} = \mathbf{M}\mathbf{w}. \tag{6.19$'$}$$

If the calculation leading to $\mathbf{w}'(\mathbf{w})$ has been done correctly, then $ad - bc = 1$. Comparing coefficients through (4.2*b*) we find diretly the generating polynomial f_2.

[22] We indicate by f_{2k}^{ϱ} the polynomials producing to the *root* transformation. For a discussion on how to find the polynomials given expansions of the form (6.4), (6.6), and (6.8), see reference [6], sections 2 and 3. Those calculations were done using the symbolic manipulation programs MACSYMA and SMP.

4.6.4.2 The third-order aberration polynomials

Next come the fourth-order terms in the polynomial f_4 which, applied to terms linear in $\mathbf{p}$ and $\mathbf{q}$, generates terms of orders three, five, seven, etc., due to (3.10). The terms of order *three* define this aberration to be *third*-order. The general form of f_{2k} is

$$f_{2k} = \sum_{k_+k_0k_-=k} R_{k_+k_0k_-} M_{k_+k_0k_-}, \tag{6.20}$$

with the monomials $M_{k_+k_0k_-}$ given by (6.10).

As in equation (4.18), we write $\mathfrak{R}_{2k} = \exp(:f^{\rho}_{2k}:)$, $\mathbf{w} = \begin{pmatrix}\mathbf{p}\\ \mathbf{q}\end{pmatrix}$, and consider the expansion of the exponential series to some aberration order, say *seven*. The *paraxial* transformation (first order) was found to be

$$\mathbf{w}_1 = \mathfrak{R}_2\mathbf{w} = \begin{pmatrix}1 & \varsigma_2 n\\ 0 & 1\end{pmatrix}\begin{pmatrix}\mathbf{p}\\ \mathbf{q}\end{pmatrix} = \begin{pmatrix}\mathbf{p}+\varsigma_2\mathbf{q}\\ \mathbf{q}\end{pmatrix}. \tag{6.21}$$

Now, to the aberration order we consider polynomials up to order *eight*:

$$\begin{aligned}\mathfrak{R}\mathbf{w} &= \cdots\mathfrak{R}_8\mathfrak{R}_6\mathfrak{R}_4\mathfrak{R}_2\mathbf{w} = \cdots\mathfrak{R}_8\mathfrak{R}_6\mathfrak{R}_4\mathbf{w}_1\\ &= [1+:f^{\rho}_8:][1+:f^{\rho}_6:][1+:f^{\rho}_4:+\tfrac{1}{2!}(:f^{\rho}_4:)^2+\tfrac{1}{3!}(:f^{\rho}_4:)^3]\mathbf{w}_1 + O_9\\ &= \mathbf{w}_1 + :f^{\rho}_4:\mathbf{w}_1 + [:f^{\rho}_6: + \tfrac{1}{2!}(:f^{\rho}_4:)^2]\mathbf{w}_1 + [:f^{\rho}_8: + :f^{\rho}_6::f^{\rho}_4: + \tfrac{1}{3!}(:f^{\rho}_4:)^3]\mathbf{w}_1 + O_9\\ &=: \mathbf{w}_1 + \mathbf{w}_3 + \mathbf{w}_5 + \mathbf{w}_7 + O_9,\end{aligned} \tag{6.22}$$

where the brackets above group the operator by their resulting order in $\mathbf{p}$ and $\mathbf{q}$, and the last line defines the two-vectors $\mathbf{w}_{2k-1}$ as being of order $2k-1$ in the components of $\mathbf{p}$ and $\mathbf{q}$.

For third order, thus, with the aid of (6.15) we may calculate the Poisson brackets $\{M_{k_+k_0k_-}, \mathbf{p}\}$ and $\{M_{k_+k_0k_-}, \mathbf{q}\}$ for $k_+ + k_0 + k_- = 2$, and compare the resulting third-order polynomial (with the unknown coefficients $R_{k_+k_0k_-}$), with the third order terms in (6.4) and (6.6). We thus find the *third-order aberration coefficients* and thereby determine f_4.

4.6.4.3 The 'direct' solution

The comparison of the general transformation $\mathbf{w}'(\mathbf{w})$ with the series (6.22) is a problem in which an efficient algorithm may save much work. Both Forest [19] and Navarro Saad[23] used the algorithm that is direct, but computationally somewhat cumbersome. It is the following.

In the expression (6.22), the first Poisson bracket of f_6 with $\mathbf{p}$ and $\mathbf{q}$ now yields *fifth*-order terms. From this we have to *subtract* the fifth-order *remanent* of f_4 (*i.e.*, $\frac{1}{2!}(:f^{\rho}_4:)^2\mathbf{w}_1$) and compare it with the fifth-order terms of the transformation (6.4)–(6.6). Actually, only a *subset* of comparisons are necessary. Take, for instance, all coefficients in (6.4) and the coefficient of the $(\mathbf{q}^2)^2\mathbf{q}$-term in (6.6). This yields the fifth-order aberration polynomial f_6. The seventh order terms (6.4–6) are now compared with $\mathbf{w}_7$. To find the coefficients of f_8 we must subtract the seventh-order remanent $[:f^{\rho}_6::f^{\rho}_4: + \frac{1}{3!}(:f^{\rho}_4:)^3]\,\mathbf{w}_1$ that comes from the *compounding* of lower-order aberrations.

Each successive factor $\mathfrak{M}_{2k} = \exp(:f_{2k}:)$ yields its logarithm f_{2k} by comparison with the aberration order $2k-1$ terms of $\overline{\mathbf{p}}$, $\overline{\mathbf{q}}$, once the remanent of the compounded lower-order aberrations has been subtracted. This remanent subtraction produces, successively, the expressions for $\exp(:f_2:)\mathbf{p}$, $\exp(:f_4:)\exp(:f_2:)\mathbf{p}$, $\exp(:f_6:)\exp(:f_4:)\exp(:f_2:)\mathbf{p}$, etc. The comparison takes place with $\{f_{2k}, \mathbf{p}\}$, and similarly for $\mathbf{q}$. In principle thus, the problem is solved. The rest is clever computing.

[23]Unpublished.

4.6.4.4 The solution by quadrature

In order to find iteratively the aberration polynomials (6.20) generating an explicitly computed transformation, there seems to be a definite computational advantage in the following iterative solution method.[24]

We assume we know all polynomials f_j up to some order k. (The arguments hold also for non-axissymmetric transformations, where j and k may be half-integers.) The factorization (4.17) of $\mathfrak{M}$ may be written as

$$\mathfrak{M} = \mathfrak{M}_{>2k}\, \mathfrak{M}_{2k}\, \mathfrak{M}_{<2k}\, \mathfrak{M}_2,$$

where $\mathfrak{M}_{<2k} = \mathfrak{M}_{2k-2} \cdots \mathfrak{M}_4$ is known explicitly as a transformation and we wish to find $\mathfrak{M}_{2k} = \exp{:}f_{2k}{:}$. We compute the following:

$$\begin{aligned} \mathfrak{M}_2^{-1}\, \mathfrak{M}_{<2k}^{-1} \mathbf{w}' &= \mathfrak{M}_2^{-1}\, \mathfrak{M}_{<2k}^{-1}\, \mathfrak{M}_{>2k}\, \mathfrak{M}_{2k}\, \mathfrak{M}_{<2k}\, \mathfrak{M}_2 \mathbf{w} \\ &= \mathfrak{M}'_{>2k}\, \mathfrak{M}'_{2k} \mathbf{w} = \mathbf{w} + \mathbf{W}_{2k-1}(\mathbf{w}) + O_{2k}. \end{aligned}$$

The leading uncompensated aberration $\mathbf{W}_{k-1}(\mathbf{w})$ is a four-vector with polynomial entries of order $k-1$ and

$$\begin{aligned} \mathfrak{M}'_{2k} &= \mathfrak{M}_2^{-1}\, \mathfrak{M}_{<2k}^{-1}\, \mathfrak{M}_{2k}\, \mathfrak{M}_{<2k}\, \mathfrak{M}_2 \\ &= \exp{:}\mathfrak{M}_{<2k}^{-1}\, f_{2k}(\mathfrak{M}_2^{-1}\mathbf{w}){:} = 1 + {:}f_{2k}(\mathfrak{M}_2^{-1}\mathbf{w}){:} + {:}O_{2k+2}(\mathbf{w}){:}. \end{aligned}$$

Since we know $\mathbf{W}_{2k-1}(\mathbf{w})$, the task is to find the scalars $h_{2k}(\mathbf{w}_1) = f_{2k}(\mathbf{w})$ from

$$\{h_{2k}(\mathbf{w}), \mathbf{w}\} = \mathbf{W}_{2k-1}(\mathbf{w}), \quad \textit{i.e.} \quad \frac{\partial h_{2k}}{\partial q_i} = W_{2k-1}^{p_i}(\mathbf{w}), \quad \frac{\partial h_{2k}}{\partial p_i} = -W_{2k-1}^{q_i}(\mathbf{w}),$$

where we have written the expression in components (p_i, q_i) and indicated the corresponding components of $\mathbf{W}$ as W^{p_i} and W^{q_i}. The problem has been thus reduced to a quadrature.

The uniqueness of h_{2k} is insured since $\partial h/\partial w_n$ is an exact derivative due to the symplectic condition applied to $\mathbf{w}' = \mathbf{w} + \mathbf{W}(\mathbf{w})$ [19]. A convenient path is to integrate first for $\mathbf{q}$ at $\mathbf{p} = \mathbf{0}$ and then integrate over $\mathbf{p}$.

The results for the root transformation (6.4) are[25]

$$g_2^{\rho} = n\varsigma_2 \mathbf{q}^2, \tag{6.23a}$$

$$g_4^{\rho} = n(\varsigma_4 - 2\varsigma_2^3)(\mathbf{q}^2)^2 + 2\varsigma_2^2 \mathbf{q}^2 \mathbf{p}\cdot\mathbf{q} - (1/2n)\varsigma_2 \mathbf{p}^2 \mathbf{q}^2, \tag{6.23b}$$

$$\begin{aligned} g_6^{\rho} = {} & n(\varsigma_6 - 6\varsigma_2^2\varsigma_4 + 2\varsigma_2^5)(\mathbf{q}^2)^3 + (4\varsigma_2\varsigma_4 - 2\varsigma_2^4)(\mathbf{q}^2)^2 \mathbf{p}\cdot\mathbf{q} \\ & - (1/2n)\varsigma_4 (\mathbf{q}^2)^2 \mathbf{p}^2 + (1/2n^2)\varsigma_2^2 \mathbf{q}^2 \mathbf{p}\cdot\mathbf{q}\, \mathbf{p}^2 - (1/8n^3)\varsigma_2 \mathbf{q}^2 (\mathbf{p}^2)^2, \end{aligned} \tag{6.23c}$$

$$\begin{aligned} g_8^{\rho} = {} & n[\varsigma_8 - 2\varsigma_2^2\varsigma_6 - \tfrac{40}{3}\varsigma_2\varsigma_4^2 + \tfrac{10}{3}\varsigma_2^4\varsigma_4 - \tfrac{4}{3}\varsigma_2^7](\mathbf{q}^2)^4 \\ & + [2\varsigma_2\varsigma_6 + 4\varsigma_4^2 + \tfrac{8}{3}\varsigma_2^3\varsigma_4 - \tfrac{4}{3}\varsigma_2^6](\mathbf{q}^2)^3 \mathbf{p}\cdot\mathbf{q} + (1/n)[-\tfrac{1}{2}\varsigma_6 - \varsigma_2^2\varsigma_4 + \varsigma_2^5](\mathbf{q}^2)^3 \mathbf{p}^2 \\ & + (1/n)[-\tfrac{14}{3}\varsigma_2^2\varsigma_4 + 4\varsigma_2^5](\mathbf{q}^2)^2 (\mathbf{p}\cdot\mathbf{q})^2 + [2/n^2](\varsigma_2\varsigma_4 - \varsigma_2^4)(\mathbf{q}^2)^2 \mathbf{p}\cdot\mathbf{q}\, \mathbf{p}^2 \\ & - \tfrac{4}{3}(1/n^2)\varsigma_2^4 \mathbf{q}^2 (\mathbf{p}\cdot\mathbf{q})^3 + (1/n^3)[\tfrac{1}{12}\varsigma_2^3 - \tfrac{1}{8}\varsigma_4](\mathbf{q}^2)^2(\mathbf{p}^2)^2 \\ & + \tfrac{1}{3}(1/n^3)\varsigma_2^3 \mathbf{q}^2 (\mathbf{p}\cdot\mathbf{q})^2 \mathbf{p}^2 + \tfrac{1}{4}(1/n^4)\varsigma_2^2 \mathbf{q}^2 \mathbf{p}\cdot\mathbf{q}(\mathbf{p}^2)^2 - \tfrac{1}{16}(1/n^5)\varsigma_2 \mathbf{q}^2 (\mathbf{p}^2)^3. \end{aligned} \tag{6.23d}$$

---

[24]This subsection was added in proofs by Wolf, from observations made by Forest.

[25]Equations (6.23) and (6.24) have been calculated by Dragt and Forest in the factorization order (3.20), rather than in the order (3.14) in which we hoped to uniform the results of this chapter. The results obtained by Navarro Saad following (3.14) have been put in doubt by the first two authors. It was not possible to recalculate Navarro Saad's results in the short time afforded by the publication deadline: Miguel has left IIMAS for a better-paid job in a Mexican software company and Bernardo has been left in a quandry with a diskful of computer programs.

A similar computation for the surface transformation $\mathfrak{S}$ that produces (6.8), yields the *surface* aberration polynomials given by

$$g_2^\sigma = \varsigma_2(n-n')\mathbf{q}^2, \tag{6.24a}$$

$$\begin{aligned} g_4^\sigma = & -\varsigma_2^3[(n-n')/n]\{n[2-(\varsigma_4/\varsigma_2^3)]-2n'\}(\mathbf{q}^2)^2 \\ & + 2\varsigma_2^2[(n-n')/n]\mathbf{q}^2\mathbf{p}\cdot\mathbf{q} + \varsigma_2[(n-n')/2nn']\mathbf{q}^2\mathbf{p}^2, \end{aligned} \tag{6.24b}$$

$$\begin{aligned} g_6^\sigma = & [1/n^3][(\varsigma_6-6\varsigma_2^2\varsigma_4+2\varsigma_2^5)n^4+(-\varsigma_6+12\varsigma_2^2\varsigma_4-4\varsigma_2^5)n^3n'-6\varsigma_2^2\varsigma_4n^2n'^2+4\varsigma_2^5nn'^3-2\varsigma_2^5n'^4](\mathbf{q}^2)^3 \\ & +[1/n^3n'][(2\varsigma_2\varsigma_4-4\varsigma_2^4)n^4+(2\varsigma_2\varsigma_4+6\varsigma_2^4)n^3n'-(4\varsigma_2\varsigma_4+4\varsigma_2^4)n^2n'^2+6\varsigma_2^4nn'^3-4\varsigma_2^4n'^4](\mathbf{q}^2)^2\mathbf{p}\cdot\mathbf{q} \\ & + [1/2n^3n'][\varsigma_4n^3-\varsigma_4n^2n'+2\varsigma_2^3nn'^2-2\varsigma_2^3n'^3](\mathbf{q}^2)^2\mathbf{p}^2 \\ & + [2/n^3n']\varsigma_2^3[n^3-n^2n'+nn'^2-n'^3]\mathbf{q}^2(\mathbf{p}\cdot\mathbf{q})^2 \\ & + [1/2n^3n'^2]\varsigma_2^2[n^3+nn'^2-2n'^3]\mathbf{q}^2\mathbf{p}\cdot\mathbf{q}\,\mathbf{p}^2 \\ & + [1/8n^3n'^3]\varsigma_2[n^3-n'^3]\mathbf{q}^2(\mathbf{p}^2)^2, \end{aligned} \tag{6.23c}$$

$$\begin{aligned} g_8^\sigma = & -[1/3n^5n'][(8\varsigma_2\varsigma_4^2-32\varsigma_2^4\varsigma_4+32\varsigma_2^7)n^7+(-3\varsigma_8+6\varsigma_2^2\varsigma_6+24\varsigma_2\varsigma_4^2+86\varsigma_2^4\varsigma_4-124\varsigma_2^7)n^6n' \\ & \quad +(3\varsigma_8-12\varsigma_2^2\varsigma_6-72\varsigma_2\varsigma_4^2-120\varsigma_2^4\varsigma_4+200\varsigma_2^7)n^5n'^2+(6\varsigma_2^2\varsigma_6+40\varsigma_2\varsigma_4^2+164\varsigma_2^4\varsigma_4-188\varsigma_2^7)n^4n'^3 \\ & \quad +(122\varsigma_2^4-152\varsigma_2^4\varsigma_4)n^3n'^4+(54\varsigma_2^4\varsigma_4-20\varsigma_2^7)n^2n'^5-24\varsigma_2^7nn'^6+12\varsigma_2^7n'^7](\mathbf{q}^2)^4 \\ & +[1/3n^5n'][(18\varsigma_2\varsigma_6-116\varsigma_2^3\varsigma_4+52\varsigma_2^6)n^6+(-12\varsigma_2\varsigma_6+12\varsigma_4^2+240\varsigma_2^3\varsigma_4-124\varsigma_2^6)n^5n' \\ & \quad +(-6\varsigma_2\varsigma_6-12\varsigma_4^2-216\varsigma_2^3\varsigma_4+108\varsigma_2^6)n^4n'^2+(176\varsigma_2^3\varsigma_4-64\varsigma_2^6)n^3n'^3 \\ & \quad +(4\varsigma_2^6-84\varsigma_2^3\varsigma_4)n^2n'^4+60\varsigma_2^6nn'^5-36\varsigma_2^6n'^6](\mathbf{q}^2)^3\mathbf{p}\cdot\mathbf{q} \\ & +[1/2n^5n'][\varsigma_6n^5+(-\varsigma_6-2\varsigma_2^2\varsigma_4+2\varsigma_2^5)n^4n'+(12\varsigma_2^2\varsigma_4-8\varsigma_2^5)n^3n'^2 \\ & \quad +(4\varsigma_2^5-10\varsigma_2^2\varsigma_4)n^2n'^3+8\varsigma_2^5nn'^4-6\varsigma_2^5n'^5](\mathbf{q}^2)^3\mathbf{p}^2 \\ & -[1/3n^5n'^2][(32\varsigma_2^5-16\varsigma_2^2\varsigma_4)n^6+(-24\varsigma_2^2\varsigma_4-48\varsigma_2^5)n^5n'+(54\varsigma_2^2\varsigma_4+36\varsigma_2^5)n^4n'^2 \\ & \quad +(-44\varsigma_2^2\varsigma_4-32\varsigma_2^5)n^3n'^3+(30\varsigma_2^2\varsigma_4+24\varsigma_2^5)n^2n'^4-48\varsigma_2^5nn'^5+36\varsigma_2^5n'^6](\mathbf{q}^2)^2(\mathbf{p}\cdot\mathbf{q})^2 \\ & +[1/n^5n'^2][2\varsigma_2\varsigma_4n^5-\varsigma_2\varsigma_4n^4n'+2\varsigma_2\varsigma_4n^3n'^2-3\varsigma_2\varsigma_4n^2n'^3+6\varsigma_2^4nn'^4-6\varsigma_2^4n'^5](\mathbf{q}^2)^2\mathbf{p}\cdot\mathbf{q}\,\mathbf{p}^2 \\ & +[1/24n^5n'^3][(3\varsigma_4+4\varsigma_2^3)n^5+(2\varsigma_2^3-3\varsigma_4)n^2n'^3+12\varsigma_2^3nn'^4-18\varsigma_2^3n'^5](\mathbf{q}^2)^2(\mathbf{p}^2)^2 \\ & +[1/3n^5n'^2]\varsigma_2^4[8n^5-12n^4n'+12n^3n'^2-8n^2n'^3+12nn'^4-12n'^5]\mathbf{q}^2(\mathbf{p}\cdot\mathbf{q})^3 \\ & +[1/3n^5n'^3]\varsigma_2^3[2n^5+n^2n'^3+6nn'^4-9n'^5]\mathbf{q}^2(\mathbf{p}\cdot\mathbf{q})^2(\mathbf{p}^2)^2 \\ & +[1/4n^5n'^4]\varsigma_2^2[n^5+n^2n'^3+nn'^4-3n'^5]\mathbf{q}^2\mathbf{p}\cdot\mathbf{q}\,(\mathbf{p}^2)^2 \\ & -[1/16n^5n'^5]\varsigma_2[n^5-n'^5]\mathbf{q}^2(\mathbf{p}^2)^3. \end{aligned} \tag{6.23d}$$

It is interesting to note that each g_m depends only on the ς_ℓ with $\ell \le m$. Also, the only term in the g_m that depends on ς_m is the coefficient of $(\mathbf{q}^2)^{m/2}$. Thus, the freedom to adjust the shape of an interface produces relatively little freedom in the adjustment of aberrations.

4.6.4.5 The Seidel third-order aberrations

For *third*-order aberrations, the basis of six fourth-order monomials described above is —except for the $(\mathbf{q}^2)^2$ term— the one traditionally used in the description of aberrations since Seidel [22] in 1853, and bear his name. They are given in the table next page:

COEFFICIENT	TERM	SEIDEL ABERRATION
$A = R_{2,0,0}$	$(\mathbf{p}^2)^2$	SPHERICAL ABERRATION
$B = R_{1,1,0}$	$\mathbf{p}^2\mathbf{p}\cdot\mathbf{q}$	COMA
$C = R_{0,2,0}$	$(\mathbf{p}\cdot\mathbf{q})^2$	ASTIGMATISM
$D = R_{1,0,1}$	$\mathbf{p}^2\mathbf{q}^2$	CURVATURE OF FIELD
$E = R_{0,1,1}$	$\mathbf{p}\cdot\mathbf{q}\,\mathbf{q}^2$	DISTORTION
$F = R_{0,0,2}$	$(\mathbf{q}^2)^2$	POCUS

Table. The Seidel aberrations. The first five A–E are the traditional ones; *pocus*, F, is added to the list. These are all third-order aberrations.

The separate effects of the first five Seidel aberrations may be found in various classic volumes such as [8] and [23,24]. The first four (A–D) entail a loss of focus, with coma particularly deleterious for image quality. Distortion does not unfocus, and its effect may be compensated (in astronomical plates, for example), if the corresponding aberration coefficient E is known. The last term, $F(\mathbf{q}^2)^2$, has no effect on the quality of an image. It does, however, affect the arrival *direction* of a ray at the image plane, and therefore may be important if the optical system under study is used in conjunction with free propagation: it decreases the depth of field at the corners of an image plate. It does not seem to have a name, so Wolf used *pocus* (for its **p**-*unfocusing* property) in reference [15]. For the aberrations of astigmatism and curvature of field, we shall present an alternative classification in the next section.

There are several features of the foregoing table which deserve comment. The first one, implicit in the hamiltonian formulation of optics, is the behavior under *Fourier conjugation*: $(\mathbf{p},\mathbf{q}) \mapsto (\mathbf{q},-\mathbf{p})$. This is a map between the terms in every f_m which represents a quarter 2π rotation in the symplectic group $Sp(2,\Re)$.[26] Under this transformation, $M_{k_+k_0k_-}$ becomes $(-1)^{k_0}M_{k_-k_0k_+}$. Thus in the table spherical aberration and coma are Fourier conjugate to pocus and distortion, respectively. Both astigmatism and curvature of field are self-conjugate. But the general monomial $M_{k_+k_0k_-}$ has no general symmetry under this operation.

We should also notice that these aberrations are counted to third order. If we are in seventh aberration order, for example, we may expect to see the fifth- and seventh-order effects of the third-order Seidel monomials; there would also be aberrations particular to fifth order, and compound effects of third and fifth order, plus the aberrations particular to seventh order. When we investigate the effect of each of the third-order Seidel aberrations on phase space, it may be economic to *disregard* the self-compounded remanents present in higher orders of **p** and **q**. The mappings (6.4)–(6.6), when cut to any finite number of terms, are no longer strictly canonical, but only canonical ***to the aberration order***. Yet the coefficients in the mapping are, as we saw, sufficient —and necessesary– to determine the *aberration* coefficients. In this way we may nest the action of third-order aberration optical systems $\mathfrak{M} = \mathfrak{M}_4\,\mathfrak{M}_2$ on the space of *linear* and *cubic* polynomials in **p** and **q**. We need only consider those cubic monomials which are produced out of **p** and **q** through the action of the operator $:f_4:$. These monomials are

$$\mathbf{p}^2\mathbf{p},\quad \mathbf{p}^2\mathbf{q},\quad \mathbf{p}\cdot\mathbf{q}\,\mathbf{p},\quad \mathbf{p}\cdot\mathbf{q}\,\mathbf{q},\quad \mathbf{q}^2\mathbf{p},\quad \mathbf{q}^2\mathbf{q}. \qquad (6.25)$$

[26] It becomes an *eighth* root of the identity element in the *metaplectic* group $Mp(2,\Re)$. The latter applies to *wave* optics, though.

Let $\mathbf{v}_1$ be the space of linear monomials, and $\mathbf{v}_3$ that of cubic monomials. The action of $\mathfrak{M} = \exp(:f_4:)\exp(:f_2:)$ can thus be written in $(2+6)\times(2+6)$ block-matrix form as

$$\begin{aligned}\mathfrak{M}_4\mathfrak{M}_2\begin{pmatrix}\mathbf{v}_1\\ \mathbf{v}_3\end{pmatrix} &= \mathfrak{M}_4\begin{pmatrix}\mathbf{D}_2 & \mathbf{0}\\ \mathbf{0} & \mathbf{D}_6\end{pmatrix}\begin{pmatrix}\mathbf{v}_1\\ \mathbf{v}_3\end{pmatrix} = \begin{pmatrix}\mathbf{D}_2 & \mathbf{0}\\ \mathbf{0} & \mathbf{D}_6\end{pmatrix}\begin{pmatrix}\mathfrak{M}_4\mathbf{v}_1\\ \mathfrak{M}_4\mathbf{v}_3\end{pmatrix}\\ &= \begin{pmatrix}\mathbf{D}_2 & \mathbf{0}\\ \mathbf{0} & \mathbf{D}_6\end{pmatrix}\begin{pmatrix}\mathbf{1} & \mathbf{F}\\ \mathbf{0} & \mathbf{1}\end{pmatrix}\begin{pmatrix}\mathbf{v}_1\\ \mathbf{v}_3\end{pmatrix} = \begin{pmatrix}\mathbf{D}_2 & \mathbf{D}_2\mathbf{F}\\ \mathbf{0} & \mathbf{D}_6\end{pmatrix}\begin{pmatrix}\mathbf{v}_1\\ \mathbf{v}_3\end{pmatrix}.\end{aligned} \tag{6.26}$$

Here $\mathbf{D}_2$ is the two-dimensional representation of *Sp(2,ℜ)* in (4.23), $\mathbf{D}_6$ is a *six*-dimensional representation obtained from the linear combination coefficients in $\mathbf{v}_3$ due to the linear transformations of $\mathbf{v}_1$, and $\mathbf{F}$ is a 2×6 matrix involving (linearly) the coefficients of f_4. The **0**'s and **1**'s are adequate rectangular and square matrices.

This construction is simplified group-theoretically when we recognize that the six-dimensional representation of *Sp(2,ℜ)*, $\mathbf{D}_6$, is *reducible.* It may be block-diagonalized to the 2×2 and the 4×4 representations of the gaussian symplectic group. These are finite-dimensional, *non*-unitary representations of the group.[27] This matrix representation of third-order aberrations $\mathfrak{M}$'s and its reduction to irreducible parts was developed in reference [15], and will be fundamented in the next section.

Mappings of four-dimensional phase space are hard to visualize. One helpful device is to consider a one-dimensional optical world (or a system with *cilyndrical* lenses), where phase-space is two-dimensional. First-order optical systems may produce (almost) all linear transformations of the plane. If the q-axis is vertical and p horizontal, then spherical aberration moves phase-space points mantaining their p-coordinate constant. No change in ray direction comes from spherical aberration, this is a purely *geometrical* aberration. Now, horizontal lines $q =$ constant in the phase-space diagram represent pencils of rays converging to the image point q. These horizontal lines are bent into parallel cubic curves by the p^4 term in f_4, since, witness: $\exp(A{:}p^4{:})q = q + A{:}p^4{:}q + O_5 = q + Ap^3 + O_5$. Pocus acts in a similar way producing a $\frac{1}{2}\pi$-rotated version of this ***phase-space diagram.*** Coma bends horizontals into quadratics and shifts verticals; distortion bends verticals and shifts horizontals, keeping thus focused images focused, but *displaced* by an amount proportional to q^2.

Another device for picturing the influence of aberrations on ray pencils is to imagine the object to be mapped as composed by a rectangular grid of point light sources. We may examine the image on the screen finding first the image points of the rays in the pencil which are parallel to the optical axis, then those of rays in succesive cones around those lines, representing some isophote distribution in the source direction. This yields a ***spot diagram*** in the image.

The spot diagram of spherical aberration is appears as a succesion of concentrical circles around each source point, their size independent of the location of the point. This indicates unfocusing takes place evenly across the screen. Coma displays the familiar ***comet*** shape which gives the aberration its name: the centers of the circles displace as the radius grows, in such a way that the envolvents are a pair of lines with a 60° angle. The two *meridional* rays of each cone (*i.e.*, those lying in a plane with the optical axis) fall onto the *same* image point in the direction of $\mathbf{q}$, and the two *saggital* rays (*i.e.*, those perpendicular to the meridional plane) onto the diametrically opposed point of the image. There is thus a two-to-one mapping of the $\mathbf{q}$-projection of phase space, which is the visible image.

---

[27] See Appendix B. (Editor's note.)

Astigmatism affects only meridional rays, and the issuing pencil cones map onto straight segments of the image, centered and directed along $\mathbf{q}$. Curvature of field produces concentric circles of radii proportional to $\mathbf{q}^2|\mathbf{p}|$. Distortion lets all circles fall onto the same $\mathbf{q}$-space image point but, as we saw, displaces the point. Pocus does apparently nothing to the image in $\mathbf{q}$-space, but spherically aberrates $\mathbf{p}$-space.

This description of the effects of Seidel aberrations is basically known in the optics literature (see, for instance, reference [8]), albeit in a more geometrical construction. From the point of view of Lie methods, it is interesting to note, morever, that the set transformations of phase space *modulo* third order, constitutes a ***group***. Its structure is that of a *semidirect product* —evidenced by the type of block-triangular matrices in (6.22)— between the ***first order*** transformation group $Sp(2,\Re)$, and a ***pure aberration*** group of "translations". The former is three-dimensional and the second, six dimensional (f_4 has six independent coefficients). The normal subgroup is the latter. As we shall see in the next section, the action of $Sp(2,\Re)$ on the aberration *multiplet* is reducible to a 5×5 and the trivial representations of the group. The total number of parameters is thus *nine*.

The nine-dimensional ***aberration*** group A^3 was introduced in [13], developed in [15,16], and appears below in Subsect. **4.7.4**. The ***basic*** representation (6.22) describes its action on phase space to third order. Of course, each element has also its representation as $\exp(:f_4:)\exp(:f_2:)$, acting to *all* orders on $(\mathbf{p},\mathbf{q})$-phase space. But then, products of such elements will produce compound aberrations of order higher than third. By remaining within A^3 and its representation (6.22), we are describing matters essentially and exclusively to third order.

As will be seen in **4.8.4**, this nine-parameter group A^3 posseses a *presentation*[28] which is more confortable and economic to work with than the 8×8 matrix form in (6.22). The *product*, in particular, involves up to 5×5 matrices acting on 5-vectors. For higher-order aberrations the results have still not been put in a satisfactorily compact form.

4.7 The symplectic classification of aberrations

After the explicit calculations for axis-symmetric optical systems in the last section, it should be clear that the aberration monomials $M_{k_+k_0k_-}$ given in (6.10) transform under $Sp(2,\Re)$ as *cartesian* tensors, with symmetric indices in its 3×3 representation (4.28). The latter are not general 3×3 matrices, but are in fact conjugate to $SO(2,1)$ matrices acting on the three-dimensional vector space of componentes $\mathbf{p}^2$, $\mathbf{p}\cdot\mathbf{q}$, $\mathbf{q}^2$.

4.7.1 A three-dimensional space for axis-symmetric systems

We introduce the following three-dimensional space with coordinates $\{\xi_i\}_{i=1,2,3}$ or $\{\xi_\sigma\}_{\sigma=+,0,-}$, in terms of the phase space coordinates of axis-symmetric systems:

$$-\tfrac{1}{\sqrt{2}}(\xi_1+i\xi_2)=\xi_+=\tfrac{1}{\sqrt{2}}\mathbf{p}^2, \tag{7.1a}$$

$$\xi_3=\xi_0=\mathbf{p}\cdot\mathbf{q}, \tag{7.1b}$$

$$\tfrac{1}{\sqrt{2}}(\xi_1-i\xi_2)=\xi_-=\tfrac{1}{\sqrt{2}}\mathbf{q}^2. \tag{7.1c}$$

Paraxial optical transformations act on this $\boldsymbol{\xi}$-space and *rotate* it. The (square) radius of a sphere on the origin of this space is

$$\boldsymbol{\xi}^2=\xi_1^2+\xi_2^2+\xi_3^2=\xi_0^2-2\xi_+\xi_-=(\mathbf{p}\cdot\mathbf{q})^2-\mathbf{p}^2\mathbf{q}^2=-(\mathbf{p}\times\mathbf{q})^2=-\Pi, \tag{7.2}$$

[28] See Chapter 2 in this volume. (Editor's note.)

i.e., *minus* the Petzval invariant Π in (3.19).

We now define the Lie operators $\{\pounds_\sigma\}_{\sigma=+,0,-}$ written in terms of phase space quantities as

$$\pounds_+ = -\tfrac{1}{2}{:}\mathbf{p}^2{:}, \tag{7.3a}$$
$$\pounds_0 = \tfrac{1}{2}{:}\mathbf{p}\cdot\mathbf{q}{:}, \tag{7.3b}$$
$$\pounds_- = \tfrac{1}{2}{:}\mathbf{q}^2{:}. \tag{7.3c}$$

Notice the *minus* sign here of $\pounds_+$ *vs.* the plus sign of K_+ in (3.16*a*). The following *so(3)* commutation rules [10, Eq.(3.12)] hold:

$$[\pounds_0, \pounds_+] = \pounds_+, \tag{7.4a}$$
$$[\pounds_0, \pounds_-] = -\pounds_-, \tag{7.4b}$$
$$[\pounds_+, \pounds_-] = 2\pounds_0. \tag{7.4c}$$

Since these *so(3)* generators act on $\boldsymbol{\xi}$-space, they may be realized as differential operators on the space of functions $f(\boldsymbol{\xi})$ as

$$-K_+ = \pounds_1 + i\pounds_2 = \pounds_+ = \sqrt{2}\left(\xi_+\frac{\partial}{\partial\xi_0} + \xi_0\frac{\partial}{\partial\xi_-}\right), \tag{7.5a}$$
$$K_0 = \pounds_3 = \pounds_0 = \xi_+\frac{\partial}{\partial\xi_+} - \xi_-\frac{\partial}{\partial\xi_-}, \tag{7.5b}$$
$$K_- = \pounds_1 - i\pounds_2 = \pounds_- = \sqrt{2}\left(\xi_0\frac{\partial}{\partial\xi_+} + \xi_-\frac{\partial}{\partial\xi_0}\right). \tag{7.5c}$$

These $\{\pounds_\sigma\}$, $\sigma = +, 0, -$, are the usual raising, weight, and lowering operators, and the $\{\pounds_i\}$, $i = 1, 2, 3$, are the generators of rotations around the ith axis. The latter are $\mathcal{L}^2(\Re^3)$–skew adjoint on three-space $\boldsymbol{\xi}$ with the measure $d\xi_1\, d\xi_2\, d\xi_3$.

The weight operator $\pounds_0 = K_0$ acting on any monomial $M_{k_+k_0k_-}(\xi_+, \xi_0, \xi_-)$ will count the k_+ power of ξ_+ (*i.e.*, of $\mathbf{p}^2$) and subtract the k_- power of ξ_- (*i.e.*, of $\mathbf{q}^2$), yielding the *weight* of the monomial as $k_+ - k_-$. The raising operator $\pounds_+$ substitutes ξ_0's and ξ_-'s for ξ_+'s and ξ_0's, raising thus the weight by one unit, and $\pounds_-$ does likewise down the multiplet ladder. There is the Casimir operator

$$\begin{aligned}\pounds^2 &= \pounds_1^2 + \pounds_2^2 + \pounds_3^2\\ &= \tfrac{1}{2}(\pounds_+\pounds_- + \pounds_-\pounds_+ + 2\pounds_0^2)\\ &= -\tfrac{1}{8}[{:}\mathbf{p}^2{:}{:}\mathbf{q}^2{:} + {:}\mathbf{q}^2{:}{:}\mathbf{p}^2{:} - 2({:}\mathbf{p}\cdot\mathbf{q}{:})^2],\end{aligned} \tag{7.6a}$$

and there is also the *number* operator, $\boldsymbol{\xi}\cdot\nabla_{\boldsymbol{\xi}}$,

$$\mathcal{N} = \xi_+\frac{\partial}{\partial\xi_+} + \xi_0\frac{\partial}{\partial\xi_0} + \xi_-\frac{\partial}{\partial\xi_-}, \tag{7.6b}$$

which counts the order of $M_{k_+k_0k_-}(\xi_+, \xi_0, \xi_-)$ in $\boldsymbol{\xi}$ as $k = k_+ + k_0 + k_-$. The number operator $\mathcal{N}$ *commutes* with $\pounds^2$ and with $\pounds_0$.

4.7.2 Decomposition of aberration polynomials in multiplets

With what has been developed in the last subsection, we are squarely within angular momentum theory [10]. In this context, the choice of $\mathcal{N}$, $\pounds^2$, and $\pounds_0$ is natural and provides an eigenbasis for the space of functions $f(\boldsymbol{\xi})$. Indeed, the factorization of the optical transfer map $\mathfrak{M}$ into factors of $\exp(:f_{2k}:)$, $k = 1, 2, 3, \ldots$, has already determined the concept of aberration *order* $N = 2k - 1$. These are eigenspaces of $\mathcal{N}$ with positive integer eigenvalues $k = k_+ + k_0 + k_- = \frac{1}{2}(N + 1)$.

The highest-weight monomial for a given k is clearly $M_{k00} = (\mathbf{p}^2)^k \approx \xi_+^k$, and through the lowering operator $\mathcal{L}_-$ we may generate the rest of a unique multiplet of 'angular momentum' k. The space of all monomials of order k in $\boldsymbol{\xi}$, $M_{k_+k_0k_-}(\xi_+, \xi_0, \xi_-)$ (order $2k$ in $\mathbf{p}$, $\mathbf{q}$) is of dimension $\frac{1}{2}(k+1)(k+2)$, which is *larger* than the $(2k+1)$-dimensional multiplet k alone, so there are other multiplets. The space of monomials of order k may be *decomposed* according to the eigenvalues of $\mathcal{L}^2$ through the standard construction of Wigner–Raccah algebra, which we now proceed to develop.

We have the lemma that for the polynomials f_{2k} responsible for ***primary*** (*i.e.*, non-compound) aberration of order $N = 2k - 1$, the following decomposition holds:

$$f_{2k}(\boldsymbol{\xi}) = \sum_{\substack{j+2\nu=k \\ j,\nu=0,1,2,\ldots}} \sum_{m=-j}^{j} (\boldsymbol{\xi}^2)^\nu \, {}^k f^j_m \, \mathcal{Y}^j_m(\boldsymbol{\xi}), \tag{7.7}$$

$$(\boldsymbol{\xi}^2)^k \, {}^k f^j_m = \int_{S_2} d\Omega_{\boldsymbol{\xi}} \, (\boldsymbol{\xi}^2)^\nu \, f_{2k}(\boldsymbol{\xi}) \, \mathcal{Y}^j_m(\boldsymbol{\xi}), \qquad j + 2\nu = k, \tag{7.8}$$

where $\Omega_{\boldsymbol{\xi}}$ are the coordinates of the two-sphere, $\boldsymbol{\xi}^2$ is given in (7.2) and is simply related to the Petzval invariant; $\mathcal{Y}^j_m(\boldsymbol{\xi})$ is the ***solid spherical harmonic*** [10, Eq.(3.153)], a harmonic homogeneous polynomial of order j in $\boldsymbol{\xi}$ (*i.e.*, of order $2j$ in the components of $\mathbf{p}$, $\mathbf{q}$) and of *weight* m. The solid spherical harmonics have eigenvalue k under $\mathcal{N}$ and $j(j+1)$ under $\mathcal{L}^2$, where j is the maximum value m can attain in its multiplet. The sum of orders of $\boldsymbol{\xi}$ in $(\boldsymbol{\xi}^2)^\nu \mathcal{Y}^j_m(\boldsymbol{\xi})$ is $2\nu + j = k$, the order of f_{2k} in $\boldsymbol{\xi}$.

Hence, the j-content of f_{2k} is

$$j = k,\ k-2,\ k-4,\ \cdots, \begin{cases} 1 & k \text{ odd}, \\ 0 & k \text{ even}. \end{cases} \tag{7.9}$$

First order (paraxial) optical transformations are generated by $:f_2:$ which only contains a $j = 1$ triplet. Third order (primary) aberrations are generated by $:f_4:$, *i.e.*, $k = 2$, a six-dimensional space; this decomposes under *so(3)* (and *sp(2,ℜ)*) into a $j = 2$ ***quintuplet*** $\{\mathcal{Y}^2_m\}^2_{m=-2}$ and a $j = 0$ ***singlet*** $\boldsymbol{\xi}^2 = -\Pi$, the Petzval invariant.

Fifth-order aberration polynomials decompose into a $j = 3$ septuplet $\{\mathcal{Y}^3_m\}^3_{m=-3}$ and a $j = 1$ triplet $\{\boldsymbol{\xi}^2\mathcal{Y}^1_m\}^1_{m=-1}$; seventh-order ones into a $j = 4$ nonuplet $\{\mathcal{Y}^4_m\}^4_{m=-4}$, a $j = 2$ quintuplet $\{\boldsymbol{\xi}^2\mathcal{Y}^2_m\}^2_{m=-2}$ and a $j = 0$ singlet $\boldsymbol{\xi}^4 = \Pi^2$. For aberration order $N = 2k - 1$, there are thus the integer part of $\frac{1}{2}k + 1$ multiplets, the highest one is $\{\mathcal{Y}^j_m\}^j_{m=-j}$ for $j = k$, and next come the smaller-j multiplets spaced by two units, down to 1 or 0.

Since for meridional rays $\Pi = 0$, the only nonvanishing multiplet is the highest $j = k$ one; one-dimensional optics (or optics with cilyndric lenses) thus needs only the $j = k$ multiplet. There is a total of $\frac{1}{8}(N+3)(N+5) = \frac{1}{2}(k+1)(k+2)$ *aberration coefficients* ${}^k f^j_m$ in (7.7–8), classified here in the ***symplectic*** basis.

4.7.3 The symplectic aberration polynomials ${}^k\mathcal{Y}^j_m$

We define the symplectic aberration polynomials ${}^k\mathcal{Y}^j_m$, of ***order*** k, ***angular momentum*** j, and ***weight*** m as the polynomial in $\boldsymbol{\xi}$ (or $\mathbf{p}^2$, $\mathbf{p}\cdot\mathbf{q}$, and $\mathbf{q}^2$) that satisfies

$$\mathcal{L}_0 \, {}^k\mathcal{Y}^j_m = m \, {}^k\mathcal{Y}^j_m, \qquad \mathcal{L}^2 \, {}^k\mathcal{Y}^j_m = j(j+1) \, {}^k\mathcal{Y}^j_m, \qquad \mathcal{N} \, {}^k\mathcal{Y}^j_m = k \, {}^k\mathcal{Y}^j_m. \tag{7.9a, b, c}$$

We choose the normalization so that it be written in terms of the solid spherical harmonics as

$$ {}^k\mathcal{Y}^j_m(\boldsymbol{\xi}) = \xi^{k-j}\,\mathcal{Y}^j_m(\boldsymbol{\xi}), \qquad (k-j \text{ even}). \tag{7.10a}$$

The solid harmonics, we remind the reader, are defined in terms of the ordinary spherical harmonics $Y^j_m(\theta,\phi)$ on the coordinates $(\theta,\phi)=\Omega_\xi$ of the sphere $-\Pi=\boldsymbol{\xi}^2=$ constant [10, Eq.(3.152)],

$$ \mathcal{Y}^j_m(\boldsymbol{\xi}) = \xi^j Y^j_m(\Omega_\xi), \quad \text{so} \quad {}^k\mathcal{Y}^j_m(\boldsymbol{\xi}) = \xi^k Y^j_m(\Omega_\xi). \tag{7.10b}$$

The explicit expressions for the symplectic aberration polynomials may be found from [10, Eq. (3.153), Table 4 on p.655]. They are given by (7.10a), $\boldsymbol{\xi}^2$ given in (7.2), and

$$\begin{aligned} \mathcal{Y}^j_m(\boldsymbol{\xi}) &= \sqrt{\frac{(2j+1)(j+m)!\,(j-m)!}{4\pi}}\sum_n \frac{1}{2^{m/2+n}}\frac{\xi_+^{m+n}}{(m+n)!}\frac{\xi_0^{j-m-2n}}{(j-m-2n)!}\frac{\xi_-^n}{n!} \\ &= \sqrt{\frac{(2j+1)(j+m)!\,(j-m)!}{4\pi}}\sum_n \frac{1}{2^{m+2n}}\frac{(\mathbf{p}^2)^{m+n}}{(m+n)!}\frac{(\mathbf{p}\cdot\mathbf{q})^{j-m-2n}}{(j-m-2n)!}\frac{(\mathbf{q}^2)^n}{n!}. \end{aligned} \tag{7.11}$$

The normalization of the spherical harmonics is such that its absolute square, integrated over the unit sphere in $\boldsymbol{\xi}$-space be unity.[29]

Thus we have

$$\begin{aligned} \pounds_+\,{}^k\mathcal{Y}^j_m &= \sqrt{(j-m)(j+m+1)}\,{}^k\mathcal{Y}^j_{m+1}, && (7.12a)\\ \pounds_0\,{}^k\mathcal{Y}^j_m &= m\,{}^k\mathcal{Y}^j_{m+1}, && (7.12b)\\ \pounds_-\,{}^k\mathcal{Y}^j_m &= \sqrt{(j+m)(j-m+1)}\,{}^k\mathcal{Y}^j_{m-1}. && (7.12c) \end{aligned}$$

In particular,

$$ {}^k\mathcal{Y}^j_j = \xi^{k-j}\sqrt{\frac{(2j+1)!}{4\pi}}\,\frac{(\mathbf{p}^2)^j}{2^j\,j!} \qquad (k-j \text{ even}). \tag{7.12d}$$

Comparison with (3.16–17) shows that each multiplet $\{{}^k\mathcal{Y}^j_m\}^j_{m=-j}$ transforms under the *group* $Sp(2,\Re)$ as a finite-dimensional irreducible representation[30] $\mathcal{D}^j$. Symplectic aberration multiplets hence *will not mix* amongst each other under the transformations of paraxial optics. Their compounding, as we shall see below, involves Wigner coefficients, and is analogous —but *not* identical— to ordinary angular momentum coupling in quantum mechanics.

Finally, we would like to present an alternative normalization of the symplectic aberration polynomials [15,16] which does away with the square-root factors, and may thus be somewhat simpler for algorithmic computation. The idea is to start with the monomials

$$\begin{aligned} {}^k\chi^k_k &= (\mathbf{p}^2)^k \quad \text{or} \quad {}^k\chi^k_{-k} = (\mathbf{q}^2)^k, \\ {}^k\chi^j_m &= [\mathbf{p}^2\mathbf{q}^2-(\mathbf{p}\cdot\mathbf{q})^2]^{(k-j)/2}\,{}^j\chi^j_m \end{aligned} \tag{7.13a}$$

---

[29] Integration over the phase space of rays with the same value of the Petzval invariant does not seem to have a role in our construction as yet; perhaps wave optics will have some use for it. We keep the normalization constant for reasons of uniformity with the extant literature, although we offer one variant below.

[30] See Appendix B, equations (17)–(18). (Editor's note.)

[where $\Pi = \mathbf{p}^2\mathbf{q}^2 - (\mathbf{p}\cdot\mathbf{q})^2 \geq 0$ is the Petzval invariant (3.19) and $k - j$ is even], and lower with $\mathcal{K}_- = \pounds_- = \frac{1}{2}{:}\mathbf{q}^2{:}$, or raise it with $\mathcal{K}_+ = -\pounds_+ = \frac{1}{2}{:}\mathbf{p}^2{:}$,

$$\mathcal{K}_+ \, {}^k\mathcal{X}^j_m = (m-j)\, {}^k\mathcal{X}^j_{m+1}, \tag{7.13b}$$

$$\mathcal{K}_- \, {}^k\mathcal{X}^j_m = (m+j)\, {}^k\mathcal{X}^j_{m-1}. \tag{7.13c}$$

Explicitly, they are

$$\begin{aligned} {}^k\mathcal{X}^k_m &= \sqrt{\frac{4\pi(2k+1)(k+m)!\,(k-m)!}{(2k-1)!!}}\; {}^k\mathcal{Y}^k_m \\ &= \frac{(k+m)!\,(k-m)!}{(2k-1)!!} \sum_n \frac{(\frac{1}{2}\mathbf{p}^2)^{m+n}}{(m+n)!} \frac{(\mathbf{p}\cdot\mathbf{q})^{k-m-2n}}{(k-m-2n)!} \frac{(\frac{1}{2}\mathbf{q}^2)^n}{n!}, \end{aligned} \tag{7.14}$$

where we recall the relation $(2k)! = 2^k k!\,(2k-1)!!$.

Below we give explicitly the first few symplectic aberration polynomials. For brevity, we employ the notation $(\mathbf{p}^2)^n = p^{2n}$, $(\mathbf{q}^2)^n = q^{2n}$. The ${}^k\mathcal{X}^j_m$ have been used in references [15,16,18,20] with the Greek letter χ (*chi*), rather than *script* X as here. The difference is due to the fact that in those references, the Petzval invariant was taken to be $\Xi := \frac{1}{2}[\mathbf{p}^2\mathbf{q}^2 - (\mathbf{p}\cdot\mathbf{q})^2] = -\frac{1}{2}\xi^2 \geq 0$, and is used as ${}^{2k}\chi^j_m = \Xi^\nu\, {}^{2j}\chi^j_m$, $j + 2\nu = k$, to build the higher-order multiplets of a given angular momentum j. The relation between the two is thus ${}^k\mathcal{X}^j_m = 2^\nu\, {}^{2k}\chi^j_m$. The aberration coefficients for spherical surfaces tabulated in reference [18] must be thus adjusted by the inverse factor, $2^{-(k-j)/2}$, to compare with those provided in the last section.

The following results hold for $j = k = 1, 2, 3, 4$:

$$2\sqrt{\pi}\, {}^0\mathcal{Y}^0_0 = {}^0\mathcal{X}^0_0 = 1; \tag{7.15a}$$

$$\begin{aligned} 2\sqrt{2\pi/3}\, {}^1\mathcal{Y}^1_1 &= {}^1\mathcal{X}^1_1 = p^2, \\ 2\sqrt{\frac{\pi}{3}}\, {}^1\mathcal{Y}^1_0 &= {}^1\mathcal{X}^1_0 = (\mathbf{p}\cdot\mathbf{q}), \\ 2\sqrt{2\pi/3}\, {}^1\mathcal{Y}^1_{-1} &= {}^1\mathcal{X}^1_{-1} = q^2; \end{aligned} \tag{7.15b}$$

$$\begin{aligned} 4\sqrt{2\pi/5}\, {}^2\mathcal{Y}^2_2 &= {}^2\mathcal{X}^2_2 = p^4, \\ 2\sqrt{2\pi/5}\, {}^2\mathcal{Y}^2_1 &= {}^2\mathcal{X}^2_1 = p^2(\mathbf{p}\cdot\mathbf{q}), \\ \tfrac{4}{3}\sqrt{\pi/5}\, {}^2\mathcal{Y}^2_0 &= {}^2\mathcal{X}^2_0 = \tfrac{1}{3}\left[p^2q^2 + 2(\mathbf{p}\cdot\mathbf{q})^2\right], \\ 2\sqrt{2\pi/5}\, {}^2\mathcal{Y}^2_{-1} &= {}^2\mathcal{X}^2_{-1} = (\mathbf{p}\cdot\mathbf{q})q^2, \\ 4\sqrt{2\pi/5}\, {}^2\mathcal{Y}^2_{-2} &= {}^2\mathcal{X}^2_{-2} = q^4; \end{aligned} \tag{7.16a}$$

$$-2\sqrt{\pi}\, {}^2\mathcal{Y}^0_0 = {}^2\mathcal{X}^0_0 = p^2q^2 - (\mathbf{p}\cdot\mathbf{q})^2; \tag{7.16b}$$

$$\begin{aligned}
8\sqrt{\pi/35}\;{}^3\mathcal{Y}^3_3 &= {}^3\mathcal{X}^3_3 = p^6,\\
4\sqrt{2\pi/105}\;{}^3\mathcal{Y}^3_2 &= {}^3\mathcal{X}^3_2 = p^4(\mathbf{p}\cdot\mathbf{q}),\\
\tfrac{8}{5}\sqrt{\pi/21}\;{}^3\mathcal{Y}^3_1 &= {}^3\mathcal{X}^3_1 = \tfrac{1}{5}\,[p^4q^2 + 4p^2(\mathbf{p}\cdot\mathbf{q})^2],\\
\tfrac{4}{5}\sqrt{\pi/7}\;{}^3\mathcal{Y}^3_0 &= {}^3\mathcal{X}^3_0 = \tfrac{1}{5}\,[3p^2(\mathbf{p}\cdot\mathbf{q})q^2 + 2(\mathbf{p}\cdot\mathbf{q})^3],\\
\tfrac{8}{5}\sqrt{\pi/21}\;{}^3\mathcal{Y}^3_{-1} &= {}^3\mathcal{X}^3_{-1} = \tfrac{1}{5}\,[p^2q^4 + 4(\mathbf{p}\cdot\mathbf{q})^2q^2],\\
4\sqrt{2\pi/105}\;{}^3\mathcal{Y}^3_{-2} &= {}^3\mathcal{X}^3_{-2} = (\mathbf{p}\cdot\mathbf{q})q^4,\\
8\sqrt{\pi/35}\;{}^3\mathcal{Y}^3_{-3} &= {}^3\mathcal{X}^3_{-3} = q^6;
\end{aligned} \tag{7.17}$$

$$\begin{aligned}
\tfrac{16}{3}\sqrt{2\pi/35}\;{}^4\mathcal{Y}^4_4 &= {}^4\mathcal{X}^4_4 = p^8,\\
\tfrac{8}{3}\sqrt{\pi/35}\;{}^4\mathcal{Y}^4_3 &= {}^4\mathcal{X}^4_3 = p^6(\mathbf{p}\cdot\mathbf{q}),\\
\tfrac{8}{21}\sqrt{2\pi/5}\;{}^4\mathcal{Y}^4_2 &= {}^4\mathcal{X}^4_2 = \tfrac{1}{7}\,[p^6q^2 + 6p^4(\mathbf{p}\cdot\mathbf{q})^2],\\
\tfrac{8}{21}\sqrt{\pi/5}\;{}^4\mathcal{Y}^4_1 &= {}^4\mathcal{X}^4_1 = \tfrac{1}{7}\,[3p^4(\mathbf{p}\cdot\mathbf{q})q^2 + 4p^2(\mathbf{p}\cdot\mathbf{q})^3],\\
\tfrac{16}{105}\sqrt{\pi}\;{}^4\mathcal{Y}^4_0 &= {}^4\mathcal{X}^4_0 = \tfrac{1}{35}\,[3p^4q^4 + 24p^2(\mathbf{p}\cdot\mathbf{q})^2q^2 + 8(\mathbf{p}\cdot\mathbf{q})^4],\\
\tfrac{8}{21}\sqrt{\pi/5}\;{}^4\mathcal{Y}^4_{-1} &= {}^4\mathcal{X}^4_{-1} = \tfrac{1}{7}\,[3p^2(\mathbf{p}\cdot\mathbf{q})q^4 + 4(\mathbf{p}\cdot\mathbf{q})^3q^2],\\
\tfrac{8}{21}\sqrt{2\pi/5}\;{}^4\mathcal{Y}^4_{-2} &= {}^4\mathcal{X}^4_{-2} = \tfrac{1}{7}\,[p^2q^6 + 6(\mathbf{p}\cdot\mathbf{q})^2q^4],\\
\tfrac{8}{3}\sqrt{\pi/35}\;{}^4\mathcal{Y}^4_{-3} &= {}^4\mathcal{X}^4_{-3} = (\mathbf{p}\cdot\mathbf{q})q^6,\\
\tfrac{16}{3}\sqrt{2\pi/35}\;{}^4\mathcal{Y}^4_{-4} &= {}^4\mathcal{X}^4_{-4} = q^8.
\end{aligned} \tag{7.18}$$

We should note (as a visual check on results) that in the ${}^k\mathcal{X}^k_m$'s the monomial coefficients sum to unity. Other normalization conventions are clearly possible. We may start with (7.13*a*) and define the normalization of the rest of the multiplet asking for (7.12*c*) to hold.

To stress what has been said before, from the viewpoint of Lie methodology, the advantages of the *symplectic* classification of aberrations over the basically cartesian classification that is traditional in optics [22–25], are:

- Each symplectic aberration polynomial multiplet $\{{}^k\mathcal{Y}^j_m\}^j_{m=-j}$ transforms amongst itself under the paraxial part $\mathfrak{M}_2$ of any optical map $\mathfrak{M}$, *i.e.*, matrices are as *block-diagonal* as possible.
- Under metaxial transformations $\cdots\mathfrak{M}_6\mathfrak{M}_4$, the compounding of aberrations (3.27) requires the Poisson bracket $\{{}^{k_1}\mathcal{Y}^{j_1}_{m_1}, {}^{k_2}\mathcal{Y}^{j_2}_{m_2}\}$. This will be analyzed in terms of Wigner coefficients below. It will yield *selection rules* for the aberration compounding process.

Regarding the first point, the transformations (3.22–27) become explicitly

$$({}^k\mathcal{Y}^j_m)^{\mathrm{tr}} = \mathfrak{M}_2^{\left(\begin{smallmatrix} a & b\\ c & d\end{smallmatrix}\right)}\;{}^k\mathcal{Y}^j_m = \sum_{m'=-j}^{j} D^j_{m,m'}\begin{pmatrix} a & b\\ c & d\end{pmatrix}^{-1}\;{}^k\mathcal{Y}^j_{m'}, \tag{7.19}$$

where the representation matrix elements are given in Appendix B, equation (18). The reason for the appearance of the *inverse* matrix $(\because)^{-1}$ is that the $\mathfrak{M}$'s must satisfy (3.11):

$$\begin{aligned}
\mathfrak{M}_2(\mathbf{M})\,\mathfrak{M}_2(\mathbf{N})\,f(\mathbf{w}) &= \mathfrak{M}_2(\mathbf{M})\,f(\mathbf{N}^{-1}\mathbf{w}) = f(\mathbf{N}^{-1}\mathfrak{M}_2(\mathbf{M})\mathbf{w})\\
&= f(\mathbf{N}^{-1}\mathbf{M}^{-1}\mathbf{w}) = f((\mathbf{M}\mathbf{N})^{-1}\mathbf{w}) = \mathfrak{M}_2(\mathbf{M}\mathbf{N})\,f(\mathbf{w}).
\end{aligned} \tag{7.20}$$

Regarding the second point we may see, without further computation, that some selection rules apply, namely

$$\{{}^{k_1}y^{j_1}_{m_1}, {}^{k_2}y^{j_2}_{m_2}\} = \sum_{j_3=|m_1+m_2|}^{j_1+j_2-1} P^{j_1,j_2;j_3}_{m_1,m_2}\ {}^{k_1+k_2-1}y^{j_3}_{m_1+m_2}, \tag{7.21}$$

since (3.10) holds, $\{\mathbf{p}\times\mathbf{q}, {}^k y^j_m\} = 0$, and $\pounds_0\{f,g\} = \{\pounds_0 f, g\} + \{f, \pounds_0 g\}$ due to the Jacobi identity.

4.7.4 The third-order aberration algebra and group

We shall detail here the way in which we may use the symplectic classification of aberrations to simplify —actually to perform explicitly— the computation of third-order aberrations [15,16]. Following these references, we shall use here the aberration polynomials ${}^k\chi^j_m$ in (7.15–16).

We start by taking *all* ${}^k\chi^j_m$, as generators of an infinite-dimensional Lie algebra $\mathcal{A}$ under Poisson brackets, recalling (3.10). From that equation we see that all aberration polynomials of orders $k > k_N := \frac{1}{2}(N-1)$ (N is the *aberration order* we wish to consider), form an (infinite-dimensional) *invariant ideal* and hence a normal subalgebra $\mathcal{A}^{>N}$. We may therefore build the *quotient* algebra $\mathcal{A}^{\leq N} := \mathcal{A}/\mathcal{A}^{>N}$ with a Lie bracket $\{\cdot,\cdot\}_N$. The latter has generators ${}^k\chi^j_m$, $1 \leq k \leq k_N$. Thus, all Poisson brackets of polynomials $\{f_n, f_m\}$ which in $\mathcal{A}$ yield polynomials f_{n+m-2} in $\mathcal{A}^{>N}$, are replaced in $\mathcal{A}^{\leq N}$ by a Lie bracket yielding *zero* [*c.f.* Eq. (3.10)], *i.e.*

$$\{f_n, f_m\}_N = \begin{cases} f_{n+m-2}, & n+m-2 \leq 2k_N, \\ 0, & n+m-2 > 2k_N. \end{cases} \tag{7.22}$$

The Lie algebra defined by this Poisson bracket will be called the ***N*^th^-order aberration algebra**. We henceforth drop the subindex N in $\{\cdot,\cdot\}_N$. The connected Lie group generated by $f_n \in \mathcal{A}^{\leq N}$, $n \leq 2k_N$, will be called the ***aberration group*** of order N, denoted by A^N.

For axially symmetric systems in third aberration order, we thus consider the Lie algebra generated by the *nine* symplectic polynomials of orders one and two:

$${}^1\chi^1_1,\ {}^1\chi^1_0,\ {}^1\chi^1_{-1};\quad {}^2\chi^2_2,\ {}^2\chi^2_1,\ {}^2\chi^2_0,\ {}^2\chi^2_{-1},\ {}^2\chi^2_{-2};\quad {}^2\chi^0_0.$$

The basic aberration group *presentation* $A\{\mathbf{v}, w; \mathbf{M}\}$ is given by the *gaussian* or *paraxial* part

$$A\left\{\mathbf{0}, 0; \begin{pmatrix} a(\mathbf{u}) & b(\mathbf{u}) \\ c(\mathbf{u}) & d(\mathbf{u}) \end{pmatrix}\right\} := \exp\left(:\sum_{m=-1}^{1} u_m\, {}^1\chi^1_m:\right), \tag{7.23}$$

[where the connection between the $\{u_m\}^1_{m=-1}$ and the elements of the matrix are given by (4.26–27)], and the *aberration* part

$$A\{\mathbf{v}, w; \mathbf{1}\} := \exp\left(:\sum_{m=-2}^{2} v_m\, {}^2\chi^2_m + w\, {}^2\chi^0_0:\right), \tag{7.23}$$

parametrized by a five-dimensional *row* vector $\mathbf{v} = \{v_m\}^2_{m=-2}$ (the *quintuplet* aberration coefficients), and a scalar w (the *singlet* coefficient). The general group element is then defined as

$$A\{\mathbf{v}, w; \mathbf{M}\} := A\{\mathbf{v}, w; \mathbf{1}\}\, A\{\mathbf{0}, 0; \mathbf{M}\}. \tag{7.25}$$

We note that since the ${}^2\mathcal{X}$'s commute amongst themselves in the $\mathcal{A} \leq 3$ Lie bracket, they generate an invariant *abelian* subgroup of A^3, behaving under paraxial transformations as quadrupole+ monopole operators under the rotation group. The group composition law is thus ascertained to be

$$A\{\mathbf{v}_1, w_1; \mathbf{M}_1\}\, A\{\mathbf{v}_2, w_2; \mathbf{M}_2\} = A\{\mathbf{v}_1 + \mathbf{v}_2 \mathcal{D}^{(2)}(\mathbf{M}_1^{-1}), w_1 + w_2; \mathbf{M}_1\mathbf{M}_2\}. \tag{7.26}$$

Here, $\mathbf{M}_1\mathbf{M}_2$ is the ordinary 2×2 matrix multiplication, the singlet coefficients w *simply add*, and we have the 5×5 '*no-square root*' representation $\mathcal{D}^{(j=2)}$ of $Sp(2,\Re)$, given by

$$\mathcal{D}^{(2)}\begin{pmatrix} \alpha & \beta \\ \gamma & \delta \end{pmatrix} = \begin{pmatrix} \alpha^4 & 4\alpha^3\beta & 6\alpha^2\beta^2 & 4\alpha\beta^3 & \beta^4 \\ \alpha^3\gamma & \alpha^2(\alpha\delta + 3\beta\gamma) & 3\alpha\beta(\alpha\delta + \beta\gamma) & \beta^2(3\alpha\delta + \beta\gamma) & \beta^3\delta \\ \alpha^2\gamma^2 & 2\alpha\gamma(\alpha\delta + \beta\gamma) & \alpha^2\delta^2 + 4\alpha\beta\gamma\delta + \beta^2\gamma^2 & 2\beta\delta(\alpha\delta + \beta\gamma) & \beta^2\delta^2 \\ \alpha\gamma^3 & \gamma^2(3\alpha\delta + \beta\gamma) & 3\gamma\delta(\alpha\delta + \beta\gamma) & \delta^2(\alpha\delta + 3\beta\gamma) & \beta\delta^3 \\ \gamma^4 & 4\gamma^3\delta & 6\gamma^2\delta^2 & 4\gamma\delta^3 & \delta^4 \end{pmatrix}. \tag{7.27}$$

The unit element is $A\{\mathbf{0}, 0; \mathbf{1}\}$ and the inverse of $A\{\mathbf{v}, w; \mathbf{M}\}$ is $A\{-\mathbf{v}\mathcal{D}^{(2)}(\mathbf{M}), -w; \mathbf{M}^{-1}\}$.

In this third-order scheme, the **free propagation** mapping through length ℓ, (3.15*a*)–(4.3), is

$$\mathfrak{M}^{(3)}(\ell) = A\left\{\left(-\frac{z}{8n^3}, 0, 0, 0, 0\right), 0; \begin{pmatrix} 1 & 0 \\ -\ell/n & 1 \end{pmatrix}\right\}, \tag{7.28}$$

the **root** transformation (6.22) is[31]

$$\mathfrak{R}_4\mathfrak{R}_2 = \mathfrak{R}^{(3)} = A\left\{\left(0, 0, -\frac{\varsigma_2}{2n}, 0, n\varsigma_4\right), -\frac{\varsigma_2}{3n}; \begin{pmatrix} 1 & -2n\varsigma_2 \\ 0 & 1 \end{pmatrix}\right\}, \tag{7.29}$$

and the refracting surface transformation producing (2.15b)–(6.8) to third order is

$$\mathfrak{S}^{(3)} = A\left\{\left(0, 0, \frac{\varsigma_2}{2}\left[\frac{1}{n'} - \frac{1}{n}\right], \frac{2\varsigma_2^2}{n'}[n - n'], \frac{2\varsigma_2^3}{n'}[n - n']^2 + \varsigma_4[n - n']\right), -\frac{\varsigma_2}{3}\left[\frac{1}{n} - \frac{1}{n'}\right]; \begin{pmatrix} 1 & -2\varsigma_2[n - n'] \\ 0 & 1 \end{pmatrix}\right\}. \tag{7.30}$$

The exponentiation of the general element of the algebra $\mathcal{A}^3$ may be given in closed form. In reference [16] it is shown that

$$\exp\left(z\!:\sum_{m=-1}^{1} u_m\, {}^1\mathcal{X}^1_m + \sum_{m=-2}^{2} v_m\, {}^2\mathcal{X}^2_m + w\, {}^2\mathcal{X}^0_0\!:\right) = A\{z\mathbf{v}\, \mathcal{E}(z\mathbf{u}), zw; \mathbf{M}(z\mathbf{u})\}, \tag{7.31}$$

where the $Sp(2,\Re)$ part $\mathbf{M}(z\mathbf{u})$ is given in (4.26–27) with $(\alpha, \beta, \gamma) = (zu_1, zu_0, zu_{-1})$. For the aberration part,

$$\mathcal{E}(z\mathbf{u}) = \mathcal{D}^{(2)}(\mathbf{B})\, \mathcal{E}^d(z\omega)\, \mathcal{D}^{(2)}(\mathbf{B}^{-1}), \tag{7.32a}$$

$$\mathbf{B} = \mathbf{B}(\mathbf{u}) = \frac{1}{\sqrt{2}}\begin{pmatrix} (\omega + u_0)/\omega & (\omega - u_0)/2u_1 \\ -2u_1/w & 1 \end{pmatrix}, \tag{7.32b}$$

$$\mathcal{E}^d(x) = \operatorname{diag}\left(\frac{e^{4x} - 1}{4x}, \frac{e^{2x} - 1}{2x}, 1, \frac{1 - e^{-2x}}{2x}, \frac{1 - e^{-4x}}{4x}\right), \tag{7.32c}$$

[31] Comparison with [15, Eq.(6.2)] shows that the $\Pi \leftrightarrow \Xi$ correspondence makes $w^{\text{here}} = \frac{1}{2}w^{\text{there}}$; *cf.* Eq. (7.24).

where $\omega(\mathbf{u})$ is as in (4.27), $\mathbf{B}$ is the *Bargmann* matrix which diagonalizes $\mathbf{M}(z\mathbf{u}) = \mathbf{B}\mathbf{M}^d(z\omega)\mathbf{B}^{-1}$ to $\mathbf{M}^d(z\omega) = \operatorname{diag}(e^{-z\omega}, e^{z\omega})$, and $\mathcal{D}^{(2)}(\mathbf{B})$ is the corresponding 5×5 matrix given by (7.27).

The matrix $\mathcal{E}(z\mathbf{u})$ disentangles the paraxial and third-order aberration parts in closed form. This expression serves to find the symplectic map due to a z-homogeneous optical fiber with quartic index profile (4.11); there we need $\exp(z{:}H{:})$ with the Hamiltonian (4.12-13*b*). The calculation of (4.35–36) was done this way in reference [16].

When the matrix $\mathbf{M}(z\mathbf{u})$ is triangular and hence not diagonalizable, $\mathcal{E}(0,0,zu_{-1})$ still exists and is upper-triangular. Its matrix elments may be obtained through a limit, and are

$$\mathcal{E}(0,0,x) = \begin{cases} \begin{pmatrix} 2+m \\ 2+m' \end{pmatrix} \dfrac{(2x)^{m-m'}}{m-m'+1}, & m \geq m', \\ 0, & m < m'. \end{cases} \tag{7.33}$$

This allowed us, for example, to give the *logarithm* of the root and refracting surface transformations (7.29–30) at the end of reference [14]. The disentangling matrix $\mathcal{E}$ has a form similar to (7.32) when the exponentiated operator has the *simple* form $\sum_{m=-1}^{1} u_m\, {}^1\mathcal{X}^1_m + \sum_{m=-k}^{k} v_m\, {}^k\mathcal{X}^k_m$ [26], but has yet to be found for the general case of *compounding* aberrations of order higher than three. This is equivalent to the Baker-Campbell-Hausdorff formula.

We shall not dwell here upon other constructions which, to third order, are developed in [15,16]. It is worth mentioning, however, *spinor* symplectic harmonics. The idea is to define the *two*-vectors ${}^{1/2}\mathcal{X}^{1/2}_{1/2} := \mathbf{p}$, ${}^{1/2}\mathcal{X}^{1/2}_{-1/2} := \mathbf{q}$, and those that are obtained from these through Poisson brackets by the scalar ${}^k\mathcal{X}^j_m$'s to yield ${}^k\mathcal{X}^j_m$'s with k half-integer. Notice that $\mathbf{p}\times\mathbf{q}$ rotates the two components of each $\mathbf{p}$ and $\mathbf{q}$, while ${}^2\mathcal{X}^0_0 = (\mathbf{p}\times\mathbf{q})^2$ produces what we may define to be ${}^{3/2}\mathcal{X}^{3/2}_m$, where ${}^{3/2}\mathcal{X}^{3/2}_{3/2}$ is, quite clearly, $\mathbf{p}^2\mathbf{p}$. In this way, after generating with (7.3) or (7.13) —according to the normalization scheme or taste– there are six ${}^{3/2}\mathcal{X}^j_m$'s: a doublet $j = \frac{1}{2}$ and a quadruplet $j = \frac{3}{2}$. Together with the two ${}^{1/2}\mathcal{X}^{1/2}_m$'s we have an eight-dimensional homogeneous space for A^3. Transformations in this space are not canonical in general[32] since we are cutting the $\mathbf{p}$–$\mathbf{q}$ transformation to third order.

4.7.5 Applications of the symplectic multiplet decomposition

Equation (7.19) may or may not be computationally more economic than (3.22). There are efficient methods for computing f^{tr}_m numerically; they make use of (3.11) and have been implemented in the computer code **MARYLIE**. See references [27–30]. Nevertheless, it *does* show clearly what contributes to *what*. In this subsection we shall concentrate on one Lie algebraic fact: the separation of astigmatism and curvature of field into a member of a symplectic quintuplet and a singlet, —***curvatism*** and ***astigmature***, as they were playfully called in reference [15]. This has to do with the conservation of the *Petzval*, as we shall see now.

The monomials $M_{020} = (\mathbf{p}\cdot\mathbf{q})^2$ and $M_{101} = \mathbf{p}^2\mathbf{q}^2$, both have weight zero, and are linear combinations of the symplectic quintuplet ${}^2\mathcal{Y}^2_0 = \frac{3}{4}\sqrt{5/\pi}\,{}^2\mathcal{X}^2_0$ and singlet ${}^2\mathcal{Y}^0_0 = -1/2\sqrt{\pi}\,{}^2\mathcal{X}^0_0$, namely

[32] Only those are canonical which are special cases of *Cremona* maps. See reference [19].

$$\begin{aligned} \tfrac{4}{3}\sqrt{\pi/5}\;{}^2\mathcal{Y}^2_0 = \quad {}^2\mathcal{X}^2_0 &= \tfrac{1}{3}\,[p^2q^2 + 2(\mathbf{p}\cdot\mathbf{q})^2], \\ -2\sqrt{\pi}\;{}^2\mathcal{Y}^0_0 = \quad {}^2\mathcal{X}^0_0 &= p^2q^2 - (\mathbf{p}\cdot\mathbf{q})^2; \end{aligned} \tag{7.34a}$$

hence

$$v_0\,{}^2\mathcal{X}^2_0 + w\,{}^2\mathcal{X}^0_0 = C(\mathbf{p}\cdot\mathbf{q})^2 + D\mathbf{p}^2\mathbf{q}^2 \tag{7.34b}$$

yields

$$C = \tfrac{2}{3}v_0 - w, \qquad v_0 = C + D, \qquad (7.34c,d)$$
$$D = \tfrac{1}{3}v_0 + w, \qquad 3w = 2D - C. \qquad (7.34e,f)$$

The coefficient C is that of *astigmatism* and D is *curvature of field*. The particular linear combination (7.34f), $2D - C = 3w$ is called ***the Petzval*** [8, Sect. 5.5.3; 9, Sect. 42.6].[33]

The first observation, implicit in (3.23) and explicit to third order in the product (7.26), is that the *singlet* contributions to the aberration polynomial, are *purely additive*. Singlets exist in f_{4m}, $m = 1, 2, 3, \ldots$. Thus, *the Petzval of a compound optical system is simply the sum of the Petzvals of its individual elements.* Free propagation in a homogeneous medium has *zero* Petzval, so all contributions come through refracting surfaces. These can be read off from (7.30), and are proportional to $n'^{-1} - n^{-1}$ and the quadratic surface coefficient ς_2; this is simply the Petzval of the root transformation (7.29) of the 'front' surface, minus that of the 'back' surface.

The existence and properties of the Petzval are closely related to the presence of axial symmetry or, equivalently, invariance under rotations about the optical z-axis. Rotations about the z axis are generated by the Lie operator $J_z = :\mathbf{p}\times\mathbf{q}: = :q_x p_y - q_y p_x:$. It is easily checked that J_z commutes with $\mathfrak{L}_\pm$ and $\mathfrak{L}_0$, because they are constructed from rotationally invariant quantities. It also annihilates all ${}^k\mathcal{Y}^j_m$ since they are by construction all rotationally invariant. Since ${}^2\mathcal{Y}^0_0$ is made from $\mathbf{p}\times\mathbf{q}$ —see (7.16b)— it follows that ${}^2\mathcal{Y}^0_0$ should be annihilated by $\mathfrak{L}_\pm$ and $\mathfrak{L}_0$. Finally, comparison of (7.6a) for the Casimir operator $\mathfrak{L}^2$ and the expression (7.16b) for ${}^2\mathcal{Y}^0_0$ show that they have a very similar structure, and the fact that $\mathfrak{L}_\pm$ and $\mathfrak{L}_0$ commute with $\mathfrak{L}^2$ can also be viewed as a consequence of rotational invariance.

This close relation between ${}^2\mathcal{Y}^0_0$ and $\mathfrak{L}^2$ also appears to raise a paradox. Note that, apart from normalization, ${}^2\mathcal{Y}^0_0$ can be obtained from $\mathfrak{L}^2$ simply by removing the colons. Now consider the symplectic Lie algebra *sp(4,ℜ)* generated by *all* quadratic polynomials in the components of **p** and **q** when rotational symmetry is *not* enforced. This Lie algebra also has a quadratic Casimir operator, and one might think that removing the colons from this operator would produce an invariant *quartic* polynomial from *sp(4,ℜ)* analogous to ${}^2\mathcal{Y}^0_0$. On the other hand, it is known that there is *no* invariant quartic polynomial for *sp(4,ℜ)*. The resolution of this apparent paradox is that when the Casimir operator for *sp(4,ℜ)* is constructed and the colons removed, the quartic polynomial that is produced *vanishes* identically. The various monomials that are produced by removing the colons in the Casimir operator, all cancel.

Let us now examine some other consequences of the invariance of the Petzval. In **4.5**, we saw the propagation of rays in optical fibers where the refraction index is $(\mathbf{q}, z)$-dependent. We may express H_4 in (4.13c) in the symplectic basis, and find

$$\begin{aligned} H_4 &= \frac{1}{8\alpha_0^3}(\mathbf{p}^2)^2 - \frac{\alpha_2}{2\alpha_0^2}\mathbf{p}^2\mathbf{q}^2 - \alpha_4(\mathbf{q}^2)^2 \\ &= \frac{1}{8\alpha_0^3}\,{}^2\mathcal{X}^2_2 - \frac{\alpha_2}{2\alpha_0^2}({}^2\mathcal{X}^2_0 + \tfrac{2}{3}\,{}^2\mathcal{X}^0_0) - \alpha_4\,{}^2\mathcal{X}^2_{-2} \\ &= \frac{1}{\alpha_0^3}\sqrt{\frac{\pi}{10}}\,{}^2\mathcal{Y}^2_2 - \frac{2\alpha_2}{3\alpha_0^2}\sqrt{\frac{\pi}{5}}({}^2\mathcal{Y}^2_0 - \sqrt{5}\,{}^2\mathcal{Y}^0_0) - 4\alpha_4\sqrt{\frac{2\pi}{5}}\,{}^2\mathcal{Y}^2_{-2}. \end{aligned} \tag{7.35}$$

[33]Note, however that in both references the coefficients C and D differ in normalization from those used in this chapter.

When $H_4^{\rm int}$ is computed using (4.22), the ${}^2\mathcal{Y}_0^0$ component is unaffected by the paraxial transformation $\mathfrak{M}_2$, while the ${}^2\mathcal{Y}_m^2$ components are transformed among themselves.

Thus we have

$$H_4^{\rm int} = \frac{2\alpha_2\sqrt{\pi}}{3\alpha_0^2}\,{}^2\mathcal{Y}_0^0 + \text{quintuplet content.} \tag{7.36}$$

Finally, when $H_4^{\rm int}$ is inserted into (4.21*a*) to compute the third-order aberration polynomial for the fiber, we find the relation

$$f_4 = -\frac{2\alpha_2\sqrt{\pi}}{3\alpha_0^2}\,{}^2\mathcal{Y}_0^0 \int_{z^{\rm i}}^{z} dz'\,\frac{\alpha_2}{\alpha_0^2} + \text{quintuplet content.} \tag{7.37}$$

Thus, the singlet or Petzval content of f_4 is given by a simple integral involving only $\alpha_0(z)$ and $\alpha_2(z)$. The Petzval of the fiber is thus

$$2D - C = \int_{z^{\rm i}}^{z} dz'\,\frac{\alpha_2}{\alpha_0^2}, \tag{7.38}$$

as may be verified from (4.30*c*, *d*).

As another application of the Petzval conservation to the treatment of lenses, suppose $n(\mathbf{r})$ is position-dependent *only* as a result of a curved interface between two media having constant refraction indices n and n'. Let the interface be given by the axially-symmetric polynomial form (6.2), and introduce some smooth approximation to the Heaviside step function,

$$\Theta_\epsilon(x) := \begin{cases} 0, & x < -\epsilon, \\ 1, & x > \epsilon > 0. \end{cases} \tag{7.39}$$

Then $n(\mathbf{r})$ may be written in the form

$$n(\mathbf{r}) = n + (n' - n)\,\Theta_\epsilon\big(z - \varsigma_2\mathbf{q}^2 - \varsigma_4(\mathbf{q}^2)^2 - \cdots\big). \tag{7.40}$$

Suppose this is expanded in Taylor series. Then one finds the result

$$n(\mathbf{r}) = n + (n' - n)\,\Theta_\epsilon(z) + (n - n')\varsigma_2\mathbf{q}^2\Theta_\epsilon'(z) + \cdots. \tag{7.41}$$

Comparison with (4.11) gives the relations

$$\alpha_0(z) = n + (n' - n)\Theta_\epsilon(z), \tag{7.42a}$$
$$\alpha_2(z) = -(n' - n)\varsigma_2\Theta_\epsilon'(z). \tag{7.42b}$$

These expressions for α_0 and α_2 can be substituted into (7.38) to give the contribution to the Petzval *coming from the interface.* We find the result

$$\begin{aligned} 2D - C &= \varsigma_2 \int_{-\epsilon}^{\epsilon} dz'\,\frac{(n - n')\Theta_\epsilon'(z)}{[n + (n' - n)\Theta_\epsilon(z')]^2} \\ &= \left.\frac{\varsigma_2}{n + (n' - n)\Theta_\epsilon(z')}\right|_{z'=-\epsilon}^{\epsilon} \\ &= \varsigma_2\left(\frac{1}{n'} - \frac{1}{n}\right). \end{aligned} \tag{7.43}$$

This we verify in (7.30), remembering from (7.34c) that $w = \frac{1}{3}(2D - C)$. We observe that ς_2 is related to the radius of curvature, r, of the interface by $\varsigma_2 = -1/2r$. Thus, (7.43) is just the expected contribution to the Petzval for an interface [8, Sect. 5.5.3; 9, Sect. 42.6].

The seventh-order symplectic aberration singlet,

$$ {}^4\chi_0^0 = [\mathbf{p}^2\mathbf{q}^2 - (\mathbf{p}\cdot\mathbf{q})^2]^2 = M_{040} - 2M_{121} + M_{202}, \tag{7.44}$$

may be subject to similar analysis. The corresponding aberration coefficient for a spherical surface (6.2) is [18];[34]

$$ w^{(4)} = \frac{2}{15\,r^3}\left(\frac{1}{n^3} - \frac{1}{n'^3}\right). \tag{7.45}$$

Still to be made, is an investigation into the relation between the various symplectic aberration multiplets for the surface and —simpler— the root transformations. For spherical surfaces [18] we may expect a *dynamical* Lie algebra approach for optics, such as appears for the quantum harmonic oscillator. Other revolution conic surfaces should lead to similar considerations.

4.7.6 Compounding aberrations through Wigner coefficients

In equations (3.27), we gave the expressions for the compounding of two symplectic maps in terms of Poisson brackets (3.1) between the aberration polynomials. These polynomials may be expanded in the monomials (6.10) or in symplectic polynomials (7.10), so we need the Poisson bracket between any two functions f, g of ξ_+, ξ_0, and ξ_- given in (7.1). To this end, the chain rule provides the result

$$\begin{aligned}\{f,g\} &= \frac{\partial f}{\partial q_1}\frac{\partial g}{\partial p_1} + \frac{\partial f}{\partial q_2}\frac{\partial g}{\partial p_2} - \frac{\partial f}{\partial p_1}\frac{\partial g}{\partial q_1} - \frac{\partial f}{\partial p_2}\frac{\partial g}{\partial q_2} \\ &= 2\xi_+\left(\frac{\partial f}{\partial \xi_0}\frac{\partial g}{\partial \xi_+} - \frac{\partial f}{\partial \xi_+}\frac{\partial g}{\partial \xi_0}\right) + 2\xi_0\left(\frac{\partial f}{\partial \xi_-}\frac{\partial g}{\partial \xi_+} - \frac{\partial f}{\partial \xi_+}\frac{\partial g}{\partial \xi_-}\right) + 2\xi_-\left(\frac{\partial f}{\partial \xi_-}\frac{\partial g}{\partial \xi_0} - \frac{\partial f}{\partial \xi_0}\frac{\partial g}{\partial \xi_-}\right) \\ &= \left(\sqrt{2}\frac{\partial f}{\partial \xi_+}K_+ + 2\frac{\partial f}{\partial \xi_0}K_0 + \sqrt{2}\frac{\partial f}{\partial \xi_-}K_-\right)g.\end{aligned} \tag{7.46}$$

In the last expression[35] we have used the $sp(2,\Re)$ generators K_σ given in (3.16) and (7.5), and we recall again that their relations to the $so(3)$ generators $\mathcal{L}_\sigma$ are $K_+ = -\mathcal{L}_+$, $K_0 = \mathcal{L}_0$, and $K_- = \mathcal{L}_-$.

For the monomials M_{k_+,k_0,k_-} defined in (6.10), from (6.15) or (7.46), we find the closed form

$$\begin{aligned}\{M_{k_+,k_0,k_-}, M_{\ell_+,\ell_0,\ell_-}\} = {}& 2[k_0(\ell_+ - \ell_-) - (k_+ - k_-)\ell_0]\, M_{k_++\ell_+,k_0+\ell_0-1,k_-+\ell_-} \\ &+ 4[k_-\ell_+ - k_+\ell_-]\, M_{k_++\ell_+-1,k_0+\ell_0+1,k_-+\ell_--1}\,.\end{aligned} \tag{7.47}$$

Let us consider now the Poisson bracket between two normalized symplectic polynomials ${}^k\mathcal{Y}^j_m$ as defined in (7.10). We first notice that $\mathcal{L}_0\{f,g\} = \{\mathcal{L}_0 f, g\} + \{f, \mathcal{L}_0 g\}$, due to the Jacobi identity. That means if f and g are eigenfunctions of $\mathcal{L}_0$ with weights m_1 and m_2, $\{f,g\}$ will be an eigenfunction of the same with weight $m_1 + m_2$. Second, from the middle line of (7.46), we see that if f and g are homogeneous of order k_1 and k_2 in the components of $\boldsymbol{\xi}$, then $\{f,g\}$ will be homogeneous of order $k = k_1 + k_2 - 1$. Lastly, since $\{\xi^2, f\} = 0$, the factors $\xi^{k_i - j_i}$ of ${}^{k_i}\mathcal{Y}^{j_i}_{m_i}$ in (7.10) may be extracted from

[34]Because of reasons given earlier, the values are related to the notation in that reference through $v^7_{00} = \frac{1}{4}w^{(4)}$.

[35]This expression defines the phase-space Lie operator $:f:$ as a *Lie field* on $sp(2,\Re)$ and $so(3)$.

the Poisson bracket. We may thus state [*cf.* Eq. (7.21)]

$$\{{}^{k_1}y^{j_1}_{m_1}, {}^{k_2}y^{j_2}_{m_2}\} = \xi^{k_1+k_2-j_1-j_2}\{y^{j_1}_{m_1}, y^{j_2}_{m_2}\} = \sum_{\substack{j=|m_1+m_2| \\ k_1+k_2+j \text{ odd}}}^{j_1+j_2-1} P^{j_1,j_2;j}_{m_1,m_2}\ {}^{k_1+k_2-1}y^{j}_{m_1+m_2}. \tag{7.48}$$

The last line makes use of the first two observations; the upper limit in the sum is a consecuence of the third, *i.e.*, that ${}^k y^j_m$ exists only for order $k \geq j$, the lower limit from $|m| \leq j$ and an argument to be given below. The task at hand is to find the Poisson bracket coefficients $P^{j_1,j_2;j}_{m_1,m_2}$ in terms of the Wigner coupling coefficients.

In calculating (7.48) through (7.46) we use (7.12) for the action of $\mathcal{K}_\sigma$, and the following formulae which are easily established from (7.11):

$$\frac{\partial}{\partial \xi_+} y^j_m(\xi) = \sqrt{\frac{(2j+1)(j+m)(j+m-1)}{2(2j-1)}}\, y^{j-1}_{m-1}(\xi), \tag{7.49a}$$

$$\frac{\partial}{\partial \xi_0} y^j_m(\xi) = \sqrt{\frac{(2j+1)(j+m)(j-m)}{2j-1}}\, y^{j-1}_{m}(\xi), \tag{7.49b}$$

$$\frac{\partial}{\partial \xi_-} y^j_m(\xi) = \sqrt{\frac{(2j+1)(j-m)(j-m-1)}{2(2j-1)}}\, y^{j-1}_{m+1}(\xi), \tag{7.49c}$$

Hence the second Poisson bracket in (7.48) is

$$\begin{aligned}\{y^{j_1}_{m_1}, y^{j_2}_{m_2}\} = \sqrt{\frac{2j_1+1}{2j_1-1}}\Big(&-\sqrt{(j_1+m_1)(j_1+m_1-1)(j_2-m_2)(j_2+m_2+1)}\, y^{j_1-1}_{m_1-1}\, y^{j_2}_{m_2+1} \\ &+2\sqrt{(j_1+m_1)(j_1-m_1)}\, m_2\, y^{j_1-1}_{m_1}\, y^{j_2}_{m_2} \\ &+\sqrt{(j_1-m_1)(j_1-m_1-1)(j_2+m_2)(j_2-m_2+1)}\, y^{j_1-1}_{m_1+1}\, y^{j_2}_{m_2-1}\Big).\end{aligned} \tag{7.50}$$

We now recall the product of spherical harmonics [10, Eqs. (3.138,189)], expressed in terms of the Wigner or $3jm$ coefficients, *viz.*

$$Y^{\ell_1}_{m_1}(\Omega_\xi) Y^{\ell_2}_{m_2}(\Omega_\xi) = \sum_{\ell=|\ell_1-\ell_2|}^{\ell_1+\ell_2} \sqrt{\frac{(2\ell_1+1)(2\ell_2+1)}{4\pi(2\ell+1)}}\, C^{\ell_1,\ell_2,\ell}_{0,0,0} C^{\ell_1,\ell_2,\ell}_{m_1,m_2,m_1+m_2} Y^{\ell}_{m}(\Omega_\xi), \tag{7.51}$$

with $\ell_1 = j_1 - 1$, $\ell_2 = j_2$, and $\ell = j$. We use this to reduce the products of solid spherical harmonics in (7.50) to *single* spherical harmonics, times $\xi^{j_1+j_2-1}$. We then pull the sum over j to the left, and thus find the Poisson bracket coefficients $P^{j_1,j_2;j}_{m_1,m_2}$ in (7.48) to be

$$
\begin{aligned}
P^{j_1,j_2;j}_{m_1,m_2} = &\sqrt{\frac{(2j_1+1)(2j_2+1)}{4\pi(2j+1)}}\, C^{j_1-1,j_2,j}_{0,0,0} \\
&\times\Big(-\sqrt{(j_1+m_1)(j_1+m_1-1)(j_2-m_2)(j_2+m_2+1)}\, C^{j_1-1,j_2,j}_{m_1-1,m_2+1,m_1+m_2} \\
&\qquad +2\sqrt{(j_1+m_1)(j_1-m_1)}\, m_2\, C^{j_1-1,j_2,j}_{m_1,m_2,m_1+m_2} \\
&\quad +\sqrt{(j_1-m_1)(j_1-m_1-1)(j_2+m_2)(j_2-m_2+1)}\, C^{j_1-1,j_2,j}_{m_1+1,m_2-1,m_1+m_2}\Big).
\end{aligned}
\tag{7.52}
$$

We note that because of the reduced Wigner coefficient, $P^{j_1,j_2;j}_{m_1,m_2}$ in (7.48) is zero when j_1+j_2+j is *even* [10, Eq. (3.194)]; since k_1-j_1 and k_2-j_2 are both even, then k_1+k_2+j must be *odd* for the coefficient to be *nonzero*. The sum over j in (7.48) thus extends over j's spaced by two. This keeps ξ^2 to integer powers of $k-j$, $k=k_1+k_2-1$. One property which is visible in the form (7.52) is that $P^{j_1,j_2;j}_{j_1,j_2}=0$, $P^{j_1,j_2;j}_{-j_1,-j_2}=0$.

The Poisson-Wigner coefficients $P^{j_1,j_2;j}_{m_1,m_2}$ as presented in (7.52) are not yet completely satisfactory since they are not manifestly antisymetric under the exchange $(j_1,m_1)\leftrightarrow(j_2,m_2)$. It would also be interesting to investigate whether these coefficients can be brought to a more compact form. Several known recursion relations for Wigner coefficients fail *by signs* to achieve this end, probably because the algebra at hand is *sp(2,ℜ)*, rather than *so(3)* as usual. The Poisson bracket (7.48) or (7.50) is an operation involving spherical harmonics which does not seem to have been treated in the literature. It is relevant, we saw, for calculations on the compounding of optical aberrations.

4.8 Concluding discussion

We have presented a new method for characterizing optical systems and computing aberrations. It was shown that every optical system gives rise to, and is characterized by, an optical symplectic map, and that symplectic maps can be written as products of Lie transformations. These Lie transformations can be related to paraxial optics and the description of aberrations; symmetric systems can be classified using the group *Sp(2,ℜ)*, and may be applied to the problem of combining aberrations for continuous systems and for discontinuous interfaces.

4.8.1 The code MARYLIE

The discussion in this chapter has been limited mostly to systems symmetric under rotations around the optical axis. It is also possible, however, to treat systems with no particular symmetry, and systems with alignment and positioning errors.

Indeed, a charged particle optics computer code, based on Lie algebraic methods, has been developed at the University of Maryland. This code, called MARYLIE, is intended to be used in the design of accelerators and storage rings. It works through third order with no assumptions about any particular symmetry. That is, it works with symplectic maps $\mathfrak{M}$ characterized by polynomials f_2, f_3, and f_4:

$$\mathfrak{M} = \exp(:f_2:)\exp(:f_3:)\exp(:f_4:). \tag{8.1}$$

Also, a version of **MARYLIE** is under development that will work with maps $\mathfrak{M}$ characterized by polynomials f_1, f_2, f_3, and f_4:

$$\mathfrak{M} = \exp(:f_1:)\exp(:f_2:)\exp(:f_3:)\exp(:f_4:). \tag{8.2}$$

This code will treat systems with *misaligned* and *misplaced* elements [7]. Finally, a Lie algebraic code called **MICROLIE** is under consideration for the design of electron microscopes [31]. It would work with maps charaterized by polynomials f_1 through f_6.

Experience with **MARYLIE** and preliminary versions of **MICROLIE** indicates that Lie algebraic methods are very efficient for ray tracing and for the calculation and control of aberrations. Using the methods of this chapter it should be possible to construct a Lie algebraic code, tentatively called **RAYLIE**, for the treatment of geometrical light optics. It could treat graded index systems through at least fifth-order aberrations, and constant index systems through seventh-order aberrations; if desired, it may be extended to even higher-order aberrations.

Whether or not such a Lie algebraic code would be of use requires further investigation. There are already high-speed computer ray-tracing programs for the case of systems composed of elements with constant refraction indices. Their high speed results from the fact that rays can be traced very rapidly through these elements simply by employing Snell's law at the various interfaces. These codes work by tracing large numbers of rays and searching for lens spacing and parameter values that optimize various features of the ray pattern. However, despite their usefulness, search codes tend to run rather blindly. One of the advantages of the Lie algebraic approach seems to be in providing additional insight concerning the sources of various aberrations, and to what extent they may be best corrected by bending lenses, using special elements, and imposing various symmetries. Finally, all the present codes for treating graded-index systems, trace rays by direct numerical integration. For such systems **RAYLIE** would be orders of magnitude faster, and would have the further advantage of providing explicit information about the optical transfer map.

4.8.2 Aberration ideals

A final comment is in order about an underlying group-theoretical reason for the success of the Lie algebraic method of treating aberrations. As exemplified by equation (3.27), the key ingredient in combining maps is the Baker-Campbell-Hausdorff formula for exponents. This formula says that a Lie operators appearing as exponents can be combined according to a certain calculus, and that the result depends only on single and multiple commutators of these operators. Furthermore, equation (3.9) says that these commutators can be evaluated in terms of Poisson brackets. Finally, equation (3.10) relates the degree of a Poisson braket to the degrees of its constituents.

Suppose we consider all possible polynomials f_n, with $n \geq 2$. Then these polynomials form an infinite dimensional Lie algebra under Poisson brackets, and exponentiation of this Lie algebra gives the infinite-dimensional group of symplectic maps.[36]

---

[36] For the definition of a Lie algebra and the relation between a Lie algebra and a Lie group, see sections 2.4, 3.1, and 5.2 of reference [5].

Next, suppose we want to replace all polynomials f_n, with $n > N$, by zero. For example, if $N = 5$, then any f_6 would be counted as modulo zero, and so would $\{f_4, f_5\}$. If we do this, as briefly discussed at the beginning of Subsect. 7.4, then the result is a *quotient* Lie algebra. This Lie algebra is *finite-dimensional*. Finally, exponentiation of this finite-dimensional Lie algebra produces a finite-dimensional group of symplectic maps.[37]

In the context of optics, the polynomials f_n describe aberrations of order $n-1$. The decision to ignore all polynomials with $n > N$ corresponds to neglecting all aberrations of order greater than $N-1$. What the above discussion has shown is that the neglect of aberrations beyond any fixed order is consistent with the underlying group structure of symplectic maps.

4.8.3 Misaligned and misplaced optical elements

Suppose now that the polynomials $f_1 = ap_1 + bp_2 + cq_1 + dq_2$ are included as well. This is necessary if we wish to have a Lie algebraic treatment of misaligned (in $\mathbf{p}$) and misplaced (in $\mathbf{q}$) elements. At first sight, this would seem to spoil our above considerations because, according to (3.10), Poisson bracketing an f_n with an f_1 produces an f_{n-1}. That is, the order of f_n is *lowered* by elements within the algebra. Thus, for the example above with $N = 5$, Poisson bracketing an f_1 with an f_6, which was supposed to be neglected because any f_6 *should* be neglected, produces an f_5 that should *not* be neglected. The nested subalgebra structure breaks down.

This apparent dilemma in the use of Lie algebraic methods to treat misaligned and misplaced elements can be resolved as follows: suppose we consider quantities f_{mn} of the form

$$f_{mn} = \epsilon^m f_n. \tag{8.3}$$

Here f_n is a homogeneous polynomial of order n as before, and ϵ is a small parameter. Next, we assign to each f_{mn} an integer, called the ***rank*** of f_{mn}, by the rule

$$\operatorname{rank} f_{mn} := m + n. \tag{8.4}$$

Then, it is easily verified that we have the relation

$$\operatorname{rank}\{f_{k\ell}, f_{mn}\} = \operatorname{rank} f_{k\ell} + \operatorname{rank} f_{mn} - 2. \tag{8.5}$$

Comparison of (8.5) with (3.10) shows that the concept of *rank* plays a role similar to that of order.

Now we may consider all f_{mn} with rank greater or equal to two. These quantities also form an infinite dimensional Lie algebra under Poisson brackets. Also, all quantitites with *rank* greater than some number N form an *ideal*. Thus, it is possible to consistently neglect such elements to again produce a finite-dimensional quotient Lie algebra. For example, consider the case $N = 4$. There we have the

[37] This ideal is an invariant Lie subalgebra. The process of discarding or neglecting an ideal is equivalent to forming a *quotient* Lie algebra. Exponentiation of this Lie algebra gives a quotient Lie group, the aberration group [15]. See also [32, Part I, Sect.6].

quantities

$$\text{rank}\,2:\ \epsilon f_1,\ f_2, \tag{8.6a}$$
$$\text{rank}\,3:\ \epsilon^2 f_1,\ \epsilon f_2,\ f_3, \tag{8.6b}$$
$$\text{rank}\,4:\ \epsilon^3 f_1,\ \epsilon^2 f_2,\ \epsilon f_3,\ f_4. \tag{8.6c}$$

These form a finite dimensional Lie algebra under Poisson brackets modulo quantities of rank greater than four.

What does this have to do with optics? Suppose we make the natural assumption that we are interested only in optical systems for which misalignment and misplacement errors are *small*. Then, it is natural to consider the quantitity ϵf_1, rather than f_1, where ϵ is an indication of the smallness of the errors under consideration. Now all Lie algebraic calculations can proceed as before with all entitites having rank greater or equal to two. The only difference is that decisions as to what to neglect and what to retain are made on the basis of rank rather than order. Finally, when the calculation is complete, we recognize that the use of the parameter ϵ was, in fact, merely an ordering technique for making expansions both in aberration order and in powers of the errors. We then set $\epsilon = 1$. When this is done, the burden of smallness is borne by the various polynomials f_1 themselves. The conclusion to be drawn from these considerations is that it is also possible to have a self-consistent Lie algebraic treatment of optical systems with errors, provided these errors are small [7].

Acknowledgements: The first two authors are grateful to the United States Department of Energy (contract AS05-80ER10666) for its support of this work. The second author is also grateful to the National Sciences and Engineering Research Council of Canada, for fellowship support.

References

[1] A.J. Dragt, Lie-algebraic theory of geometrical optics and optical aberrations. *J. Opt. Soc. America* **72**, 372–379 (1982).

[2] K. Halbach, *Am. J. Phys.* **32**, 90 (1964); M. Klein, *Optics* (Wiley, New York, 1970), p. 84; W. Brower, *Matrix Methods in Optical Instrument Design* (Benjamin, New York, 1964).

[3] A. Ghatak and K. Thyagarajan, *Contemporary Optics* (Plenum, New York, 1978), p. 15.

[4] M. Hertzberger, *Trans. Am. Math. Soc.* **53**, 218 (1943); O. Stavroudis, *The Optics of Rays, Wavefronts, and Caustics* (Academic Press, New York, 1978), p. 245.

[5] A.J. Dragt, *Lectures in Nonlinear Orbit Dynamics* (American Institute of Physics, Conference Proceedings, Vol. 87, 1982).

[6] A.J. Dragt and J. Finn, Lie series and invariant functions for analytic symplectic maps, *J. Math. Phys.* **17**, 2215–2227 (1976).

[7] L.M. Healy, University of Maryland, Department of Physics and Astronomy, Ph.D. Thesis.

[8] M. Born and E. Wolf, *Principles of Optics*, 2nd Ed., (Macmillan, New York, 1964).

[9] R.K. Luneberg, *Mathematical Theory of Optics* (University of California Press, 1964).

[10] L.C. Biedenharn and J.D. Louck, *Angular Momentum in Quantum Physics*, Encyclopedia of Mathematics, Vol. 8, Ed. by G.-C. Rota (Addison-Wesley, Reading Mass., 1981).

[11] A.J. Dragt and E. Forest, Computation of nonlinear behavior of hamiltonian systems using Lie algebraic methods, *J. Math. Phys.* **24**, 2734–2744 (1983).

[12] V. Guillemin and S. Sternberg, *Symplectic Techniques in Physics* (Cambridge University Press, 1984).

[13] K.B. Wolf, A group-theoretical model for gaussian optics and third order aberrations, in *Proceedings of the XII International Colloquium on Group-theoretical Methods in Physics*, Trieste, 1983, (Lecture Notes in Physics, Vol. 201, Springer Verlag, 1984), pp. 133–136.

[14] M. Navarro-Saad and K.B. Wolf, Factorization of the phase-space transformation produced by an arbitrary refracting surface. Preprint CINVESTAV, Mexico (March 1984); *J. Opt. Soc. Am.* (in press).

[15] M. Navarro-Saad and K.B. Wolf, The group theoretical treatment of aberrating systems. I. Aligned lens systems in third aberration order. Comunicaciones Técnicas IIMAS, preprint N° 363 (1984); *J. Math. Phys.* (in press).

[16] K.B. Wolf, The group theoretical treatment of aberrating systems. II. Axis-symmetric inhomogeneous systems and fiber optics in third aberration order. Comunicaciones Técnicas IIMAS, preprint N° 366 (1984); *J. Math. Phys.* (in press).

[17] H. Goldstein, *Classical Mechanics*, 2nd ed. (Addison-Wesley, Reading Mass., 1980).

[18] K.B. Wolf, Symmetry in Lie optics. Reporte de investigación, Departamento de Matemáticas, Universidad Autónoma Metropolitana, preprint n° 3, 1985 (submitted for publication).

[19] E. Forest, *Lie algebraic methods for charged particle beams and light optics*, University of Maryland, Department of Physics and Astronomy, Ph. D. Thesis (1984).

[20] M. Navarro-Saad, *Cálculo de aberraciones en sistemas ópticos con teoría de grupos*, Universidad Nacional Autónoma de México, Facultad de Ciencias, B. Sc. Thesis (1985).

[21] M. Navarro-Saad and K.B. Wolf, Applications of a factorization theorem for ninth-order aberration optics, *J. Symbolic Computation* **1**, 235–239 (1985).

[22] L. Seidel, Zur Dioptik, *Astr. Nachr.* № **871**, 105–120 (1853).

[23] H. Buchdahl, *Optical Aberration Coefficients* (Dover, New York, 1968).

[24] H. Buchdahl, *An Introduction to Hamiltonian Optics* (Cambridge University Press, 1970).

[25] H. Buchdahl, *J. Opt. Soc. Am.* **62**, 1314 (1972); *ib. Optik* **37**, 571 (1973); **40**, 460 (1974); **46**, 287, 393 (1976); **48**, 53 (1977).

[26] K.B. Wolf, Approximate canonical transformations and the treatment of aberrations. I. One dimensional simple N^{th} order aberrations in optical systems (preliminary version). Comunicaciones Técnicas IIMAS N° 352 (1983) (unpublished).

[27] D.R. Douglas, University of Maryland, Department of Physics and Astronomy, Ph. D. Thesis (1982).

[28] A. Dragt and D. Douglas, *IEEE Trans. Nucl. Sci.* **NS-30** (1983), p. 2442.

[29] A Dragt, L. Healy *et al.*, MARYLIE 3.0 —A program for nonlinear analysis of accelerator and beamlike lattices. To appear in *IEEE Trans. Nucl. Sci.* (1985).

[30] A. Dragt, R. Ryne, *et al.*, `MARYLIE 3.0` —A program for charged particle beam transport based on Lie algebraic methods. University of Maryland, Department of Physics and Astronomy Technical Report (1985).

[31] A. Dragt and E. Forest, Lie algebraic theory of charged particle optics and electron microscopes. Center for Theoretical Physics, University of Maryland preprint (April 1984). To appear in *Advances in Electronics and Electron Physics*, Vol. 67, P.W. Hawkes, Ed. (Academic Press, New York, 1986).

[32] M. Hausner and J. Schwartz, *Lie Groups, Lie Algebras* (Gordon and Breach, 1968).

Editor's note: This chapter contains the galley proof corrections of only two of the three authors. There was a very definite deadline, both on the part of the publishers and of the owners of the printer leased to IIMAS–UNAM, the latter for February 28, 1986. For this reason, the responsibility for any error must lie with the editor.

Chapter 5

Canonical transforms for paraxial wave optics

by OCTAVIO CASTAÑOS, ENRIQUE LÓPEZ–MORENO, and KURT BERNARDO WOLF

ABSTRACT. Paraxial geometric optics in N dimensions is well known to be described by the inhomogeneous symplectic group $I_{2N} \wedge Sp(2N,\Re)$. This applies to wave optics when we choose a particular (ray) representation of this group, corresponding to a true representation of its central extension and twofold cover $\tilde{\Gamma}_N = W_N \wedge Mp(2N,\Re)$. for wave optics, the representation distinguished by Nature is the *oscillator* one. There applies the theory of canonical integral transforms built in quantum mechanics. We translate the treatament of coherent states and other wave packets to lens and pupil systems. Some remarks are added on various topics, including a fundamental euclidean algebra and group for metaxial optics.

5.1 Introduction and basic results

In writing this chapter, we have had the benefit of reading the contributions of other participants in the León workshop. This allows us to avoid repeating the introductory material, and build upon the definitions of the Heisenberg (-Weyl) group W_N presented in Chapters 1 and 2, and the Lie-theoretical treatment of geometrical paraxial optics through the symplectic group $Sp(2N,\Re)$ in Chapters 3 and 4. Indeed, the peculiarity of W_N is that its group of automorphisms is parametrically larger than its group of *inner* automorphisms. We shall recount this fact and its consequences below. We shall work first in N dimensions and then particularize to $N = 2$ and 1, for actual and axis-symmetric systems.

5.1.1 The Heisenberg–Weyl and the real symplectic groups

The N-dimensional, $(2N+1)$–parameter, nilpotent Heisenberg-Weyl group, W_N, has for basic *presentation* ω the following product rule [*cf.* Section 3 of Chapter 2 in this volume]:

$$\omega(\mathbf{v}_1, \varsigma_1)\,\omega(\mathbf{v}_2, \varsigma_2) = \omega(\mathbf{v}_1 + \mathbf{v}_2, \varsigma_1 + \varsigma_2 + \tfrac{1}{2}\mathbf{v}_1^\top \mathbf{\Omega} \mathbf{v}_2), \tag{1.1}$$

$$\mathbf{v}_i \in \Re^{2N}, \quad \varsigma_i \in \Re, \qquad \mathbf{\Omega} = \begin{pmatrix} \mathbf{0} & \mathbf{1}_N \\ -\mathbf{1}_N & \mathbf{0} \end{pmatrix} = -\mathbf{\Omega}^\top, \tag{1.2a, b}$$

where $^\top$ means transpose and $\mathbf{v}$ is a $2N$–dimensional *column* vector.

The N–dimensional semisimple, real symplectic group *Sp(2N, ℜ)* may be *presented* (see Appendix A) through $2N \times 2N$ real matrices $\mathbf{M}$, such that

$$\mathbf{M}\boldsymbol{\Omega}\mathbf{M}^\top = \boldsymbol{\Omega}, \qquad \det \mathbf{M} = 1, \tag{1.3}$$

multiplying through ordinary matrix product. In $N \times N$ block form, $\mathbf{M} = \begin{pmatrix} \mathbf{A} & \mathbf{B} \\ \mathbf{C} & \mathbf{D} \end{pmatrix}$, there hold:

$$\mathbf{A}\mathbf{D}^\top - \mathbf{B}\mathbf{C}^\top = \mathbf{1}_N, \qquad \mathbf{A}^\top\mathbf{D} - \mathbf{C}^\top\mathbf{B} = \mathbf{1}_N, \tag{1.4a}$$

$$\mathbf{A}\mathbf{B}^\top = \mathbf{B}\mathbf{A}^\top, \quad \mathbf{A}^\top\mathbf{C} = \mathbf{C}^\top\mathbf{A}, \quad \mathbf{B}^\top\mathbf{D} = \mathbf{D}^\top\mathbf{B}, \quad \mathbf{C}\mathbf{D}^\top = \mathbf{D}\mathbf{C}^\top. \tag{1.4b}$$

5.1.2 The Weyl-symplectic group Γ_N

The statement that *Sp(2N, ℜ)* is an automorphism group of W_N is that $\omega_{\mathbf{M}}(\mathbf{v}, \varsigma) := \omega(\mathbf{M}\mathbf{v}, \varsigma)$ has the same composition rule (1.1) as $\omega(\mathbf{v}, \varsigma)$ because of (1.3). We may therefore build the ***Weyl-symplectic*** group Γ_N as the *semidirect product* $\Gamma_N = W_N \wedge Sp(2N, \Re)$, with W_N normal, whose elements $g\{\omega; \mathbf{M}\} = g\{\mathbf{v}, \varsigma; \mathbf{M}\}$ are parametrized by $\omega(\mathbf{v}, \varsigma) \in W_N$, and $\mathbf{M} \in Sp(2N, \Re)$. The W_N subgroup is $g\{\mathbf{v}, \varsigma; \mathbf{1}\}$ and the *Sp(2N, ℜ)* subgroup is $g\{\mathbf{0}, 0; \mathbf{M}\}$. The parameters in this presentation are selected so that $g\{\mathbf{v}, \varsigma; \mathbf{1}\}\, g\{\mathbf{0}, 0; \mathbf{M}\} =: g\{\mathbf{v}, \varsigma; \mathbf{M}\}$. The product rule is

$$g\{\mathbf{v}_1, \varsigma_1; \mathbf{M}_1\}\, g\{\mathbf{v}_2, \varsigma_2; \mathbf{M}_2\} = g\{\mathbf{v}_1 + \mathbf{M}_1^{\top -1}\mathbf{v}_2, \varsigma_1 + \varsigma_2 + \tfrac{1}{2}\mathbf{v}_1^\top \mathbf{M}_1 \boldsymbol{\Omega} \mathbf{v}_2; \mathbf{M}_1\mathbf{M}_2\}. \tag{1.5}$$

We note that $\mathbf{M}^{\top -1} = -\boldsymbol{\Omega}\mathbf{M}\boldsymbol{\Omega}$ for all matrices $\mathbf{M}$ in *Sp(2N, ℜ)*, the identity in Γ_N is $g\{\mathbf{0}, 0; \mathbf{1}\}$ and the inverse is $g\{\mathbf{v}, \varsigma; \mathbf{M}\}^{-1} = g\{-\mathbf{M}^\top\mathbf{v}, -\varsigma; \mathbf{M}^{-1}\}$.

The construction of Γ_N is very similar to that of the euclidean groups from the semidirect product of the d-dimensional translation and rotation groups, $I_d \wedge SO(d)$, except for the fact that in quantum mechanics —and in wave optics— the Heisenberg-Weyl group has a *twist*, *i.e.*, an extra central parameter ς. The $2N$ parameters $\mathbf{v}$ are translations in phase space, but in W_N they do not commute. Instead, translations on a closed curve may leave an *imprint* on the system [1]:

$$g\{\mathbf{v}_2, 0; \mathbf{1}\}^{-1}\, g\{\mathbf{v}_1, 0; \mathbf{1}\}^{-1}\, g\{\mathbf{v}_2, 0; \mathbf{1}\}\, g\{\mathbf{v}_1, 0; \mathbf{1}\} = g\{\mathbf{0}, \mathbf{v}_2^\top \boldsymbol{\Omega} \mathbf{v}_1; \mathbf{1}\}. \tag{1.6}$$

In quantum mechanics, the twist is a *phase*, $\exp(i\hbar\varsigma)$ that multiplies the wavefunction, with a fundamental scale constant $\hbar$, which is the representation label for W_N. If $\hbar$ is allowed to vanish, the result is a description of classical mechanics. There, the group Γ_N may be represented through $(2N+1) \times (2N+1)$ matrices $\mathfrak{g}$ with the following block structure:

$$\mathfrak{g}\{\mathbf{v}, \varsigma; \mathbf{M}\} = \begin{pmatrix} \mathbf{M} & \mathbf{v} \\ \mathbf{0} & 1 \end{pmatrix}^{-1} = \begin{pmatrix} \mathbf{M}^{-1} & -\mathbf{M}^{-1}\mathbf{v} \\ \mathbf{0} & 1 \end{pmatrix}. \tag{1.7}$$

Of course, this is not a *faithful* representation of Γ_N since the central parameter ς is missing in the right-hand side, but it is known that W_N *does not have* finite-dimensional unitary matrix representations. Unitarity is required to preserve norms and orthogonality in quantum wavefunctions and, in wave optics, the total energy of the beam. The classical representation $\mathfrak{g}$ of Γ_N is thus a faithful representation only of the *non-twisted* inhomogeneous symplectic group $I_{2N} \wedge Sp(2N, \Re)$.

5.1.3 The approach we follow

After this introductory presentation of the groups and notation, we shall apply this model to paraxial optics in Section 2. There are, to be sure, many approaches to the subject, some more physical, some more mathematical. Here we follow that of Moshinsky and collaborators to the subject of linear canonical transformations in quantum mechanics [2–5]. In fact, the applicability of *canonical transforms* [5, Part 4] to optics went largely unnoticed by us until we came across the work of Nazarathy and Shamir [6] who, from the point of view of Fourier optics, were investigating the composition of the optical transfer function in the paraxial wave approximation. This is presented in Section 3, in terms of canonical transform integral kernels for Γ_N. The treatment of the W_N transformations is incorporated here for the description of misaligned and/or misplaced optical elements in paraxial systems.

Dimension N really means dimension two or, if the system has cylindrical lenses, dimension one. Axis-symmetric systems also reduce to dimension one, but require different representations of the group. Symmetry reduction is treated in Section 4 with some generality to see the effect of the phase anomaly phenomena for the latter two systems. Section 5 uses the canonical transform machinery to describe the passage of gaussian, coherent and other self-reproducing wavefunctions through paraxial systems. In the concluding section we present some comments on the extension of the canonical transform technique to the Wigner function in optics, the study of aberrations on the basis of the 'wavization' of the results of the previous chapter. Also, we question one of the basic assumptions of this and other work: that quantum mechanical phase space is isomorphic to the optical phase space. This is true only for *paraxial* systems, *i.e.*, in first-order approximation. We suggest that the Heisenberg–Weyl scaffold is only a *contraction* of a *euclidean* structure that is a more exact Lie model for scalar optics.

5.2 The group $\Gamma_N = W_N \wedge Sp(2N, \Re)$ in optics

The hamiltonian formulation of geometrical optics was presented in Section 2 of the preceding chapter [8,9]. A light ray is described by the coordinates of ***position*** $\mathbf{q}$, its intersection with a reference $z = 0$ plane perpendicular to the optical axis of the system, and a ***momentum*** vector $\mathbf{p}$ in the *direction* of the ray, and of magnitude $|\mathbf{p}| = n \sin\theta$, n being the ***refraction index*** at $\mathbf{q}$ and θ the angle between the ray and the optical axis.

5.2.1 The phase space of geometrical optics

Assume the system has N dimensions of position $\mathbf{q}$-space. We build a $(2N+1)$–dimensional column vector with $\mathbf{p}$, $\mathbf{q}$, and 1, which serves as a homogeneous space for the geometrical group action of $\Gamma_N = W_N \wedge Sp(2N, \Re)$, in the representation given by (1.7), *i.e.*, in the block form

$$\mathfrak{g}\left\{\begin{pmatrix}\mathbf{x}\\ \mathbf{y}\end{pmatrix}, \varsigma; \begin{pmatrix}\mathbf{A} & \mathbf{B}\\ \mathbf{C} & \mathbf{D}\end{pmatrix}\right\}\begin{pmatrix}\mathbf{p}\\ \mathbf{q}\\ 1\end{pmatrix} = \begin{pmatrix}\begin{pmatrix}\mathbf{A} & \mathbf{B}\\ \mathbf{C} & \mathbf{D}\end{pmatrix} & \begin{pmatrix}\mathbf{x}\\ \mathbf{y}\end{pmatrix}\\ \mathbf{0} \quad \mathbf{0} & 1\end{pmatrix}^{-1}\begin{pmatrix}\mathbf{p}\\ \mathbf{q}\\ 1\end{pmatrix} = \begin{pmatrix}\begin{pmatrix}\mathbf{D}^\top & -\mathbf{B}^\top\\ -\mathbf{C}^\top & \mathbf{A}^\top\end{pmatrix}\left[\begin{pmatrix}\mathbf{p}\\ \mathbf{q}\end{pmatrix} - \begin{pmatrix}\mathbf{x}\\ \mathbf{y}\end{pmatrix}\right]\\ 1\end{pmatrix}. \qquad (2.1)$$

This representation is actually *inconsistent* with the nature of optics, because the latter requires $|\mathbf{p}| \leq n(\mathbf{q})$, while in principle $\mathbf{q} \in \Re^N$, and the former allows arbitrarily large values in $\mathbf{B}$ and $\mathbf{a}$. These may carry $\mathbf{p}$ outside its natural range. In paraxial optics we choose to *ignore* this bound and assume that $\mathbf{p}$ and $\mathbf{q}$ are small and follow (2.1) on grounds that the *linear* transformation of phase space is canonical and a good enough first-order approximation to geometrical optics.

5.2.2 The canonical transform representation of Γ_N

The formalization of paraxial wave optics we follow is, quite simply, that of quantum mechanics à la Schrödinger–Dirac [10], namely that *(i)* there exists an $\mathcal{L}^2(\Re^N)$ space of functions —***wavefunctions***— subject to the evolution produced by the elements of paraxial optical systems. The latter, in turn, are to be *(ii)* unitary transformations on $\mathcal{L}^2(\Re^N)$ belonging to the group Γ_N.

If we believe in this picture, we can use known results (see [4] and references therein) to state that the ***wave-optical*** representation $\lambda\!\!\!^-$ of Γ_N on $\mathcal{L}^2(\Re^N)$ is given, for the Heisenberg-Weyl part, by

$$\mathfrak{G}\left\{\begin{pmatrix}\mathbf{x}\\ \mathbf{y}\end{pmatrix}, \varsigma; \mathbf{1}\right\} f(\mathbf{q}) = \exp i\left(\textstyle\sum_j x_j \mathbb{P}_j + \sum_j y_j \mathbb{Q}_j + z\bar{\lambda}\mathbf{1}\right) f(\mathbf{q}) \\ = \exp(i\mathbf{y}^\top \mathbf{q}) \exp\left(i\bar{\lambda}[\varsigma + \tfrac{1}{2}\mathbf{y}^\top\mathbf{x}]\right) f(\mathbf{q} + \bar{\lambda}\mathbf{x}); \tag{2.2}$$

for the *Sp(2N, ℜ)* part we have an *integral transform* action [5,6,11]

$$\left[\mathfrak{G}\left\{\mathbf{0}, 0; \begin{pmatrix}\mathbf{A} & \mathbf{B}\\ \mathbf{C} & \mathbf{D}\end{pmatrix}\right\} f\right](\mathbf{q}) = \int_{\Re^N} d^N\mathbf{q}'\, D^o_{\mathbf{q},\mathbf{q}'}\begin{pmatrix}\mathbf{A} & \mathbf{B}\\ \mathbf{C} & \mathbf{D}\end{pmatrix} f(\mathbf{q}'), \tag{2.3a}$$

with the *oscillator* kernel[1]

$$D^o_{\mathbf{q},\mathbf{q}'}\begin{pmatrix}\mathbf{A} & \mathbf{B}\\ \mathbf{C} & \mathbf{D}\end{pmatrix} = \frac{\varphi_N}{\sqrt{(2\pi\bar{\lambda})^N|\det \mathbf{C}|}} \exp\left(\frac{-i}{\bar{\lambda}}[\tfrac{1}{2}\mathbf{q}^\top\mathbf{A}\mathbf{C}^{-1}\mathbf{q} - \mathbf{q}^\top\mathbf{C}^{-1}\mathbf{q}' + \tfrac{1}{2}\mathbf{q}'^\top\mathbf{C}^{-1}\mathbf{D}\mathbf{q}']\right), \tag{2.3b}$$

$$\varphi_N = \exp(-i\pi\tfrac{1}{4}N\textstyle\sum_i \operatorname{sgn}\gamma_i), \qquad \{\gamma_i\}_{i=1}^N \text{ eigenvalues of } \mathbf{C}. \tag{2.3c}$$

5.2.3 Units, singularities, phases, and covers

We have several comments on equations (2.3). First, the *representation* label $\bar{\lambda}$ of Γ_N plays the role of the reduced Planck constant $\hbar$ (see, *e.g.* [4]); in optics it is the ***reduced wavelength*** of the light, *i.e.*, $\bar{\lambda} = \lambda/2\pi$. Now, $\bar{\lambda}$ has units of *length*, as $\mathbf{q}$ does, while $\mathbf{p}$ is dimensionless. The dimensionalities of the various Γ_N parameters are hence determined as follows: ς and $\mathbf{y}$ have units $\bar{\lambda}^{-1}$; $\mathbf{x}$ has no units; $\mathbf{C}$ has units $\bar{\lambda}$, $\mathbf{B}$ has units $\bar{\lambda}^{-1}$, $\mathbf{A}$ and $\mathbf{D}$ are dimensionless.

Second, in the Heisenberg-Weyl group action, the parameter ς enters only through the *phase* $\exp(i\bar{\lambda}\varsigma)$, which means that we have the *effective* group action of $W_N/\mathfrak{Z}$ (*i.e.*, W_N modulo the integers in the last group parameter), since the group elements $\omega(\mathbf{x}, \mathbf{y}, \varsigma)$ and $\omega(\mathbf{x}, \mathbf{y}, \varsigma + 2\pi k/\bar{\lambda})$, $k \in \mathfrak{Z}$ produce the same transformation on the homogeneous space $\mathcal{L}^2(\Re^N)$.

Third, when $\mathbf{C}^{-1}$ exists, the kernel (2.3*b*) is a well behaved function of $\mathbf{q}$ and $\mathbf{q}'$. When $\det\mathbf{C} \to 0$, the kernel becomes *singular* on $\mathbf{q}' = \mathbf{A}^\top\mathbf{q}$ [see (2.1)], and the *integral transform* action collapses to a *Lie* action [3,5,11],

$$\mathfrak{G}\left\{\mathbf{0}, 0; \begin{pmatrix}\mathbf{A} & \mathbf{B}\\ \mathbf{0} & \mathbf{A}^{\top -1}\end{pmatrix}\right\} f(\mathbf{q}) = \sqrt{\det \mathbf{A}}\, \exp(i\mathbf{q}^\top\mathbf{B}\mathbf{A}^\top\mathbf{q}/2\bar{\lambda})\, f(\mathbf{A}^\top\mathbf{q}), \tag{2.4}$$

[1]Please note carefully that in this chapter we are taking two-vectors $\begin{pmatrix}\mathbf{p}\\ \mathbf{q}\end{pmatrix}$. In former work, especially in reference [5], the convention was $\begin{pmatrix}\mathbf{q}\\ \mathbf{p}\end{pmatrix}$, as it is for most authors. The last section of the previous chapter in this volume, shows that the present convention is consistent with having column-vector multiplets where the highest-weight state appears on top. Also, the canonical transform kernel $C_{\mathbf{M}}(\mathbf{q}, \mathbf{q}')$ in [4] is here indicated by $D^o_{\mathbf{q},\mathbf{q}'}(\mathbf{M})$, as in group representation theory.

for the subspaces of $\mathbf{q}$ where the eigenvalues γ_i of $\mathbf{C}$ have become zero.

Fourth —and perhaps most important— is *why* we choose, amongst all others, the *oscillator* representation[2] of $Sp(2N,\Re)$. Basically, it is for the same reason that we need the Schrödinger representation of quantum mechanics. In paraxial optics, this has been described in the first chapter of this volume.

The Schrödinger realization [10] chooses the well-known operators $\mathbb{Q}_j$ and $\mathbb{P}_k$ to be represented by q_j and $-i\lambda\partial/\partial q_k$, $k = 1, 2, \ldots, N$, satisfying $[\mathbb{Q}_j, \mathbb{P}_k] = i\lambda\delta_{j,k}$. It is thus the choice for $\{\mathbb{Q}_j\}_{j=1}^N$ to be *diagonal*. If we want to uphold this in the larger $\Gamma_N \supset W_N$, we must choose the diagonal operators in $sp(2N,\Re)$ such that they *commute* with the $\mathbb{Q}_j$. The subgroup of $Sp(2N,\Re)$ satisfying this property is the abelian group of upper-triangular matrices $\begin{pmatrix}\mathbf{1} & \mathbf{B}\\ \mathbf{0} & \mathbf{1}\end{pmatrix}$, with $\mathbf{B}$ symmetric. This is a *parabolic* subgroup of $Sp(2N,\Re)$.

The reduction of the representations of $Sp(2N,\Re)$ with respect to noncompact subgroups yields integral transforms where the integral runs over the spectra of the commuting diagonal generators. The operators in question are here $\mathbb{Q}_j$, and the subgroup generators are $\mathbb{Q}_i\mathbb{Q}_j$; the spectra of the former run thus over $\Re^N$. The generators of this group [12] are *second*-order differential operators $\mathbb{P}_i\mathbb{P}_j$ (producing *integral transforms*), first order $\mathbb{Q}_i\mathbb{P}_j + \mathbb{P}_j\mathbb{Q}_i$ (producing *geometric*, or *Lie* transformations), and zeroth order $\mathbb{Q}_i\mathbb{Q}_j$ (bringing in *Bargmann multipliers*), realized as in the Schrödinger representation. This, then, generates the *oscillator* representation.

It is also well known [2,5,11] that the *oscillator representation* is actually a 2:1 representation[3] of $Sp(2N,\Re)$, and a faithful representation only of its two-fold *cover*, the ***metaplectic*** group[4] $Mp(2N,\Re)$. We call $\tilde{\Gamma}_N := W_N \wedge Mp(2N,\Re)$ the corresponding two-fold cover of Γ_N. In other words, if $\mathbf{M}_1$ and $\mathbf{M}_2$ are two $Sp(2N,\Re)$ matrices and $\mathbf{M}_3 = \mathbf{M}_1\mathbf{M}_2$, then (omitting the W_N parameters),

$$\mathfrak{G}\{\mathbf{M}_1\}\,\mathfrak{G}\{\mathbf{M}_2\} = \sigma(\mathbf{M}_1,\mathbf{M}_2;\mathbf{M}_3)\,\mathfrak{G}\{\mathbf{M}_3\}, \qquad \mathbf{M}_i = \begin{pmatrix}\mathbf{A}_i & \mathbf{B}_i\\ \mathbf{C}_i & \mathbf{D}_i\end{pmatrix}, \tag{2.5a}$$

where σ is a *sign*, $+$ or $-$, expressible in terms of the eigenvalues $\gamma_{1,i}$, $\gamma_{2,i}$, and $\gamma_{3,i}$ of $\mathbf{C}_1$, $\mathbf{C}_2$, $\mathbf{C}_3$, and $\gamma_{4,i}$ of $\mathbf{C}_2^{-1}\mathbf{C}_3\mathbf{C}_1^{-1}$, thus:

$$\sigma(\mathbf{M}_1,\mathbf{M}_2;\mathbf{M}_3) = \exp[-i\tfrac{1}{4}\pi(\textstyle\sum_i \operatorname{sgn}\gamma_{1,i} + \sum_i \operatorname{sgn}\gamma_{2,i} - \sum_i \operatorname{sgn}\gamma_{3,i} + \sum_i \operatorname{sgn}\gamma_{4,i})]. \tag{2.5b}$$

In $\tilde{\Gamma}_N$, if the description of the elements is done in terms of the appropriate Bargmann parameters[4] the sign problem simply dissappears. It is known, however, that $Mp(2N,\Re)$ has *no* finite matrix representation; the inconvenience of the sign is happily borne for the facility of matrix multiplication in $Sp(2N,\Re)$, *vs.* the composition formulae[4] for $Mp(2N,\Re)$ parameters.

This ends our comments on the adoption of the oscillator representation (2.3) of Γ_N to describe wave optics. Let us now close this section displaying explicitly some of the Γ_N transformations produced by lens systems.

5.2.4 Optical elements and associated operators

[2]The representations of the algebra $sp(2,\Re)$ are described in Appendix B; these lead to representations of the *cover* of the group, $\overline{Sp(2,\Re)}$. The latter is the simply-connected universal cover of $Sp(2,\Re)$. The connectivity properties of $Sp(2N,\Re)$ are given in Appendix A,, together with the definition of the *metaplectic* group $Mp(2,\Re)$, relevant for odd-dimensional optics.

[3]This is reducible to 2^N *irreducible* representations of the type $D^+_{3/4}$ and $D^+_{1/4}$. See Appendix B.

[4]See Appendix A, equation (A.12).

5.2.4.1 Free propagation

It is well known that[5] in a homogeneous medium n = constant of length z, the evolution operator is given by

$$\mathfrak{F}_z := \exp\left[\frac{iz}{\lambda}\left(-n + \frac{1}{2n}\sum_j \mathbb{P}_j\mathbb{P}_j\right)\right] = \mathfrak{G}\left\{\mathbf{0}, -\frac{zn}{\lambda^2}; \begin{pmatrix} \mathbf{1} & \mathbf{0} \\ -z/n\,\mathbf{1} & \mathbf{1} \end{pmatrix}\right\}. \tag{2.6a}$$

The corresponding integral kernel (2.3) is

$$F_z(\mathbf{q}, \mathbf{q}') = \frac{e^{-i(nz/\lambda + \pi N/4)}}{\sqrt{(2\pi\lambda z/n)^N}} \exp\left[-\frac{in}{2\lambda z}(\mathbf{q} - \mathbf{q}')^2\right]. \tag{2.6b}$$

5.2.4.2 Quadratic refracting surface

Consider two homogeneous media with refraction indices n and n', separated by a quadratic surface

$$S(\mathbf{q}) = \mathbf{b}^\top\mathbf{q} + \mathbf{q}^\top\mathbf{B}\mathbf{q} \tag{2.7a}$$

(with *prismicity* $\mathbf{b}$ and gaussian *power* $\mathbf{B}$). This produces the Γ_N transformation[5]

$$\mathfrak{S}_{S,n-n'} := \exp\left[-\frac{i(n-n')}{\lambda}\left(\sum_j b_j\mathbb{Q}_j + \sum_{j,k}\mathbb{Q}_j B_{jk}\mathbb{Q}_k\right)\right] = \mathfrak{G}\left\{\begin{pmatrix} \mathbf{0} \\ -\frac{n-n'}{\lambda}\mathbf{b} \end{pmatrix}, 0; \begin{pmatrix} \mathbf{1} & -z(n-n')\mathbf{B} \\ \mathbf{0} & \mathbf{1} \end{pmatrix}\right\}. \tag{2.7b}$$

This is clearly analyzable into the product of two root transformations [15], one in n and the inverse one in n'.

We have shown [16] that the set generated by all products of $\mathfrak{F}_z$'s and $\mathfrak{S}_S$'s is *dense* in Γ_N.

We should note carefully that while free propagation (2.6) is the exact analogue of Schrödinger free flight in quantum mechanics, there is no real quantum analogue of a lens, except through harmonic oscillator 'jolts' (whose existence is purely theoretical). Of course, for charged particles, magnetic dipoles are used as lenses and replaced by (2.7), but these are finite approximations to a necessarily z–dependent Hamiltonian.

Finally, the time-independent harmonic oscillator has for exact analogue a quadratic index profile, z–independent optical fiber. This is the source of the success of coherent states [17] to describe light in these systems. From the point of view of the Γ_N representations, paraxial optics and quadratic-potential quantum mechanics are equivalent.

Inhomogeneous paraxial optical systems may be treated using the results of canonical transforms for time-dependent quadratic-potential quantum systems [16–19]. In this chapter, we prefer to treat optical systems as composed of *finite* elements: quadratic refracting surfaces between homogeneous media.

[5]See the preceding chapter, Section 5.

5.3 Concatenation to optical systems

5.3.1 Concatenation order

We want to associate to each element of an optical system, one *group* element such that, laid out from *left to right* (as is usual in optical diagrams), their product will yield the group element describing the system. The prototypical geometric optical system shown below:

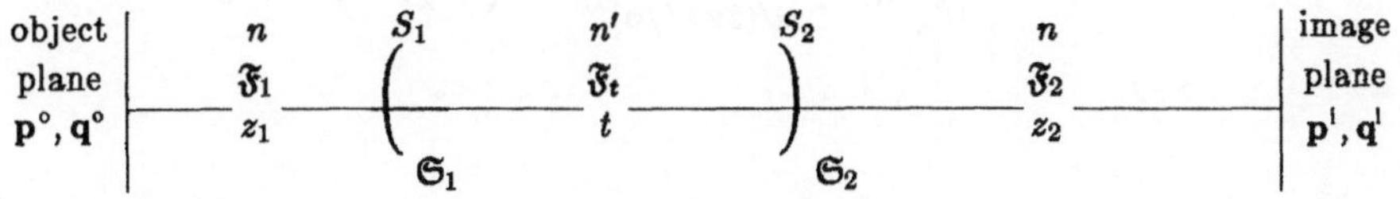

Here, $\mathfrak{F}_1$ and $\mathfrak{F}_2$ are the operators of paraxial free propagation in a homogeneous medium n after the object and before the screen, by distances z_1 and z_2; $\mathfrak{S}_1$ and $\mathfrak{S}_2$ carry the action of the two refraction surfaces, S_1 and S_2, enclosing a medium n' of thickness t, where free propagation $\mathfrak{F}_t$ applies. These operators are $\tilde{\Gamma}_N$ elements: $\mathfrak{F}_i = \mathfrak{G}\{\mathbf{0}, -kz, \mathbf{F}_i\}$ and $\mathfrak{S}_i = \mathfrak{G}\{\mathbf{0}, 0, \mathbf{S}_i\}$, where $\mathbf{F}_i$ and $\mathbf{S}_i$ are the free-propagation and lens matrices in (2.6–2.7).

The concatenation of these elements produces the Lie map [8] of the system $\mathfrak{M} = \mathfrak{F}_1\, \mathfrak{S}_1\, \mathfrak{F}_t\, \mathfrak{S}_2\, \mathfrak{F}_2$ that acts on the *object* phase space $(\mathbf{p}^\circ, \mathbf{q}^\circ)$ in geometrical optics, to produce the corresponding *image* space $(\mathbf{p}^{\mathrm{i}}, \mathbf{q}^{\mathrm{i}})$ through its representation matrix or integral kernel. These operators, we recall, are Lie (exponential) operators [9]; they hence act on linear combinations and the *arguments* of functions. [*cf.* Eq. (3.11) of the previous chapter.] For this reason in geometrical optics the matrix *inverse* to $\mathbf{M}$ appears in (1.7)–(2.1) when $\mathfrak{g}\{\mathbf{M}\}$ is acting on the $2N$-row column vector $\mathbf{w} = (\mathbf{p}, \mathbf{q})$. Thus, to the *geometrical* system of the figure there corresponds the Lie operators $\mathfrak{f}_i = \mathfrak{g}\{\mathbf{F}_1\}$, $\mathfrak{s}_i = \mathfrak{g}\{\mathbf{S}_i\}$, and their composition is $\mathfrak{s}_2\mathfrak{f}_2\mathbf{w} = \mathfrak{s}_2\mathbf{F}_2^{-1}\mathbf{w} = \mathbf{F}_2^{-1}\mathfrak{s}_2\mathbf{w} = \mathbf{F}_2^{-1}\mathbf{S}_2^{-1}\mathbf{w} = (\mathbf{S}_2\mathbf{F}_2)^{-1}\mathbf{w}$, and hence $\mathfrak{m}\mathbf{w} = \mathbf{M}^{-1}\mathbf{w}$, with $\mathbf{M} = \mathbf{F}_1\mathbf{S}_1\mathbf{F}_t\mathbf{S}_2\mathbf{F}_2$.

Similarly for the integral transforms (2.2–3) we have the composition of elements $\mathfrak{G}\{\mathbf{v}, \varsigma, \mathbf{M}\}$, specified in (1.5), with the two-factor sign (2.5); therefore $\mathfrak{M} = \mathfrak{F}_1\mathfrak{S}_1\mathfrak{F}_t\mathfrak{S}_2\mathfrak{F}_2$ too, as for their geometrical lower-case counterparts.

Let us now concatenate the transformations $\mathfrak{F}_z$ and $\mathfrak{S}_S$ in (2.6–7) to some useful optical devices.

5.3.2 Optical Fourier analyzers

An **optical Fourier analyzer** [20] is built placing an aligned ($\mathbf{b} = 0$) axis-symmetric lens, of gaussian power $\mathbf{B} = B\mathbf{1}$, between two homogeneous spaces of length z,

$$\mathfrak{G}\left\{\mathbf{0}, -\frac{zn}{\bar{\lambda}^2}, \begin{pmatrix} \mathbf{1} & \mathbf{0} \\ -z/n\,\mathbf{1} & \mathbf{1} \end{pmatrix}\right\} \mathfrak{G}\left\{\mathbf{0}, 0, \begin{pmatrix} \mathbf{1} & B\mathbf{1} \\ \mathbf{0} & \mathbf{1} \end{pmatrix}\right\} \mathfrak{G}\left\{\mathbf{0}, -\frac{zn}{\bar{\lambda}^2}, \begin{pmatrix} \mathbf{1} & \mathbf{0} \\ -z/n\,\mathbf{1} & \mathbf{1} \end{pmatrix}\right\}$$

$$= \mathfrak{G}\left\{\mathbf{0}, -\frac{2zn}{\bar{\lambda}^2}, \begin{pmatrix} (1 - Bz/n)\,\mathbf{1} & B\,\mathbf{1} \\ (z^2B/n^2 - 2z/n)\,\mathbf{1} & (1 - Bz/n)\,\mathbf{1} \end{pmatrix}\right\} \quad (3.1)$$

We may let B grow from zero (a plane) to $(n/f)\sin\beta$, $0 \leq \beta \leq \frac{1}{2}\pi$, where f is the *focal length* at $\beta = \frac{1}{2}\pi$. Then, if we choose the distances such that $z = (f/n)\tan\frac{1}{2}\beta$, $0 \leq \beta \leq \frac{1}{2}\pi$, we shall have the Γ_N transformation

$$\mathfrak{A}_\beta = \mathfrak{G}\left\{\mathbf{0}, -\frac{2zn}{\lambda^2}, \begin{pmatrix} \cos\beta\mathbf{1} & \frac{n}{f}\sin\beta\mathbf{1} \\ -\frac{f}{n}\sin\beta\mathbf{1} & \cos\beta\mathbf{1} \end{pmatrix}\right\}. \tag{3.2}$$

For $\beta = \frac{1}{2}\pi$ we have the *inverse Fourier transform* (up to a phase [5]), $\mathfrak{A}_{\pi/2}$, whose Γ_N integral kernel is the free-flight phase $\exp(-2if/\lambda)$ times

$$D^o_{\mathbf{q},\mathbf{q}'}\begin{pmatrix} 0 & 1 \\ -1 & 0 \end{pmatrix} = \frac{e^{i\pi N/4}}{\sqrt{(2\pi\lambda f/n)^N}} \exp\left(\frac{in\mathbf{q}\cdot\mathbf{q}'}{\lambda f}\right). \tag{3.3}$$

5.3.3 The phase anomaly

The optical system $\mathfrak{A}_\beta$ displays the following ***phase anomaly*** which we proceed to explain. Equation (3.3) is the inverse Fourier transform kernel [5], *except for a phase:*

$$\mathfrak{A}_{\pi/2} = e^{i\pi N/4}\mathfrak{F}^{-1}. \tag{3.4}$$

If we stack *two* Fourier analyzers together, we obtain $\mathfrak{A}_{\pi/2}\mathfrak{A}_{\pi/2} = \mathfrak{A}_\pi = e^{i\pi N/2}\mathfrak{I}_0$, where $\mathfrak{I}_0 = \mathfrak{F}^{-2} = \mathfrak{F}^2$ is the ***inversion*** operator through the origin: $(\mathfrak{I}_0 f)(\mathbf{q}) = f(-\mathbf{q})$. If we now stack *four* analyzers we obtain $\mathfrak{A}^4_{\pi/2} = \mathfrak{A}_{2\pi} = e^{i\pi N}\mathfrak{F}^4 = e^{i\pi N}\mathfrak{I}_0^2 = e^{i\pi N}$.

In geometrical optics, or in wave optics of an *even* number N of dimensions (*actual* optics is $N = 2$), $\mathfrak{A}_{2\pi}$ is the identity operator. But in wave optics in an *odd* number N of dimensions (such $N = 1$ in actual cilyndrical lenses), there is a *sign reversal* of the object wavefunction *vs.* the image wavefunction.

This phase anomaly in cylindrical Fourier analyzers is equivalent to that of plane waveguides, where the evolution operator is (3.2). Both are clear indicators of how the geometrical $Sp(2N,\Re)$ group becomes doubly covered [11] to $Mp(2N,\Re)$ in wave optics. The situation reminds us of the double cover which the spin group $SU(2)$ affords over the orbital angular momentum group $SO(3)$. The differences, however, are also striking: $Mp(2N,\Re)$ is *still* infinitely connected, and the phase anomaly vanishes in even dimensions. Yet the mathematics are close enough to allow, in the preceding chapter, the computation of symplectic polynomials through spherical harmonics for *geometric* optics. We shall comment on the situation for wave optics in the concluding section.

5.3.4 Axis-symmetric systems

The Fourier analyzer seen above is a special type of system: it is *axis*-symmetric, *i.e.*, invariant under joint $SO(N)$ rotations of $\mathbf{q}$- and $\mathbf{p}$-space, each around the optical axis of the system $\mathbf{q} = \mathbf{0}$, $\mathbf{p} = \mathbf{0}$. Free propagation (2.6) is axis-symmetric too. If we ask that the axis-symmetric lenses be aligned and well-placed [*i.e.*, $\mathbf{b} = \mathbf{0}$ in (2.7)] then the whole system has axial symmetry. This we shall use in the next section for the reduction of $\Gamma_N = W_N \wedge Sp(2N,\Re)$ to $Sp(2,\Re)$.

A ***symmetry*** of an optical system $\mathfrak{M}$ is a subgroup of the group $\tilde{\Gamma}_n$ that commutes with $\mathfrak{M}$. Lens systems of optical interest are usually axis-symmetric. (Symmetry under *boost* transformations in the q_1–q_2 plane may be of relevance for magnetic quadrupoles.) The object wavefunctions, on the other hand, are not assumed to have any particular symmetry, but may be subject to series expansion in a basis of functions which *have* such symmetry (*i.e.*, Fourier series in the azimuthal angle).

The generators of the algebra *sp(2N, ℜ)* are, as we said above, all quadratic terms in the W_N covering algebra, namely $\mathbb{Q}_j\mathbb{Q}_k$, $\mathbb{Q}_j\mathbb{P}_k$, and $\mathbb{P}_j\mathbb{P}_k$, $j,k=1,2,\ldots,N$. The rotations around the optical axis are generated by the operators

$$\mathsf{M}_{j,k} := \frac{1}{\lambda}(\mathbb{Q}_j\mathbb{P}_k - \mathbb{Q}_k\mathbb{P}_j), \qquad j,k = 1,2,\ldots N. \tag{3.5}$$

They close into an *so(N)* algebra. Their commutant in Γ_N has a vector basis

$$\mathbb{K}_+ := \frac{1}{2\lambda}\sum_j \mathbb{P}_j\mathbb{P}_j =: \frac{1}{2\lambda}\mathbb{P}^2, \tag{3.6a}$$

$$\mathbb{K}_0 := \frac{1}{4\lambda}\sum_j(\mathbb{P}_j\mathbb{Q}_j - \mathbb{Q}_j\mathbb{P}_j) =: \frac{1}{4\lambda}\{\mathbb{P},\mathbb{Q}\}_+, \tag{3.6b}$$

$$\mathbb{K}_- := \frac{1}{2\lambda}\sum_j \mathbb{Q}_j\mathbb{Q}_j =: \frac{1}{2\lambda}\mathbb{Q}^2. \tag{3.6c}$$

They may be seen to close into an *sp(2, ℜ)* algebra:

$$[\mathbb{K}_0,\mathbb{K}_\pm] = \pm i\mathbb{K}_\pm, \qquad [\mathbb{K}_+,\mathbb{K}_-] = -2i\mathbb{K}_0 \tag{3.7a, b}$$

[*cf.* preceding chapter, equation (3.17) replacing K_σ by $-i\mathbb{K}_\sigma$, and Appendix B.]

It is easy to see that the set of operators (3.5) commutes with the three operators (3.6). We have thus the *reduction* of the oscillator representation of *sp(2N, ℜ)* by its subgroup, which is [21,22] the direct sum *sp(2, ℜ)* ⊕ *so(N)*.

5.3.5 Complementarity in *Sp(2N,ℜ)*

In the oscillator representation of *sp(2N, ℜ)* [23], the subalgebra representations of *sp(2, ℜ)* and *so(N)* are not independent, but ***complementary***, *i.e.*, their Casimir operators

$$\mathbb{C}^{sp(2)} := (\mathbb{K}_0)^2 - \tfrac{1}{2}\{\mathbb{K}_+,\mathbb{K}_-\}_+, \tag{3.8a}$$

$$\mathbb{C}^{so(N)} := \tfrac{1}{2}\sum_{j,k}(\mathsf{M}_{j,k})^2, \tag{3.8b}$$

are related by

$$\mathbb{C}^{sp(2)} = -\tfrac{1}{4}\mathbb{C}^{so(N)} + \tfrac{1}{16}N(4-N). \tag{3.8c}$$

The eigenvalues of $\mathbb{C}^{so(N)}$ on the $(N-1)$–sphere are $c^{so} = \ell(\ell+N-2)$, $\ell = 0,1,2,\ldots$, and thus those of *sp(2, ℜ)* are $c^{sp} = k(1-k)$ for $k = \frac{1}{2}(\ell + \frac{1}{2}N)$, $\ell = 0,1,2,\ldots$. Here k is the *Bargmann index* for *sp(2, ℜ)* = *so(2,1)* [24]; these values fall on the lower-bound *discrete* representation series, D_k^+ [25].

We are expecially interested in the $N=2$ case of axis-symmetric optics, there, *so(2)* has a single generator $\mathsf{M}_{12} := \mathbb{Q}_1\mathbb{P}_2 - \mathbb{Q}_2\mathbb{P}_1$, and (3.8*c*) becomes

$$\mathbb{C}^{sp(2)} = -\tfrac{1}{4}(\mathsf{M}_{12})^2 + \tfrac{1}{4} = \tfrac{1}{4}[1 - \lambda^{-2}(\mathbb{Q}_1\mathbb{P}_2 - \mathbb{Q}_2\mathbb{P}_1)^2]. \tag{3.9}$$

In this case, thus, M_{12} itself may be taken diagonal as well as the Casimir operator $\mathbb{C}^{so(2)} = (\mathsf{M}_{12})^2$. The behaviour under *parity*, $\mathbb{I}_0 f(\mathbf{q}) = f(-\mathbf{q})$, may be used to distinguish between the doubly degenerate eigenvalues $c^{so(2)} = (\pm m)^2$, corresponding to integer eigenvalue $m \in \mathfrak{Z}$, under the generator of rotations around the optical axis, M_{12}. We should note that, classically, $(\mathbf{p}\times\mathbf{q})^2 = (p_1q_2 - p_2q_1)^2$ is the ***Petzval***, or ***skewness*** invariant [7] of the ray. In wave optics, thus, we may use the discrete eigenvalues of this operator M_{12} as a *row* label to reduce the $\mathcal{L}^2(\Re^2)$ representation space to a sum of *sp(2,* $\Re$*)*-irreducible $\mathcal{L}^2(\Re^+)$ spaces. For *so(2)*-eigenvalue m, the *sp(2,* $\Re$*)* Bargmann index will be

$$k = \tfrac{1}{2}(1 + |m|), \qquad m \in \mathfrak{Z}, \tag{3.10}$$

corresponding to *single*-valued representations of *sp(2,* $\Re$*)*.

The same analysis holds for $N = 1$ dimension, where only the parity operator I_0 commutes with the generators of the system, in place of the *so(N)* Casimir operator. The result is that the only *sp(2,* $\Re$*)* representations present are $k = \frac{1}{4}$ (+ parity) and $k = 3/4$ (− parity). These correspond to the same eigenvalue, $\frac{3}{16}$, of the Casimir operator and constitute the oscillator representation [25] on $\mathcal{L}^2(\Re)$. Axis-symmetric systems in one dimension therefore *respect* (*i.e.*, commute with) the parity of the object wavefunction.

5.3.6 Partial wave expansions

The conserved quantity under axially-symmetric two-dimensional gaussian systems is the expectation value of the operator M_{12} in (3.5), classicaly $\mathbf{p}\times\mathbf{q}$. The *sp(4,* $\Re$*)* generators which remain in a paraxial axially-symmetric system are hence K_σ, $\sigma = 0, \pm$ in (3.6). These generate matrices (1.3–4) which have diagonal block-matrices: $\mathbf{B} = B\mathbf{1}$, $\mathbf{C} = C\mathbf{1}$, $\mathbf{A} = A\mathbf{1}$, and $\mathbf{D} = D\mathbf{1}$. The phase parameter ς may be kept since it also commutes with *so(2)*. The *Sp(2,* $\Re$*)* group composition (2.5) between these elements holds, with $\sigma = +1$; this is so because all real diagonal 2×2 submatrices have positive determinants; the relevant group is no longer *Mp(2,* $\Re$*)*, but *Sp(2,* $\Re$*)*.

The *Sp(2,* $\Re$*)* group acts on $\mathcal{L}^2(\Re^2)$ object phase functions through the kernel

$$D_{\mathbf{q},\mathbf{q}'}(\mathbf{M}) = -\frac{i}{2\pi\lambda C}\exp\left(\frac{i}{2\lambda C}(A\mathbf{q}^2 - 2\mathbf{q}\cdot\mathbf{q}' + D\mathbf{q}'^2)\right). \tag{3.11}$$

When we parametrize in polar coordinates,

$$q_1 = q\cos\theta, \quad q_2 = q\sin\theta, \qquad q \in \Re^+, \quad \theta \in \text{circle}, \tag{3.12}$$

we may use the Fourier series expansion [21] in eigenfunctions of M_{12},

$$f(\mathbf{q}(q,\theta)) = \frac{1}{\sqrt{2\pi}}\sum_{m=-\infty}^{\infty} f_m(q)e^{im\theta}, \tag{3.13a}$$

$$f_m(q) = \frac{1}{\sqrt{2\pi}}\int_{-\pi}^{\pi} f(\mathbf{q}(q,\theta))e^{-im\theta}, \tag{3.13b}$$

reduces the $\Re^2$–integral transform action (2.3)–(3.11) to an $\Re^+$ integral transform of the Hankel type.

For each pair of Fourier components $\pm m$ of the object function, the radial function undergoes a ***k–radial*** canonical transform [21,22,25],

$$\left[\mathfrak{C}\left\{\begin{pmatrix} A & B \\ C & D\end{pmatrix}\right\} f_m\right](q) = \int_0^\infty dq'\, D^{k(m)}_{\mathbf{q},\mathbf{q}'}\begin{pmatrix} A & B \\ C & D\end{pmatrix} f_m(q'), \qquad k(m) = \tfrac{1}{2}(1+|m|), \tag{3.14a}$$

$$D^{k(m)}_{\mathbf{q},\mathbf{q}'}\begin{pmatrix} A & B \\ C & D\end{pmatrix} = \frac{e^{-i\pi k(m)}}{2\pi C}\exp\left(\frac{i}{2\lambda C}(Aq^2 + Dq'^2)\right) J_{|m|}\left(\frac{q\,q'}{\lambda C}\right), \tag{3.14b}$$

$$\left[\mathfrak{C}\left\{\begin{pmatrix} A & B \\ 0 & A^{-1}\end{pmatrix}\right\} f_m\right](q) = (\operatorname{sgn} A)^{2\,k(m)}\sqrt{|A|}\,\exp\left(\frac{iAB}{2\lambda}q^2\right) f_m(|A|q), \tag{3.14c}$$

An object wavefunction with only one Fourier component m present, will map into an image wavefunction with the same m.

For quadrupole lenses, a similar $sp(4,\Re) \supset sp(2,\Re) \oplus so(1,1)$ decomposition leads to a continuous partial-wave classification, where the hyperbolic-radial coordinate function transforms according to the continuous series, 'hyperbolic' canonical transforms of $Sp(2,\Re)$ [26–28]. These results on symmetry reduction were applied in reference [28] to find the scattering matrix elements of the Pöschl–Teller potential, and identifies its wavefunctions to be the $sp(2,\Re)$ Clebsch-Gordan coefficients.

5.3.7 Spots off the axis

A typical situation in optics is to have a circular light spot off the axis. On order to render (3.14) applicable, the optical axis should be placed at the center of the spot —or the other way 'round. Let $f^\circ(\mathbf{q})$ be an *object* wavefunction which, under a system $\mathfrak{M}$, specified by the matrix $\mathbf{M} \in Sp(2N,\Re)$, maps to an *image* wavefunction $f^{\mathrm{i}}(\mathbf{q}) = [\mathfrak{G}(\mathbf{M})f^\circ](\mathbf{q})$. Now, displace $f^\circ(\mathbf{q})$ by $\mathbf{x}$ to $f^\circ_{\mathbf{x}}(\mathbf{q}) := f^\circ(\mathbf{q}+\mathbf{x})$, and send it through $\mathfrak{M}$:

$$\begin{aligned}
f^\circ_{\mathbf{x}}(\mathbf{q}) &:= f^\circ(\mathbf{q}+\mathbf{x}) = \exp i(x_1\mathbb{P}_1 + x_2\mathbb{P}_2)\, f^\circ(\mathbf{q}) = \left[\mathfrak{G}\left\{\begin{pmatrix}\mathbf{x}\\ \mathbf{0}\end{pmatrix}, 0, \mathbf{1}\right\} f^\circ\right](\mathbf{q}) \\
\overset{\mathfrak{M}}{\Rightarrow}\quad f^{\mathrm{i}}_{\mathbf{x}}(\mathbf{q}) &:= [\mathfrak{M} f^\circ_{\mathbf{x}}](\mathbf{q}) = \left[\mathfrak{G}\left\{\mathbf{0}, 0, \begin{pmatrix}\mathbf{A} & \mathbf{B}\\ \mathbf{C} & \mathbf{D}\end{pmatrix}\right\} \mathfrak{G}\left\{\begin{pmatrix}\mathbf{x}\\ \mathbf{0}\end{pmatrix}, 0, \mathbf{1}\right\} f^\circ\right](\mathbf{q}) \\
&= \left[\mathfrak{G}\left\{\begin{pmatrix}\mathbf{D}\mathbf{x}\\ -\mathbf{B}\mathbf{x}\end{pmatrix}, 0, \begin{pmatrix}\mathbf{A} & \mathbf{B}\\ \mathbf{C} & \mathbf{D}\end{pmatrix}\right\} f^\circ\right](\mathbf{q}) = \left[\mathfrak{G}\left\{\begin{pmatrix}\mathbf{D}\mathbf{x}\\ -\mathbf{B}\mathbf{x}\end{pmatrix}, 0, \mathbf{1}\right\} f^{\mathrm{i}}\right](\mathbf{q}) \\
&= \exp\!\big(-i\lambda\tfrac{1}{2}(\mathbf{B}\mathbf{x})^\top \cdot (\mathbf{D}\mathbf{x})\big) \exp\!\big(-i(\mathbf{B}\mathbf{x})^\top \cdot \mathbf{q}\big)\, f^{\mathrm{i}}(\mathbf{q}+\mathbf{D}\mathbf{x}).
\end{aligned} \tag{3.15}$$

The result is an image function displaced by $\mathbf{x}^{\mathrm{i}} = \mathbf{D}\mathbf{x}$, with a ray direction $\mathbf{B}\mathbf{x}$ (given by the q-dependent phase), times a phase $\exp\big(-i\lambda\frac{1}{2}(\mathbf{B}\mathbf{x})^\top \cdot (\mathbf{D}\mathbf{x})\big)$ dependent on the system and displacement only. This is the expression for the geometric image point ($\mathbf{p} = \mathbf{D}\mathbf{x}$, $\mathbf{q} = -\mathbf{B}\mathbf{x}$) of the object point ($\mathbf{p} = \mathbf{0}$, $\mathbf{q} = \mathbf{x}$) under the matrix $\mathbf{M}$, which is the center of the new image. Thus with the transformation $f^{\mathrm{i}} \mapsto f^{\mathrm{i}}{}_{\mathbf{x}}$ in (3.15), we reduce the study of off-axis beams, to that of beams through the origin of the object plane. The same argument holds to reduce the study of beams having some inclination $\mathbf{y} \neq \mathbf{0}$ at the object plane. We tilt the object wavefunction with a prism operator $\exp(i\mathbf{y} \cdot \mathbb{Q})$, *i.e.*, we displace in $\mathbf{p}$; this acts through a factor $\exp(i\mathbf{y} \cdot \mathbf{q})$ and we obtain the image with tilt given by an equation similar to (3.15).

As a result, we may always align our optical axis such that if the object beam has a convenient axis of symmetry, this axis becomes the optical axis of the system. *Then* we may perform a decomposition in Fourier partial waves m, and use its symmetry, if any. In this way we reduce the problem to the study of the centered image point under the axially symmetric $Sp(2,\Re)$ part of the paraxial system.

If it is the *system* which turns out to be misaligned or tilted, the image wavefunction will form as before at the geometric phase space point $\mathbf{v} = (\mathbf{x}, \mathbf{y})$, centered at $\mathbf{x}$, with a factor $\exp(i\mathbf{y}^\top\mathbf{q})$ due to the tilt of the image ray, and a $\mathbf{q}$-independent phase. This *local mapping* from a neighborhood of an object point at the origin of phase space, to a neighborhood of its image point, is the effect of a paraxial system referred to its *design orbit* [9]. In the following we thus consider only axially-symmetric paraxial optical systems, aligned with the center of axial symmetry of the object wavefunctions. This will yield the image wavefunctions with the same symmetry. In particular, simple gaussian beams correspond to the $m = 0$ component, while the *discrete* coherent states occupy the higher-m eschelons. See chapter 7 in this volume.

5.4 Gaussian beams and other self-reproducing wavefunctions

An object wavefunction given by a centered Dirac δ, $f^{\circ}(\mathbf{q}) = \delta(\mathbf{q})$, entering an axis-symmetric system $\mathfrak{M}$ represented by a 4×4 diagonal-block matrix $\mathbf{M} \otimes \mathbf{1}$, $\mathbf{M} = \begin{pmatrix} A & B \\ C & D \end{pmatrix}$, will image onto

$$f^{\mathrm{i}}(\mathbf{q}) = D^{\circ}_{\mathbf{q},\mathbf{0}}(\mathbf{M} \otimes \mathbf{1}) = \frac{-i}{2\pi\lambda\!\!\!^{-} C} \exp\left(\frac{iA}{2\lambda\!\!\!^{-} C}\, \mathbf{q}^2\right) =: \frac{1}{A} G_{iC/A}(\mathbf{q}). \tag{4.1}$$

Due to (3.11) we have expressed here the result in terms of the simple Gaussian exponential function

$$G_{\omega}(\mathbf{q}) := [\mathfrak{F}_{i\omega}\delta](\mathbf{q}) = \left[\mathfrak{C}\left\{\begin{pmatrix} \mathbf{1} & \mathbf{0} \\ -i\omega\mathbf{1} & \mathbf{1} \end{pmatrix}\right\}\delta\right](\mathbf{q}) = \frac{1}{2\pi\lambda\!\!\!^{-}\omega} \exp\left(-\frac{q^2}{2\lambda\!\!\!^{-}\omega}\right) \tag{4.2}$$

(the width ω has units of $\lambda\!\!\!^{-}$ and $\mathrm{Re}\,\omega \geq 0$). For the radial image wavefunctions this will be $D^{k(m)}_{\mathbf{q},\mathbf{0}}(\mathbf{M})$, as given in (3.14); only the $m = 0$ partial wave is nonzero since the object function (the gaussian beam) is a function only of q^2. From (3.10) we note that $k(m = 0) = \frac{1}{2}$. These functions are normalized[6] so that their $\mathcal{L}^1(\Re^2)$–norm be unity: $\int_{\Re^2} d^2\mathbf{q}\, G_{\omega}(\mathbf{q}) = 1$. [Its $\mathcal{L}^2(\Re^2)$–norm is $\int_{\Re^2} d^2\mathbf{q}\, |G_{\omega}(\mathbf{q})|^2 = 1/4\pi\lambda\!\!\!^{-}\omega$.]

5.4.1 Simple gaussian beams

Gaussian exponential functions are particularly easy to analyze through the canonical transform formalism, since they are generated out of Dirac δ's through *complex* canonical transforms [12].

Gaussian exponential functions are the free-space propagation of a Dirac δ through an *imaginary* length $i\omega$. Also, Gaussian beams are a family of wavefunctions which have the minimum dispersion product allowed by the Heisenberg uncertainty relation [5, Subsect.7.6.6]. The momentum content of $G_{\omega}(\mathbf{q})$ is the Fourier transform of (4.1), namely $(\lambda\!\!\!^{-}/\omega)G_{\lambda\!\!\!^{-}/\omega}(\mathbf{p})$. A very narrow Gaussian light source has thus inevitably a large spread in direction, and vice versa. This is the Heisenberg uncertainty relation of paraxial wave optics.

Using only matrix algebra, we may calculate de effect of a paraxial system on a gaussian beam:

$$\begin{aligned} \left[\mathfrak{C}\left\{\begin{pmatrix} A & B \\ C & D \end{pmatrix}\right\} G_{\omega}\right](\mathbf{q}) &= \left[\mathfrak{C}\left\{\begin{pmatrix} A & B \\ C & D \end{pmatrix}\right\}\mathfrak{C}\left\{\begin{pmatrix} 1 & 0 \\ -i\omega & 1 \end{pmatrix}\right\}\delta\right](\mathbf{q}) \\ &= \left[\mathfrak{C}\left\{\begin{pmatrix} A - i\omega B & B \\ C - i\omega D & D \end{pmatrix}\right\}\delta\right](\mathbf{q}) \\ &= D^{k(m=0)}_{\mathbf{q},\mathbf{0}}\begin{pmatrix} A - i\omega B & B \\ C - i\omega D & D \end{pmatrix} = \frac{1}{A - i\omega B} G_{\omega^{\mathrm{M}}}(\mathbf{q}). \end{aligned} \tag{4.3a}$$

The image is also a gaussian exponential of *transformed* width

$$\omega^{\mathrm{M}} = \frac{\omega D + iC}{A - i\omega B}, \qquad \mathrm{Re}\,\frac{1}{\omega^{\mathrm{M}}} = \frac{\omega}{\omega^2 D^2 + C^2}. \tag{4.3b}$$

A gaussian object (or *input*) function thus remains Gaussian —this is the content of the ***ABCD*** law of Kogelnik [29]. The width ω, however, goes complex to ω^{M}; the *absolute value* remains a gaussian

[6] There is a *difference* of normalization with respect to the gaussian exponential functions used in Chapter 8 in this volume.

exponential of width $[\mathrm{Re}\,(1/\omega^{\mathrm{M}})]^{-1}$, and there is a phase oscillation, quadratic in q, given by the factor $\exp[i\frac{1}{2}\mathrm{Im}(1/\omega^{\mathrm{M}})\,\mathbf{q}^2]$.

Under free propagation, when $\mathfrak{M}$ is $\mathfrak{F}_z$, the Gaussian $G_\omega(\mathbf{q})$ becomes $G_{\omega^{\mathrm{M}}}(\mathbf{q})$, of width $\omega^{\mathrm{M}} = \omega + iz/n$. The modulus of this function is a Gaussian of width $\omega + z^2/\omega n^2$, so the beam has its *waist* at $z = 0$ and widens before and after. At the optical axis $\mathbf{q} = \mathbf{0}$, the function has the value $1/\omega^{\mathrm{M}}$, whose amplitude is $|1/\omega^{\mathrm{M}}| = 1/\sqrt{\omega^2 + z^2/n^2}$, which is again maximal at the waist and falls off thereafter. At the axis, $1/\omega^{\mathrm{M}} = 1/(\omega + iz/n)$ is a Breit–Wigner function with respect to the optical distance z/n from the waist. For narrow Gaussians, as z passes through zero, this Breit-Wigner factor has a rapid shift in phase by π.

We have plotted the absolute value and the phase of $G_{\omega(z)}(q)$ in the q–z plane, on a TV screen. The phase is defined modulo 2π, so $\exp(iyq)$ plots as a zig-zag factory roof. When the color allocation is rotated along this height, the TV image follows the wavefront movement (the intensity of the TV image is kept proportional to $\log|G_{\omega(z)}(q)|$). In this way we can visually follow the wavefronts through the waist. There, indeed, the wavefronts hasten through. Similar images were generated to simulate the phase singularities around the paraxial focus as a third-order spherical effect. These results were not yet ready for inclusion in this volume.[7]

5.4.2 Coherent states

Coherent states [17] may be treated by the canonical transform formalism [5, Sect.9.2] in the following way. We denote the complex *Bargmann* matrix

$$\mathbf{B} := \frac{1}{\sqrt{2}}\begin{pmatrix} 1 & -i/ƛ \\ -iƛ & 1 \end{pmatrix}. \tag{4.4}$$

Then, the ***coherent*** state $\Upsilon_{\mathbf{c}}(\mathbf{q})$ parametrized by $\mathbf{c} \in \mathfrak{C}^2$, ($\mathbf{c} = \mathbf{r} + i\mathbf{t}$, $\mathbf{r}, \mathbf{t} \in \mathfrak{R}^2$), is the *Bargmann transform* of a Dirac $\delta_{\mathbf{c}}$, sitting at $\mathbf{c}$ with units of ƛ. In $N = 2$ dimensions,

$$\begin{aligned} \Upsilon_{\mathbf{c}}(\mathbf{q}) :&= \sqrt{2\pi}[\mathfrak{G}\{\mathbf{B}\}\,\delta_{\mathbf{c}}](\mathbf{q}) = \sqrt{2\pi}D^{o}_{\mathbf{q},\mathbf{c}}(\mathbf{B}) \\ &= \pi^{-1/2}ƛ^{-1}\exp((-\tfrac{1}{2}\mathbf{q}^2 + \sqrt{2}\,\mathbf{q}\cdot\mathbf{c} - \tfrac{1}{2}\mathbf{c}^2)/ƛ^2) \\ &= \pi^{-1/2}ƛ^{-1}\exp(\mathbf{c}^2/2ƛ^2)\exp(-(\mathbf{q} - \sqrt{2}\mathbf{c})^2/2ƛ^2) \\ &= 2\sqrt{\pi}\,ƛ\,G_{ƛ}(\mathbf{q} - \sqrt{2}\mathbf{c})\exp(c^2/2ƛ^2) \\ &= \exp\big(\mathrm{Re}\,(c^2/2ƛ^2)\big)\left[\mathfrak{G}\left\{\begin{pmatrix} -\sqrt{2}\mathbf{x}/ƛ \\ \sqrt{2}\mathbf{t}/ƛ^2 \end{pmatrix}, 0; \mathbf{1}\right\}\Upsilon_{\mathbf{0}}\right](\mathbf{q}). \end{aligned} \tag{4.5}$$

These states are normalized so that $(\Upsilon_{\mathbf{c}}, \Upsilon_{\mathbf{c}'}) = \exp(\mathbf{c}^* \cdot \mathbf{c}'/ƛ^2)$.

Coherent states represent wave beams closest to geometrical optical rays. The real part of $\mathbf{c}$, $\mathbf{r} = \mathrm{Re}\,\mathbf{c}$, gives the coordinate of the *center* of the beam as $ƛ\mathbf{x} = \sqrt{2}\,\mathbf{r} = \sqrt{2}\,\mathrm{Re}\,\mathbf{c}$, and its imaginary part, $\mathbf{t} = \mathrm{Im}\,\mathbf{c}$, tells us of the *tilt* through the introduction of the phase factor $\exp(i\sqrt{2}\,\mathbf{t}\cdot\mathbf{q}/ƛ^2)$. This tilt may be compared with the parameter $\mathbf{y}$ in (2.2) and seen to correspond to a translation in momentum by $\mathbf{y} = \sqrt{2}\,\mathbf{t}/ƛ^2$ (in units of $ƛ^{-1}$). The width of the Gaussian is thus ƛ in both position and momentum space.

The *skewness* of a geometrical ray is measured by its ***Petzval*** invariant as $\mathbf{p}\times\mathbf{q}$. For coherent states (4.5) this is $\mathbf{x}\times\mathbf{y} = 2\,\mathrm{Im}(c_1 c_2^*)/ƛ^2$. For *meridional* rays (those which lie in a plane with the optical

[7] We thank Dr. Salvador Cuevas and José Fernando Barral, of the Instituto de Astronomía, UNAM, for their enthusiastic collaboration.

axis, $\mathbf{p}//\mathbf{q}$), the Petzval invariant is zero. The $\mathbf{c} = \mathbf{0}$ coherent state moves along the optical axis. Coherent states with $\mathbf{c} = \mathbf{r} + i\mathbf{t}$ will be meridional when $\arg c_1 = \arg c_2$, or $\mathbf{c} = e^{i\alpha}\mathbf{d}$, with $\mathbf{d}$ a *real* vector. When the two components of $\mathbf{c}$ have different phases, the beam is skew.

When acted upon by a paraxial system $\mathfrak{M}$, coherent states transform as

$$\begin{aligned}\mathfrak{M} : \Upsilon_{\mathbf{c}}(\mathbf{q}) &\mapsto [\mathfrak{G}\{\mathbf{M}\}\,\mathfrak{G}\{\mathbf{B}\}\,\delta_{\mathbf{c}}](\mathbf{q}) \\ &= \left[\mathfrak{G}\left\{\frac{1}{\sqrt{2}}\begin{pmatrix} A - iB\lambda & B - iA/\lambda \\ C - iD\lambda & D - iC/\lambda\end{pmatrix}\right\}\delta_{\mathbf{c}}\right](\mathbf{q}) \\ &= D^{o}_{\mathbf{q},\mathbf{c}}\begin{pmatrix} (A - iB\lambda)/\sqrt{2} & (B - iA/\lambda)/\sqrt{2} \\ (C - iD\lambda)/\sqrt{2} & (D - iC/\lambda)/\sqrt{2}\end{pmatrix}.\end{aligned} \tag{4.6}$$

Thus, in general, the $\Upsilon_{\mathbf{c}}$ *loose* their coherency, becoming simply complex Gaussians.

5.4.3 Self-reproducing wavefunctions

Coherent states remain coherent only when, in (4.6), $\mathfrak{G}\{\mathbf{M}\}\,\mathfrak{G}\{\mathbf{B}\}\,\delta_{\mathbf{c}} = \mathfrak{G}\{\mathbf{B}\}\delta_{\mathbf{c}'}$. This in turn happens only when $\mathbf{c}' = \mathbf{r}' + i\mathbf{t}' = (D + i\lambda B)\mathbf{r} + i(A - i\lambda^{-1}C)\mathbf{t}$ is some $\mathbf{M}'\mathbf{c}$, and this requirement fixes $\mathbf{M}' = (D + i\lambda B)\mathbf{1} = (A - i\lambda^{-1}C)\mathbf{1}$. Systems which mantain the coherency of the states are thus systems in the *SO(2)* subgroup of *Sp(2, ℜ)* given by $\mathfrak{A}_\beta$ in (3.2). These may be realized through a lens Fourier-type system as there, or by quadratic-profile optical fibers.

Paraxial wave optical systems in one dimension have also various classes of ***self-reproducing*** phase functions. These are the eigenfunctions of *any* element of the $w \wedge sp(2, \Re)$ algebra or, for the axially-symmetric case, of the *sp(2, ℜ)* generators (3.6) and of the symmetry generators (3.5).

Take a general element in *sp(2, ℜ)*, $\mathbb{L} = \sum_\sigma a_\sigma \mathbb{K}_\sigma$, quadratic in $\mathbb{P}$ and $\mathbb{Q}$, and suppose we know [see the previous chapter, Eq. (4.26)] its exponential $\mathfrak{G}\{\mathbf{M}(\alpha)\} = \exp(i\alpha\mathbb{L})$. Let ψ^m_λ be a complete eigenfunction set of $\mathbb{L}$ with spectrum $\{\lambda\}$, and of $\mathbb{M}_{12}$ with eigenvalues $m \in \mathfrak{Z}$. The passage of these wavefunctions through a generic axis-symmetric system $\mathfrak{M}$ is then found through *decomposing* $\mathfrak{G}\{\mathbf{M}\}$ into a product of an $\mathfrak{G}\{\mathbf{L}(\tau)\}$ belonging to the subgroup generated by $\mathbb{L}$ for some value of τ, and an element $\mathfrak{G}\{\mathbf{S}\}$, where $\mathbf{S}$ may be chosen to be an *upper triangular* matrix, *i.e.*, a *point* transformation (2.4c) representing a system *in focus* ($\mathbf{q}'$ independent of $\mathbf{p}^\circ$). Then,

$$\begin{aligned}[\mathfrak{G}\{\mathbf{M}\}\,\psi^m_\lambda](\mathbf{q}) &= [\mathfrak{G}\{\mathbf{S}\}\,\mathfrak{G}\{\mathbf{L}(\tau)\}\,\psi^m_\lambda](\mathbf{q}) \\ &= e^{i\tau\lambda}[\mathfrak{G}\{\mathbf{S}\}\,\psi^m_\lambda](\mathbf{q}) \\ &= F_M\, e^{i\tau\lambda}\,\exp(i\mathbf{q}^\top\mathbf{U}_M\mathbf{q}/\lambda)\,\psi^m_\lambda(\mathbf{V}_M\mathbf{q}),\end{aligned} \tag{4.7}$$

where τ, F_M, $\mathbf{U}_M$, and $\mathbf{V}_M$ depend on $\mathbf{M}$. The functional form of ψ^m_λ is thus *preserved* as the main factor. Self-reproducing functions for all orbits are given in reference [30]. These comprise the radial harmonic oscillator functions for $\mathbb{L} = \frac{1}{2}(\mathbb{P}^2 + \omega^2\mathbb{Q}^2)$, the radial *repulsive* oscillator functions for $\mathbb{L} = \frac{1}{2}(\mathbb{P}^2 - \omega^2\mathbb{Q}^2)$ (these are in general Whittaker functions), and Bessel functions for $\mathbb{L} = \frac{1}{2}\mathbb{P}^2$. Airy functions also appear in $\mathbb{L} = \frac{1}{2}\mathbb{P}^2 + \alpha\mathbb{Q}$, for the one-dimensional *WSp(2, ℜ)* case [5, Ch.10]. We shall not elaborate on these further, except to note that the oscillator functions are of interest as *discrete modes* in optics [18]. *Correlated* coherent states [31] may be followed through the same analysis.

We may also mention the *inverse* problem: *which* object wavefunctions reproduce up to a phase under a *given* axially-symmetric system $\mathfrak{M}$? Assume we have identified the generator $\mathbb{L}$ satisfying $\mathfrak{G}\{\mathbf{M}\} = \exp(i\tau\mathbb{L})$ for some value of τ, and we know spectrum $\{\lambda\}$ and the eigenfunctions $\omega^m_\lambda(q)$ of $\mathbb{L}$ and $\mathbb{M}_{12}$. The action of $\mathfrak{M}$ on the object phase function $\psi^m_\lambda(\mathbf{q}) = \omega^m_\lambda(q)e^{im\theta}$ will be to multiply it by the phase factor $e^{i\tau\lambda}$. Fourier systems (3.2) thus have the discrete-mode coherent states (oscillator wavefunctions) for self-reciprocal —even if the system is not in focus. Systems *in* focus have Dirac δ's for self-reciprocal —even if they are not developed in terms of eigenfunctions of $\mathbf{M}_{12}$. Free propagation has self-reciprocal Bessel functions $J_m(\lambda q)e^{im\theta}$. This corresponds to the partial wave decomposition of a plane wave, which is manifestly self-reproducing under such propagation, but for a phase $e^{i\lambda\tau}$ for any length $z = \tau$.

5.4.4 Pupils

Pupils are ***bona fide*** optical elements, not only for collimating the light beam, but for actual filtering of the image. Pupils are the optical analogue of general linear filters in parallel communication systems. Lenses are a special type of pupil: those that affect the wavefunction phase only.

Suppose that at some $z =$ constant plane we put a slide of density $|\Pi(\mathbf{q})|^2$, where $\Pi(\mathbf{q})$ is the ***pupil*** function. If the pupil affects phases, it is a complex function. A *lens* is an extreme case of a filter with pupil function $\exp(-iq^2/2\lambda\!\!\!^{-} f)$, f being the focal length. In the other extreme, a simple diaphragm between q_a and q_b is given by a function which is unity in that interval and zero elsewhere. Favourite pupils for filtering Gaussians are gaussian pupils $\exp(-q^2/2\lambda\!\!\!^{-} w)$ of width w.

Let the wavefunction arriving at the pupil plane within the apparatus to be $\psi^{\text{in}}(\mathbf{q})$. The action of the pupil is to turn this into the wavefunction

$$\psi^{\text{out}}(\mathbf{q}) = \Pi(\mathbf{q})\,\psi^{\text{in}}(\mathbf{q}). \tag{4.8}$$

Now, $\psi^{\text{out}}(\mathbf{q})$ starts its journey through the rest of the optical system, which we may characterize by a Γ_N transformation acting through the corresponding integral kernel of $\mathfrak{G}\{(\mathbf{v},\varsigma),\mathbf{M}\}$ in (2.2–2.4). The problem to solve is to find the effect of the pupil on the *final* image wavefunction

$$\psi^{\text{pup}} = \mathfrak{G}\{\omega,\mathbf{M}\}\psi^{\text{out}}, \tag{4.9a}$$

in terms of the pupil function Π and the image formed in the *absence* of pupils

$$\psi^{\text{no-pup}} = \mathfrak{G}\{\omega,\mathbf{M}\}\psi^{\text{in}} \tag{4.9b}$$

(or, conversely, $\psi^{\text{no-pup}}$ in terms of ψ^{pup} and Π). To do this, we need the canonical transform *convolution*, which we provide explicitly below for one-dimensional optics.

Consider two wavefunctions $f(q)$ and $g(q)$, and their *Sp(2,* $\mathfrak{R}$*)*-canonical transform images

$$f^M(q) = [\mathfrak{G}\{\mathbf{M}\}f](q), \qquad g^M(q) = [\mathfrak{G}\{\mathbf{M}\}g](q), \tag{4.10}$$

after some system $\mathfrak{G}\{\mathbf{0},0;\mathbf{M}\}$, $\mathbf{M} = \begin{pmatrix} a & b \\ c & d \end{pmatrix}$. Let

$$h(q) = f(q)\,g(q) =: (f \cdot g)(q), \tag{4.11}$$

and we search for $h^M(q)$. We find

$$h^M(q) = [\mathfrak{G}\{\mathbf{M}\}h](q) = [\mathfrak{G}\{\mathbf{M}\}(f \cdot g)](q) =: [f^M \overset{M}{*} g^M](q), \tag{4.12}$$

where $\overset{M}{*}$ defines the operation of **M**–***convolution***. Its general form is that of a *bilocal* product

$$[f^M \overset{M}{*} g^M](q) := \iint_{-\infty}^{\infty} dq_1\, dq_2\, f^M(q_1)\, C^{(M)}(q; q_1, q_2)\, g^M(q_2). \tag{4.13a}$$

The convolution *kernel* $C^{(M)}(q;q_1,q_2)$ is the integral of three canonical transform kernels (2.3), one direct and two inverse ones:

$$\begin{aligned} C^{(M)}(q;q_1,q_2) &= \int_{-\infty}^{\infty} dq'\, D^o_{q,q'}(\mathbf{M})\, D^o_{q_1,q'}(\mathbf{M})^*\, D^o_{q_2,q'}(\mathbf{M})^* \\ &= \frac{1}{2\pi c\sqrt{d}}\exp i\left(\frac{a}{2c}(q^2-q_1^2-q_2^2)+\frac{1}{2cd}(q-q_1-q_2)^2\right). \end{aligned} \tag{4.13b}$$

The last equality is obtained performing the gaussian integral. When $\mathbf{M}$ is the Fourier matrix $\mathbf{F}$ in (3.3), $\overset{M}{*}$ becomes the ordinary Fourier convolution $*$ as the convolution kernel (4.13) collapses to $\delta(q-q_1-q_2)$. A frequency-filter pupil on $\psi^{\rm in}$ is thus arranged through $\mathbf{M}=\mathbf{F}$. These formulae constitute the cartesian coordinates for an $\mathbf{M}$-convolution for $\mathbf{M}\in Sp(2N,\Re)$.

We may study a corresponding convolution between functions transformed by a Heisenberg–Weyl group element $\omega\{\mathbf{v},\varsigma\}\in W$. In this case the group does not have an integral transform action, but a *Lie multiplier* action given by (2.2). The transformations analogous to (4.9–13) are now the *W–convolution*

$$h^\omega(q)=(f\cdot g)^\omega(q)=[f^\omega \overset{\omega}{*} g^\omega](q):=f^\omega(q)\exp i[-yq-\lambda\!\!\!^{-}(\varsigma+\tfrac{1}{2}xy)]\,g^\omega(q). \tag{4.14}$$

A Γ–convolution is also easily built in terms of the $Sp(2,\Re)$– and W–convolutions given above, for $\gamma=(\omega;\mathbf{M})\in\Gamma$, as

$$\begin{aligned} [f^\gamma \overset{\gamma}{*} g^\gamma](q) :&= [\mathfrak{G}\{\omega;\mathbf{M}\}(f\cdot g)](q) \\ &= [\mathfrak{G}\{\omega;\mathbf{1}\}(f^M \overset{M}{*} g^M)](q) \\ &= \exp i[yq+\lambda\!\!\!^{-}(z+\tfrac{1}{2}xy)](f^M \overset{M}{*} g^M)(q+\lambda\!\!\!^{-}x). \end{aligned} \tag{4.15}$$

Substitutions lead us to find the *γ–convolution* of two functions f, g, as an analogue of (4.13), except for the kernel, which is here

$$C^{(\gamma)}(q;q_1,q_2)=\exp i[y(q-q_1-q_2)-\lambda\!\!\!^{-}(z-\tfrac{3}{2}xy)]\,C^{(M)}(q+\lambda\!\!\!^{-}x;q_1+\lambda\!\!\!^{-}x,q_2+\lambda\!\!\!^{-}x). \tag{4.16}$$

The action of the pupil $\mathbf{\Pi}$ in a system $\mathfrak{M}=\mathfrak{G}\{\gamma\}$ on the no-pupil image wavefunction is thus

$$\begin{aligned} \mathbf{\Pi}^\gamma:\ \psi^{\text{no-pup}}\mapsto\psi^{\text{pup}} &= \mathfrak{G}\{\omega;\mathbf{M}\}\psi^{\rm out}=\mathfrak{G}\{\omega;\mathbf{M}\}(\Pi\cdot\psi^{\rm in}) \\ &= (\mathfrak{G}\{\omega;\mathbf{M}\}\Pi)\overset{\gamma}{*}(\mathfrak{G}\{\omega;\mathbf{M}\}\psi^{\rm in}) \\ &= \Pi^\gamma \overset{\gamma}{*}\psi^{\text{no-pup}}=\mathbf{\Pi}^\gamma\,\psi^{\text{no-pup}}, \end{aligned} \tag{4.17}$$

i.e., a γ–convolution between the no-pupil image and the γ–image of the pupil function.

The *pupil* is thus a general linear *filter*[8] Filters in Fourier theory are operators which in the Fourier *basis* are represented by diagonal generalized matrices (including integral kernels). Here the filter $\mathbf{\Pi}$ is an operator diagonal in the *q–basis*, *cf.* (4.10) *at the plane of the pupil.* Subsequent γ–evolution turns $\mathbf{\Pi}$ into $\mathbf{\Pi}^\gamma$, whose action (4.17) we may now describe through an integral transform kernel

$$\mathbf{\Pi}^\gamma\!:\!\psi^{\text{no-pup}}=\psi^{\text{pup}}(q)=\int_{-\infty}^{\infty}dq'\,\Pi^\gamma(q,q')\,\psi^{\text{no-pup}}(q'); \tag{4.18a}$$

the kernel is

$$\Pi^\gamma(q,q')=\int_{-\infty}^{\infty}dq''\,D^o_{q,q''}(\gamma)\,\Pi(q'')\,D^o_{q'',q'}(\gamma^{-1}). \tag{4.18b}$$

In this form we use the unitarity of the integral transform kernels, $D^o_{q,q''}(\gamma)^*=D^o_{q,q''}(\gamma^{-1})$ for $\gamma\in\Gamma$.

---

[8]Filters in communication theory are usually required to be *causal* when the signals are fed to it in *time*. Here we have a signal *across* the optical axis whose data points are to be processed *simultaneously*.

5.5 Some further questions

There are presently under development several other themes connected with paraxial wave optics and beyond, but for which we have only partial or preliminary results. We mention some of them below with the hope that they will further the reader's interest.

5.5.1 The Wigner distribution and Woodward's cross-ambiguity function

Fourier optics and radar detectors work very confortably with the Wigner distribution $W(\mathbf{p},\mathbf{q})$ associated to a wavefunction $\psi(\mathbf{q})$. This is essentially the double Fourier transform (on $\mathbf{p}$ and $\mathbf{q}$) of the matrix elements of the Heisenberg-Weyl group. The latter is the *Woodward cross-ambiguity* function $H(\mathbf{x},\mathbf{y})$, that may be defined for two different functions, f and g, as

$$H(\mathbf{x},\mathbf{y},z;f,g) := \left(f, \mathfrak{G}\left\{\begin{pmatrix}\mathbf{x}\\ \mathbf{y}\end{pmatrix}, z, \mathbf{1}\right\} g\right) = \int_{\Re^N} d^N\mathbf{q}\, f(\mathbf{q}-\tfrac{1}{2}\lambda\!\!\!^{-}\mathbf{x})^* e^{i(\mathbf{q}\cdot\mathbf{y}+\lambda\!\!\!^{-} z)} g(\mathbf{q}+\tfrac{1}{2}\lambda\!\!\!^{-}\mathbf{x}). \tag{5.1}$$

This was discussed in the first chapter of this volume; it tells us how 'peaked' the beam is, and how to improve resolution for range and velocity detection [32–34]. If the system undergoes a paraxial Γ_N transformation $\mathfrak{G}\{\gamma\}$, $\psi(\mathbf{q})$ generally transforms with an integral kernel, while the Wigner and Woodward functions map *geometrically* to $W(\mathbf{g}\mathbf{p},\mathbf{g}\mathbf{q})$ with the linear transformation (2.1). This fact was recognized independently by Bastiaans [36] and by García-Calderón and Moshinsky [37]. Their result is actually more general, and may be stated as follows.

Suppose the optical system $\mathfrak{M}$ is *not* paraxial (but linear on the wavefunctions), and its effect is determined through a kernel $\mathcal{M}(\mathbf{q},\mathbf{q}')$ on the wavefunctions, *i.e.*,

$$f^{\mathrm{i}}(\mathbf{q}) = [\mathfrak{M} f^{\circ}](\mathbf{q}) = \int_{\Re^N} d^N\mathbf{q}'\, \mathcal{M}(\mathbf{q},\mathbf{q}')\, f^{\circ}(\mathbf{q}'), \tag{5.2}$$

(f^{i} = image, f° = object). Then, the Woodward ambiguity function (5.1) transforms as

$$H^{\mathrm{i}}(\mathbf{x},\mathbf{y},z;\mathfrak{M}f,\mathfrak{M}g) = \iint_{\Re^{2N}} d\mathbf{x}'\, d\mathbf{y}'\, \tilde{\mathcal{M}}(\mathbf{x},\mathbf{y};\mathbf{x}',\mathbf{y}')\, H(\mathbf{x}',\mathbf{y}',z;f,g), \tag{5.3a}$$

i.e., through an integral transform on $\Re^{2N}$ with kernel

$$\tilde{\mathcal{M}}(\mathbf{x},\mathbf{y};\mathbf{x}',\mathbf{y}') := \lambda\!\!\!^{-} \iint_{\Re^{2N}} d\mathbf{q}\, d\mathbf{q}'\, \mathcal{M}(\mathbf{q}-\tfrac{1}{2}\lambda\!\!\!^{-}\mathbf{x}, \mathbf{q}'-\tfrac{1}{2}\lambda\!\!\!^{-}\mathbf{x}')^*\, \exp i(\mathbf{y}^{\top}\mathbf{q}-\mathbf{y}'^{\top}\mathbf{q}')\, \mathcal{M}(\mathbf{q}+\tfrac{1}{2}\lambda\!\!\!^{-}\mathbf{x}, \mathbf{q}'+\tfrac{1}{2}\lambda\!\!\!^{-}\mathbf{x}'). \tag{5.3b}$$

When the kernel $\mathcal{M}(\mathbf{q},\mathbf{q}')$ is a canonical transform kernel (2.3), then (5.3*b*) collapses to a Dirac δ and the transformations in the Heisenberg–Weyl arguments of the Woodward function (5.1) —and those of the Wigner function— are *linear.* García-Calderón and Moshinsky [37] also considered, in their quantum mechanical framework, the Schrödinger time evolution under *free fall*, *i.e.*, the Hamiltonian $\frac{1}{2}\mathbb{P}^2 + \nu\mathbb{Q}$. The integral kernel is then an Airy function. This corresponds in one-dimensional optics to a z-homogeneous *graded*-index medium $n(q) = n_0 + \nu q$, where the beam falls into the denser region. When $\mathfrak{M}$ is an operator more general than a paraxial transformation, this formalism does not seem hindered. How is this extended, then, to describe aberrations? Part of the answer is developed in the following subsection.

5.5.2 Schrödinger wavization for aberrations

Paraxial optics is a very ***symmetric*** theory, much as angular momentum or the harmonic oscillator in quantum mechanics. This statement also includes the structure that may be built around it to describe covariant *departures* from rotational symmetry or harmonic oscillator expansions for other potentials. There are good indications that wave optics with *aberration* may be described in terms of the basic tools presented here, applied to the results of geometric optics of last chapter. In particular, the last subsection there defines the symplectic polynomials ${}^{\kappa}\mathcal{X}^{j}_{m}(\boldsymbol{\xi}(\mathbf{p},\mathbf{q}))$. These are built for aberration order $N = 2\kappa - 1$, and are of order 2κ in the components of $\mathbf{p}$ and $\mathbf{q}$. How should we 'wavize' such functions of geometrical phase space? Common quantum-mechanical sense would follow the Schrödinger way. This has been done successfully for the *paraxial* approximation since *quadratic* expressions in the components of $\mathbf{p}$ and $\mathbf{q}$ *have a unique quantization* [4, Sect.IV.B]; not so all higher-order polynomials, except for the special classes $pf(q) + g(q)$ and $qf(p) + g(p)$.[9]

Consider the example of p^2q^2 (in one dimension). This relates directly to the curvature of field/astigmatism phenomenon in meridional rays for two dimensions. The quantum operator corresponding to p^2q^2 is, *in the standard order* (all $\mathbb{Q}$'s to the left, all $\mathbb{P}$'s to the right), $\mathbb{Q}^2\mathbb{P}^2 - 2i\lambda\!\!\!{}^{-}\mathbb{Q}\mathbb{P} + \lambda\!\!\!{}^{-2}\phi\mathbb{I}$, where ϕ is a number which depends on the *quantization scheme* [4, Sect.4], through the second derivative of the Cohen ordering function, evaluated at the origin. This 'constant' is $-\frac{2}{3}$ for the Born–Jordan rule, $\frac{1}{2}$ for the Weyl–Mc Coy rule, $+1$ for the symmetrization scheme, etc. The quartic term is the expected one![10] The second-order term seems to indicate that we may have a concomitant paraxial magnification effect of order $\lambda\!\!\!{}^{-}$. The last term, although a constant phase and of order $\lambda\!\!\!{}^{-2}$, depends on the quantization scheme of the theorist. Is there a privileged scheme chosen by Nature? If we believe in Lie algebras, we may want to insist that the paraxial transformation properties of the symplectic operators ${}^{\kappa}\mathbb{X}^{j}_{m}$ [*i.e.*, the putative operator corresponding to ${}^{\kappa}\mathcal{X}^{j}_{m}(\boldsymbol{\xi}(\mathbf{p},\mathbf{q}))$] be the same as for their geometrical counterparts —see Eqs. (7.13) of the previous chapter, namely

$$[{}^{1}\mathbb{X}^{1}_{\pm}, {}^{\kappa}\mathbb{X}^{j}_{m}] = \lambda\!\!\!{}^{-}(m \mp j)\,{}^{\kappa}\mathbb{X}^{j}_{m\pm1}. \tag{5.4}$$

In reference [38] we showed (in one dimension) that equation (5.4) could be upheld in the *Weyl* quantization scheme and *only* there. We remind the reader that in the Weyl scheme, to $p^m q^n$ corresponds the operator obtained from m $\mathbb{P}$'s and n $\mathbb{Q}$'s, summed over all their permutations as individual objects, divided by $(m+n)!$.

If the foregoing arguments about wavization are well founded, we may consider the geometrical optical system $\mathfrak{M}$ decomposed in the Dragt–Finn factorization [39] (see the preceding chapter, Subsect. **4.3.2**): $\cdots\mathfrak{M}_8\mathfrak{M}_6\mathfrak{M}_4\mathfrak{M}_2$. We have studied the $\mathfrak{M}_2$ factor in this chapter. Assume we now have $\mathfrak{M}_4 = \exp(i\sum_{j,m}\alpha^j_m\,{}^{2}\mathbb{X}^j_m)$. Then, if the α^j_m are small, we may develop arguments on the order of approximation. These are complicated by the fact that under *no* quantization scheme is the classical Poisson bracket ring isomorphic to the whole Heisenberg-Weyl ring [4].

The commutator of two ${}^{\kappa_i}\mathbb{X}^{j_i}_{m_i}$'s yields a highest-order term $\lambda\!\!\!{}^{-}\,{}^{\kappa}\mathbb{X}^{j}_{m}$, *plus* terms of order $\lambda\!\!\!{}^{-2}$, $\lambda\!\!\!{}^{-3}$, ..., that depend on our way of writing ${}^{\kappa}\mathbb{X}^{j}_{m}$, *i.e.*, as in *standard* form (all $\mathbb{P}$'s to the right) or antistandard form (all $\mathbb{Q}$'s to the right) [40]. A firm geometrical/wave correspondence exists only for quadratic terms and several other isolated classes of operators. The former we have exploited here; among the latter are those cases where only operators in $\mathbb{P}$ appear, etc. One such instance is developed in Chapter 8, and is applied to study the effects of spherical aberration on gaussian beams. There, the operator to be exponentiated is $\mathbb{H} = -\sqrt{n^2 - \mathbb{P}^2}$.

---

[9] Results in this regard are addressed in the following chapter. (Editor's note.)

[10] It has been noted by Professor Klauder (private communication) that this is a *good* operator which leads to unitary transformations in a group.

We may refrain from searching for *iso*morphisms between geometrical and wave optics; these endeavours fail, as is well known, even between classical and quantum mechanics. Rather, we may propose that we calculate the symplectic map $\mathfrak{M}$ of a system using the *geometric* optics Lie method approach of the previous chapter, and *then* turn it into an operator throught the Weyl quantization rule and, in the decomposition $\cdots \mathfrak{M}_8 \mathfrak{M}_6 \mathfrak{M}_4 \mathfrak{M}_2$, let it act on the object wavefunction. The action of $\mathfrak{M}_2$ has been studied to all orders in reference [38] and here. Next act the

$$\mathfrak{M}_{2k} = \mathbb{1} + i \sum_{j,m} {}^{\kappa}v^j_m \, {}^{\kappa}\mathbb{X}^j_m - \frac{1}{2!}\Big(\sum_{j,m} {}^{\kappa}v^j_m \, {}^{\kappa}\mathbb{X}^j_m\Big)^2 - \cdots, \qquad \text{for } k = 2,3,4\ldots. \tag{5.5}$$

How many terms should we keep here? Consider *coma*, *i.e.*, ${}^2\mathbb{X}^2_1 = \frac{1}{2}(\mathbb{P}^2\,\mathbb{P}\cdot\mathbb{Q} + \mathbb{Q}\cdot\mathbb{P}\,\mathbb{P}^2) = \mathbb{Q}\cdot\mathbb{P}\,\mathbb{P}^2 + i\lambda\mathbb{P}^2$, a third-order differential operator. Now, ${}^2\mathbb{X}^2_1$ is the *Fourier conjugate* of *distortion*, ${}^2\mathbb{X}^2_{-1} = \frac{1}{2}(\mathbb{P}\cdot\mathbb{Q}\,\mathbb{Q}^2 + \mathbb{Q}^2\,\mathbb{Q}\cdot\mathbb{P}) = \mathbb{Q}^2\,\mathbb{Q}\cdot\mathbb{P} + i\lambda\mathbb{Q}^2$ for which no quantization-scheme constant appears, since its quantization is also unique. The action of distortion is a point-to-point mapping on the wavefunction argument and a multiplier factor. Through inverse Fourier transformation, the action of $\exp(v^{\text{coma}}\,{}^2\mathbb{X}^2_1)$ can be expressed in the form of an integral transform, with a non-quadratic kernel; this integral may not be strictly unitary, may be difficult to perform, but let us gloss over *these* problems.

If the wavefunction is very smooth —a Gaussian, for example— and the coma parameter v^{coma} is small, then the image wavefunction obtained from the integral transform can be approximated by the first few terms in the exponential series as a differential operator on the object wavefunction. Third-order coma is thus a third-order differential operator, plus a sixth-order one, plus a ninth-order one, etc. If the object wavefunction is a Gaussian, this process multiplies it by polynomials of orders three, six, nine, etc. The resulting function may be plotted and should resemble the known pattern of diffraction in coma [41, Fig. 9.6]. The zeros of the polynomial become the dark fringes around the main image. Some very preliminary graphs obtained for the third-order differential operator in the series indicates that weak coma is indeed reproduced![11] The study of diffraction in aberration may be also performed on the Wigner or Woodward functions (5.1), with a similar differential operator expansion. This corresponds to results reported in [42] from the perspective of Fourier optics.

5.5.3 A basic euclidean algebra for optics

There is one further open topic we would like to discuss, based on reference [43] for one-dimensional optics. It bears on the foundations of the Heisenberg-Weyl description of optics. As we said in the introductory section, the paraxial approximation to optics assumes that the momentum operator $\mathbb{P}$ has eigenvalues ranging over the real line. This is implicit in the Schrödinger formulation through the Stone–von Neumann theorem [44]. In optics, the spectrum of "$\mathbb{P}$" cannot extend beyond the interval $[-n, n]$. The geometrical optics phase space has the connectivity properties of a *cylinder*: ray *direction* ranges over a *circle*.

This last observation may be put to work in equation (2.2) expressing the action of the Heisenberg-Weyl group on a space of functions [4, Sect. V-A; 43]. Assume the tilt parameter, y, is counted modulo $2\pi/\lambda$ [or, to account for multivaluation possibilities, $y \equiv y \pmod{2\pi M/\lambda}$, $M \in \mathfrak{Z}^+$]. The factor $\exp(iyq)$ in (2.2) must be unity, and this is satisfied only when $q = \lambda k$, $k \in \mathfrak{Z}$, *i.e.*, at the price of replacing the original configuration space by a *lattice* of equidistant points spaced by λ. This in turn specifies that the only possible translations are those with integer parameter[12] x. This is consistent with counting the third parameter, ς, modulo $2\pi/\lambda$ [*cf.* Eq. (2.2)]. We should be interested, therefore,

[11] Work by W. Lassner, M. Navarro-Saad, and K.B. Wolf (stopped with the IIMAS computer breakdown in January '85) still awaits completion.

[12] Note that our definition here exchanges $x \leftrightarrow y$ with respect to the notation used in reference [43].

in the *proper* subgroup of $W_1(\Re^3)$ that we obtain when we let x run over the integers $\mathfrak{Z}$, and y and z over the circle $\mathfrak{S}$:

$$W(\mathfrak{Z}\,\mathfrak{S}^2) := \{\omega(x,y,z) \in W(\Re^3) \mid x \in \mathfrak{Z},\ y,z \in \Re \ (\mathrm{mod}\ 2\pi/\lambda)\}. \tag{5.6}$$

This is a *mixed* Lie group. It has two continuous parameters y and z with infinitesimal generators $\mathbb{Q} = q\cdot$ and $\mathbb{I} = \lambda$, and one *finite* translation generator $\mathbb{E} := \omega\big(\binom{1}{0}, 0\big)$. Now, $\mathbb{Q}$ is essentially self-adjoint, and its spectrum is $\lambda\mathfrak{Z}$; this is also in agreement with the Whittaker-Shannon sampling theorem [20, Sect. 2.3] for a wave phenomenon with band-limited wavelengths longer than λ, corresponding to the building of wavefunctions from plane waves coming from every direction in the circle. But $\mathbb{E}$ is a *unitary* operator with spectrum $e^{i\theta}$, θ ranging over the circle. Wavefunctions are thus functions in $\ell^2(\mathfrak{Z})$, the Hilbert space of square-summable sequences. There *is no* operator '$\mathbb{P}$' generating $\mathbb{E}$; we may build, however, difference or averaging operators *out* of $\mathbb{E}$ [4, Sect. VI-B].

For the same reason as in quantum mechanics, observables should correspond to essentially self-adjoint operators, that have real eigenvalues. Out of $\mathbb{E}$ we may form two such linear operators:

$$\mathring{\mathbb{P}} := -i\tfrac{1}{2}n(\mathbb{E} - \mathbb{E}^\dagger), \qquad \text{spectrum:} \quad n\sin\theta, \qquad \theta \in (-\pi, \pi], \tag{5.7a}$$

$$\mathring{\mathbb{H}} := -\tfrac{1}{2}n(\mathbb{E} + \mathbb{E}^\dagger), \qquad \text{spectrum:} \ -n\cos\theta. \tag{5.7b}$$

We see that $\mathring{\mathbb{P}}$ has the spectrum of the geometric optics momentum operator and $\mathring{\mathbb{H}}$ that of the Hamiltonian for homogeneous media [*cf.* Chapter 4, Eq. (2.7–2.8)]. Fourier series analysis from $q \in \lambda\mathfrak{Z}$ to $\theta \in (-\pi, \pi]$ confirms this association. Yet, *unlike* the quantum mechanical $\mathbb{Q}$, $\mathbb{P}$, and $\mathbb{I}$, under commutation $\mathbb{Q}$, $\mathring{\mathbb{P}}$, and $\mathring{\mathbb{H}}$ close here into the following Lie algebra:

$$[\mathbb{Q}, \mathring{\mathbb{P}}] = -i\lambda\mathring{\mathbb{H}}, \qquad [\mathbb{Q}, \mathring{\mathbb{H}}] = i\lambda\mathring{\mathbb{P}}, \qquad [\mathring{\mathbb{P}}, \mathring{\mathbb{H}}] = 0. \tag{5.8a, b, c}$$

This is *not* the Heisenberg-Weyl algebra (*cf.* $[\mathbb{Q}, \mathbb{P}] = i\lambda\mathbb{I}$, $[\mathbb{Q}, \mathbb{I}] = 0$, $[\mathbb{P}, \mathbb{I}] = 0$), but the *euclidean* algebra *iso(2)*, with $\mathring{\mathbb{P}}$ and $\mathring{\mathbb{H}}$ generators of translations, and $\mathbb{Q}$ of rotations.

Free propagation in a homogeneous medium is produced by the operator

$$\mathring{\mathfrak{F}}_z := \exp(-iz\mathring{\mathbb{H}}), \tag{5.9a}$$

[*cf.* Eq. (2.6a)], with the Kirchhof–Fresnel integral kernel given by

$$\begin{aligned}\mathring{F}_z(k, k') &= \frac{1}{2\pi}\int_{-\pi}^{\pi} d\theta \int_{-\pi}^{\pi} d\theta'\,\delta(\theta - \theta')\exp i\big(zn\cos\theta - k\theta + k'\theta'\big)\\ &= e^{i\pi(k'-k)/2}\,J_{k'-k}(zn),\end{aligned} \tag{5.9b}$$

where the position-space points are $q = \lambda k$, $q' = \lambda k'$, $k, k' \in \mathfrak{Z}$, and $J_m(x)$ is a Bessel function of the first kind.

Out of the above *propagator* we may define a ***gaussian function*** for *this* system, of width ω, as the wavefunction obtained from a Kronecker–δ initial condition through *diffusion*, *i.e.*, through *imaginary time* translation $w = -it$ [43]. This is

$$\mathring{G}_w(q) := \mathring{F}_{iw}(k = q/\lambda, 0) = I_k(wn), \tag{5.10a}$$

where I_k is the modified Bessel function [45, Fig.9.9].[13] The momentum distribution of this Gaussian function is the Fourier synthesis of the above equation, namely

$$\tilde{\mathring{G}}_w(\theta) = \exp(wn\cos\theta), \tag{5.10b}$$

[13] Notice that the *integer* points on the k–axis of the figure fall, indeed, on points which resemble a intuitive gaussian decreasing exponential function, in spite of the increasingly violent oscillations of $I_k(x)$ in the negative direction $k \to -\infty$.

which is clearly a function concentrated in the forward direction, with a width proportional to $1/w$.

In the 'quantum mechanical' version of paraxial wave optics studied in this chapter, the relevant operators are $x\mathbb{P} + y\mathbb{Q} + z\mathbb{I}$, elements of a Heisenberg–Weyl algebra W_1. The proposal in this subsection is that the correct operators are $x\mathring{\mathbb{P}} + y\mathbb{Q} + z\mathring{\mathbb{H}}$, elements of the *euclidean* algebra ***iso(2)***. This extends to the corresponding groups and rings [4].

In homogeneous media, $\mathring{\mathbb{P}}^2 + \mathring{\mathbb{H}}^2 = n^2$, so we may *deform* [24,25,46] ***iso(2)*** to a continuous-series multiplier representation of ***so(2,1)*** with generators

$$\mathbb{L}_0 = \mathbb{Q}, \qquad \mathbb{L}_1 = \tfrac{1}{2}(\mathbb{Q}\mathring{\mathbb{H}} + \mathring{\mathbb{H}}\mathbb{Q}) + c\mathring{\mathbb{P}}, \qquad \mathbb{L}_2 = \tfrac{1}{2}(\mathbb{Q}\mathring{\mathbb{P}} + \mathring{\mathbb{P}}\mathbb{Q}) - c\mathring{\mathbb{H}}, \tag{5.11a, b, c}$$

where where the first operator generates rotations in $\mathring{\mathbb{P}}$–$\mathring{\mathbb{H}}$ and the two others are *boosts* which *deform* θ to $\arctan 2(e^{\beta}\tan\frac{1}{2}\theta)$ and produce a multiplier factor on functions thereof [24]. The commutation relations between the $\mathbb{L}_\sigma$ are (3.7) replacing $\mathbb{K}_\sigma \mapsto \mathbb{L}_\sigma$. The irreducible representations to which these operators belong in $\mathcal{L}^2(\mathfrak{S})$ have been studied in reference [47].

We may also try to build the analogue of the *paraxial* (second-order) ***sp(2, ℜ)*** algebra generated by $\mathbb{P}^2$, $\frac{1}{2}(\mathbb{P}\mathbb{Q} + \mathbb{Q}\mathbb{P})$, and $\mathbb{Q}^2$ in the enveloping algebra of W_1. A first chouice for corresponding operators in the enveloping algebra of ***iso(2)*** could be $\mathring{\mathbb{H}}$, $\frac{1}{2}(\mathbb{P}\mathbb{Q} + \mathbb{Q}\mathbb{P})$, and $\mathbb{Q}^2$. However, these do *not* close into a finite-dimensional Lie algebra, *viz.*

$$[\mathring{\mathbb{H}}, \mathbb{Q}^2] = -i\lambda\!\!\!^{-}(\mathbb{Q}\mathring{\mathbb{P}} + \mathring{\mathbb{P}}\mathbb{Q}), \tag{5.12a}$$

$$[\mathring{\mathbb{H}}, \mathbb{Q}\mathring{\mathbb{P}} + \mathring{\mathbb{P}}\mathbb{Q}] = 2i\lambda\!\!\!^{-}\mathring{\mathbb{P}}^2 = 2i\lambda\!\!\!^{-}(n^2 - \mathring{\mathbb{H}}^2), \tag{5.12b}$$

$$[\mathbb{Q}^2, \mathbb{Q}\mathring{\mathbb{P}} + \mathring{\mathbb{P}}\mathbb{Q}] = -i\lambda\!\!\!^{-}(\mathbb{Q}^2\mathring{\mathbb{H}} + 2\mathbb{Q}\mathring{\mathbb{H}}\mathbb{Q} + \mathring{\mathbb{H}}\mathbb{Q}^2). \tag{5.12c}$$

We may perform a *paraxial approximation* on (5.8), (5.11), and (5.12), regaining, from the first, the Heisenberg–Weyl algebra, and from the last, the ***sp(2, ℜ)*** algebra. This we do considering the matrix elements between *forward-concentrated* wavefunctions $\psi(q = \lambda\!\!\!^{-} k)$, *i.e.*, such that $\tilde{\psi}(\theta)$ be significantly different from zero only in a small neighborhood $|\theta| \ll 1$. There, the expectation value of $\mathring{\mathbb{H}}$ is close to that of $n - \mathring{\mathbb{P}}^2/2n$. To first order in θ, the matrix elements of (5.8) become those of the Heisenberg–Weyl algebra; to second order, (5.12) reproduces the ***sp(2, ℜ)*** algebra.

The applicability in optics of this euclidean 'correction' to the Heisenberg–Weyl algebra is still speculative. It needs to be applied to inhomogeneous media $n(q)$ or $n(q, z)$; in particular, lenses should be described at least as clearly as in the paraxial W–model ($p, q \in \mathfrak{R}$), including the metaxial aberrations built upon it in the last chapter. The treatment of lenses as multiplication by pure-phase pupil functions with its ensuing Γ–convolution, discussed at the end of last section, may provide a lead.

There are several further extensions we may contemplate. A first case in point refers to *reflection* phenomena, for which our euclidean algebra seems to be able to describe left-moving as well as right-moving waves. Next, we may consider *two* dimensional optics with a corresponding algebra. Ray direction there ranges over a *sphere* rather than a circle. A third extension —a *restriction*, rather— should take into account discrete *periodic* wavefunctions, such as may model TV screens or N-point radar arrays. A fourth one applies to both Γ_N and ***euclidean*** models and should remove the restriction to monochromatic waves implicit in the consideration of a fixed $\lambda\!\!\!^{-}$, through Fourier analysis over this representation label. The Lie structure of wave optics is clearly very rich and deserves to be studied.

Added in proof, one subject which certainly deserves mention in this chapter is the treatment of *polarization* in wave optics. This line of research has been pursued by Mukunda, Simon, and Sudarshan [48–50] for paraxial optics. It has come to our attention only during the preparation of this volume, and so it was impossible to incorporate meaningfully into the text of this review chapter. The Poincaré symmetry algebra and group of the *Maxwell* equations are the foundation for that treatment. In the paraxial approximation there is, essentially a Galilei algebra and group with a '*spin*' transformation part for rotations and beam tiltings; in one dimension this is the euclidean group mentioned above (minus the 'spin'). In future work we should like to relate these two approaches.

References

[1] H. Weyl, Quantenmechanik und Gruppentheorie, *Z. Phys.* **46**, 1–46 (1928).

[2] M. Moshinsky and C. Quesne, Oscillator systems. In *Proceedings of the XV Solvay Conference in Physics (Brussels, 1970)*. E. Prigogine Ed., (Gordon and Breach, New York, 1975).

[3] M. Moshinsky, Canonical transformations in quantum mechanics, *SIAM J. Appl. Math.* **25**, 193–212 (1973).

[4] K.B. Wolf, The Heisenberg–Weyl ring in quantum mechanics. In *Group Theory and its Applications*, Vol. 3, Ed. by E.M. Loebl (Academic Press, New York, 1975).

[5] K.B. Wolf, *Integral Transforms in Science and Engineering*, (Plenum Publ. Corp., New York, 1979); Part IV.

[6] M. Nazarathy and J. Shamir, Fourier optics described by operator algebra, *J. Opt. Soc. Am.* **70**, 150–158 (1980); *ib.* First-order optics —a canonical operator representation. I. Lossless systems. *J. Opt. Soc. Am.* **72**, 356–364 (1982).

[7] O.N. Stavroudis, *The Optics of Rays, Wavefronts, and Caustics*, (Academic Press, New York, 1972).

[8] A.J. Dragt, Lie algebraic theory of geometrical optics and optical aberrations, *J. Opt. Soc. Am.* **72**, 373–379 (1983).

[9] A.J. Dragt, *Lectures on Nonlinear Orbit Dynamics*, AIP Conference Proceedings N° 87 (American Institute of Physics, New York, 1982).

[10] P.A.M. Dirac, *The Principles of Quantum Mechanics*, (Oxford University Press, 4th Ed., 1958).

[11] G. Lions and M. Vergne, *The Weil Representations, Maslov Index, and Theta Series*, (Birkhäuser, Basel, 1980).

[12] K.B. Wolf, Canonical transforms. I. Complex linear transforms, *J. Math. Phys.* **15**, 1295–1301 (1974).

[13] V. Bargmann, Irreducible unitary representations of the Lorentz group. *Ann. Math.* **48**, 568–640 (1947).

[14] V. Bargmann, Group representations in Hilbert spaces of analytic functions. In: *Analytical Methods in Mathematical Physics*, P. Gilbert and R. G. Newton Eds. (Gordon and Breach, New York, 1970); pp. 27–63.

[15] M. Navarro-Saad and K.B. Wolf, Factorization of the phase-space transformation produced by an arbitrary refracting surface. Preprint CINVESTAV (March 1984); to appear in *J. Opt. Soc. Am.*

[16] O. Castaños, E. López-Moreno, and K.B. Wolf, The Lie-theoretical description of geometric and wave gaussian optics (manuscript in preparation).

[17] J.R. Klauder and E.C.G. Sudarshan, *Fundamentals of Quantum Optics*, (Benjamin, Reading, Mass., 1968).

[18] V.V. Dodonov, E.V. Kurmyshev, and V.I. Man'ko, Generalized uncertainty relation in correlated coherent states, *Phys. Lett.* **79A**, 150–152 (1980).

[19] K.B. Wolf, On time-dependent quadratic quantum Hamiltonians, *SIAM J. Appl. Math.* **40**, 419–431 (1981).

[20] J.W. Goodman, *Introduction to Fourier Optics* (Mc Graw–Hill, New York, 1968).

[21] M. Moshinsky, T.H. Seligman, and K.B. Wolf, Canonical transformations and the radial oscillator and Coulomb problems. *J. Math. Phys.* **13**, 1634–1638 (1972).

[22] K.B. Wolf, Canonical transforms. II. Complex radial transforms. *J. Math. Phys.* **15**, 2102–2111 (1974).

[23] A. Weil, Sur certaines groups d'operateurs unitairs, *Acta Math.* **11**, 143–211 (1963).

[24] K.B. Wolf, Recursive method for the computation of SO_n, $SO_{n,1}$, and ISO_n representation matrix elements, *J. Math. Phys.* **12**, 197–206 (1971).

[25] D. Basu and K.B. Wolf, The unitary irreducible representations of $SL(2, R)$ in all subgroup reductions, *J. Math. Phys.* **23**, 189–205 (1982).

[26] K.B. Wolf, Canonical transforms. IV. Hyperbolic transforms: continuous series of SL(2,R) representations, *J. Math. Phys.* **21**, 680–688 (1980).

[27] D. Basu and K.B. Wolf, The Clebsch-Gordan coefficients of the three-dimensional Lorentz algebra in the parabolic basis, *J. Math. Phys.* **24**,478–500 (1983).

[28] A. Frank and K.B. Wolf, Lie algebras for systems with mixed spectra. The scattering Pöschl-Teller potential. *J. Math. Phys.* **26** 973–983 (1985).

[29] H. Kogelnik, On the propagation of gaussian beams of light through lenslike media including those with a loss or gain variation. *Appl. Opt.* **4**, 1562–1569 (1965).

[30] K.B. Wolf, On self-reproducing functions under a class of integral transforms, *J. Math. Phys.* **18**, 1046–1051 (1977).

[31] V.I. Man'ko and K.B. Wolf, The influence of aberrations in the optics of gaussian beam propagation. Reporte de Investigación, Vol. **3**, # 2 (1985), Departamento de Matemáticas, Universidad Autónoma Metropolitana.[14]

[32] W. Schempp, Radar reception and nilpotent harmonic analysis. I–VI. *C. R. Math. Rep. Acad. Sci. Canada* **4**, 43–48, 139–144, 219–224 (1982); *ibid.* **5**, 121–126 (1983); **6**, 179–182 (1984).

[33] W. Schempp, On the Wigner quasi-probability distribution function. I–III. *C. R. Math. Rep. Acad. Sci. Canada* **4**, 353–358 (1982); *ibid.* **5**, 3–8, 35–40 (1983).

[34] W. Schempp, Radar ambiguity function, nilpotent harmonic analysis, and holomorphic theta series. In *Special Functions: Group Theoretical Aspects and Applications.* Ed. by R.A. Askey, T.H. Koornwinder, and W. Schempp (Reidel, Dordrecht, 1984).

[35] P.M. Woodward, *Probability and Information Theory, with Applications to Radar*, (Artech House, Dedham, Mass., 1980).

[36] M.J. Bastiaans, Wigner distribution function and its applications to first-order optics. *J. Opt. Soc. Am.* **69**, 1710–1716 (1979).

[37] G. García-Calderón and M. Moshinsky, Wigner distribution functions and the representation of canonical transformations in quantum mechanics, *J. Phys. A* **13**, L185–L188 (1980).

[38] M. García-Bullé, W. Lassner, and K.B. Wolf, The metaplectic group within the Heisenberg–Weyl ring. Reporte de Investigación, Vol. **2** # 20 (1985), Departamento de Matemáticas, Universidad Autónoma Metropolitana. To appear in *J. Math. Phys.*

[14] A summary of results is given in Chapter 8 in this volume.

[39] A.J. Dragt and J.M. Finn, Lie series and invariant functions for analytic symplectic maps. *J. Math. Phys.* **17**, 2215–2227 (1976).

[40] W. Lassner, Symbol representations of noncommutative algebras. (submitted for publication, 1985).

[41] M. Born and E. Wolf, *Principles of Optics*, (Pergamon Press, 6th Ed., 1980).

[42] J. Ojeda-Castañeda and A. Boivin, The influence of wave aberrations: an operator approach (preprint, August 1984). To appear in Canadian J. Phys.; J. Ojeda-Castañeda, Focus-error operator and related special functions, *J. Opt. Soc. Am.* **73**, 1042–1047 (1983); A.W. Lohman, J. Ojeda-Castañeda, and N. Streibl, The influence of wave aberrations on the Wigner distribution (preprint, 1984).

[43] K.B. Wolf, A euclidean algebra of hamiltonian observables in Lie optics, *Kinam* **6**, 141–156 (1985).

[44] G.W. Mackey, A theorem of Stone and von Neumann, *Duke Math. J.* **16**, 313–326 (1949).

[45] M. Abramowitz and I. E. Stegun, Eds., *Handbook of Mathematical Functions*, Applied Mathematics Series, Vol. **55** (National Bureau of Standards, Washington D.C., 1st Ed., 1964).

[46] C.P. Boyer and K.B. Wolf, The algebra and group deformations $I^m[SO(n)\otimes SO(m)] \Rightarrow SO(n,m)$, $I^m[U(n) \otimes U(m)] \Rightarrow U(n,m)$, and $I^m[Sp(n) \otimes Sp(m)] \Rightarrow Sp(n,m)$, for $1 \leq m \leq n$. *J. Math. Phys.* **15**, 2096–2100 (1974).

[47] C.P. Boyer and K.B. Wolf, Canonical transforms. III. Configuration and phase descriptions of quantum systems possessing an $sl(2,R)$ dynamical algebra, *J. Math. Phys.* **16**, 1493–1502 (1975).

[48] E.C.G. Sudarshan, R. Simon, and N. Mukunda, Paraxial-wave optics and relativistic front description. I. The scalar theory, *Phys. Rev.* **A28**, 2921–2932 (1983); *ibid.* The vector theory, *Phys. Rev.* **A28**, 2933–2942 (1983).

[49] R. Simon, E.C.G. Sudarshan, and N. Mukunda, Generalized rays in first-order optics: transformation properties of gaussian Schell-model fields, *Phys. Rev.* **A29**, 3273–3279 (1984).

[50] N. Mukunda, R. Simon, and E.C.G. Sudarshan, Fourier optics for the Maxwell field: formalism and applications, *J. Opt. Soc. Am.* **A2**, 416–426 (1985).

Chapter 6

Wave theory of imaging systems

by JOHN R. KLAUDER

ABSTRACT: The Lie-algebraic formulation of geometrical rays for imaging systems is transcribed to a form suitable for wave fields. The merits of a coherent-state wave field formulation are stressed along with a related path-integral representation suitable for a study of general aberrations.

6.1 Rays

6.1.1 Gaussian ray optics

Among the variety of problems dealt with by wave equations or by their geometrical ray approximations, are those which may be termed *imaging* problems. An imaging problem is characterized by long stretches of uniform propagation in (say) the z direction, punctuated by short intervals of sudden change such as brought about by a thin lens, etc. Such systems arise in conventional optics, electron optics and magnetic optics, for example. As a first approximation it suffices to treat linear transformations of the transverse ray coordinates $\mathbf{q}(z), \mathbf{q} = (q_x, q_y) = (x, y)$, and momenta $\mathbf{p}(z) = n\cos\theta\, d\mathbf{q}(z)/dz$, $\mathbf{p} = (p_x, p_y)$, and $p = |\mathbf{p}| = n\sin\theta$, where θ is the angle between the direction of propagation of the ray and the z axis, and n is the index of refraction. For example, the linear transformation

$$(\mathbf{p},\mathbf{q}) \mapsto (\mathbf{p}',\mathbf{q}') = (\mathbf{p}, \mathbf{q} + \ell\mathbf{p}/n), \tag{1a}$$

corresponds to uniform propagation for a distance ℓ in an isotropic, homogeneous medium of index n. The linear transformation

$$(\mathbf{p},\mathbf{q}) \mapsto (\mathbf{p}',\mathbf{q}') = \big(\mathbf{p} + (\bar{n} - n)\mathbf{q}/r, \mathbf{q}\big), \tag{1b}$$

corresponds to the change in orientation induced by a thin axially symmetric, spherical lens of index $\bar{n}$ and of radius r; a cylindrical lens changes only one component of $\mathbf{p}$. In addition the linear transformation

$$(\mathbf{p},\mathbf{q}) \mapsto (\mathbf{p}',\mathbf{q}') = (e^{-\alpha}\mathbf{p}, e^{\alpha}\mathbf{q}), \tag{1c}$$

corresponds to a telescoping and imaging system.

6.1.2 Lie algebraic formulation

An elegant classical theory of ray propagation has been developed by Dragt [1] in which z plays a role analogous to time in particle mechanics, and which makes systematic use of classical canonical transformations. Recall that a canonical transformation of coordinates is a map $p'_\alpha = p'_\alpha(\mathbf{p},\mathbf{q})$, $q'_\alpha = q'_\alpha(\mathbf{p},\mathbf{q})$ such that

$$\{q'_\alpha, p'_\beta\} = \delta_{\alpha\beta}, \qquad \alpha,\beta = x,y, \tag{2a}$$

where the Poisson brackets are defined by

$$\{f(\mathbf{p},\mathbf{q}), g(\mathbf{p},\mathbf{q})\} := \sum_\alpha \left(\frac{\partial f}{\partial q_\alpha}\frac{\partial g}{\partial p_\alpha} - \frac{\partial f}{\partial p_\alpha}\frac{\partial g}{\partial q_\alpha} \right). \tag{2b}$$

If we introduce the operator $:f:$ and its iterates defined by

$$:f{:}g := \{f, g\}, \tag{3a}$$

$$:f{:}^2 g := :f::f{:}g = \{f,\{f,g\}\}, \tag{3b}$$

etc., then it follows that every transformation of the form

$$g' = e^{:f:}g = \sum_{n=0}^{\infty} \frac{1}{n!}(:f:)^n g \tag{3c}$$

is a canonical transformation for which f serves as the generator (however, see the cautionary remark below). In particular the typically nonlinear transformation $(\mathbf{p}',\mathbf{q}') = e^{:f:}(\mathbf{p},\mathbf{q})$ leads to new canonical coordinates. For example, it readily follows that

$$(\mathbf{p}, \mathbf{q} + \ell\mathbf{p}/n) = e^{-\frac{1}{2}(\ell/n):\mathbf{p}^2:}(\mathbf{p},\mathbf{q}), \tag{4a}$$

$$(\mathbf{p} + (\overline{n} - n)\mathbf{q}/r, \mathbf{q}) = e^{\frac{1}{2}(\overline{n}-n)/r:\mathbf{q}^2:}(\mathbf{p},\mathbf{q}), \tag{4b}$$

$$(e^{-\alpha}\mathbf{p}, e^{\alpha}\mathbf{q}) = e^{-\alpha:\mathbf{p}\cdot\mathbf{q}:}(\mathbf{p},\mathbf{q}), \tag{4c}$$

which explicitly displays the canonical maps leading to the indicated linear transformations. These maps may be concatenated one after another to describe the overall linear canonical transformation induced by an ideal (gaussian) imaging system.

6.1.3 Aberrations

In practice a real lens typically induces a nonlinear canonical transformation, one for which the generator is not just a homogeneous quadratic function of the canonical coordinates. It proves especially useful to characterize a given system element by the canonical map it induces when this map is expressed in a concatenated standard representation such as

$$e^{:f_2:}e^{:f_3:}e^{:f_4:}\ldots, \tag{5}$$

where f_n denotes a homogeneous polynomial of the canonical coordinates of degree n. The terms $f_n, n \geq 3$, introduce aberrations (deviations from the linear Gaussian rays), in particular, aberrations of order $(n-1)$ [1]. For example, 2nd-order aberrations are generated by

$$f_3 = a_{\alpha\beta\gamma}p_\alpha p_\beta p_\gamma + b_{\alpha\beta\gamma}p_\alpha p_\beta q_\gamma + c_{\alpha\beta\gamma}p_\alpha q_\beta q_\gamma + d_{\alpha\beta\gamma}q_\alpha q_\beta q_\gamma, \tag{6a}$$

for suitable coefficients ($\alpha, \beta, \gamma = x, y$, with summation understood). For a system with cylindrical symmetry $f_3 = 0$ and f_4 takes the form

$$f_4 = A(\mathbf{p}^2)^2 + B\mathbf{p}^2(\mathbf{p}\cdot\mathbf{q}) + C(\mathbf{p}\cdot\mathbf{q})^2 + D\mathbf{p}^2\mathbf{q}^2 + E(\mathbf{p}\cdot\mathbf{q})\mathbf{q}^2 + F(\mathbf{q}^2)^2. \tag{6b}$$

We return subsequently to comment on several of these aberrations and the transformations they induce.

The combined action of two elements each of which is expressed in the chosen standard representation induces an overall canonical map which in turn may itself be reexpressed in terms of the standard representation by repeated application of the Lie-algebra rules for combining generators of canonical transformations. Thus it becomes possible in principle to obtain expressions for $h = \{h_2, h_3, \cdots\}$, in terms of the generators $g = \{g_n\}$ and $f = \{f_n\}$, where

$$e^{:g_2:}e^{:g_3:}e^{:g_4:}\ldots e^{:f_2:}e^{:f_3:}e^{:f_4:}\ldots = e^{:h_2:}e^{:h_3:}e^{:h_4:}\ldots. \tag{7}$$

Generally, whenever g or f involve aberrations (any nonzero generators for $n \geq 3$), then there are an infinite number of h terms. However, if $f_n, g_n, n \geq 3$ are suitably small, then h_n, $n \geq 3$, are likewise correspondingly small, and it makes sense to truncate the h series at some order. With such a truncation accepted it becomes possible to consistently approximate standard representations of canonical maps and their products to a given prescribed order. Indeed, catologs of such maps and their products accurate to the desired order become feasible using Lie algebraic methods, and catalogs up to order $n = 8$ now exist [2]. The advantage of this prescription is that it ensures that the approximate map is still a *canonical* transformation. If $\rho(\mathbf{p}, \mathbf{q})$ and $\rho'(\mathbf{p}, \mathbf{q})$ denote ray densities before and after, respectively, the canonical transformation, then it follows that the total number of rays is conserved, *i.e.*,

$$\int d\mathbf{p}\, d\mathbf{q}\, \rho(\mathbf{p}, \mathbf{q}) = \int d\mathbf{p}\, d\mathbf{q}\, \rho'(\mathbf{p}, \mathbf{q}). \tag{8}$$

This relation is an immediate consequence of the identity

$$\int d\mathbf{p}\, d\mathbf{q}\, g e^{:f:} h = \int d\mathbf{p}\, d\mathbf{q}\, h e^{-:f:} g. \tag{9}$$

For application of magnetic optics to particle accelerator rings, in which many million iterations of a single cycle's transformation are often involved, the conservation of the total number of rays (idealized noninteracting particles) is essential.

6.1.4 Singular canonical transformations

It should be appreciated that certain polynomial generators of potential interest actually induce *singular* canonical transformations. For example, consider the transformation induced by the generator $f_3 = bpq^2$, where $p = p_x, q = q_x$ (here we ignore p_y and q_y), namely

$$e^{:bpq^2:}(p, q) = \big(p(b), q(b)\big). \tag{10a}$$

The mapping

$$p(b) = p(1 + bq)^2, \tag{10b}$$

$$q(b) = \frac{q}{1 + bq}, \tag{10c}$$

defined for $|bq| < 1$ by a power series and extended beyond by analyticity, exhibits a singularity at $q = -1/b$. If no rays cross the singularity then we have a perfectly acceptable canonical transformation, while more generally the mapping is canonical almost everywhere. In most applications the singularities can be glossed over and such transformations need cause no trouble.

However, consider the mapping generated by $f_4 = c(\mathbf{p}\cdot\mathbf{q})\mathbf{q}^2$, for which

$$e^{:c(\mathbf{p}\cdot\mathbf{q})\mathbf{q}^2:}(\mathbf{p},\mathbf{q}) = (\mathbf{p}(c),\mathbf{q}(c)), \tag{11a}$$

where

$$\mathbf{p}(c) = \sqrt{1+2c\mathbf{q}^2}\,[\mathbf{p} + 2c(\mathbf{p}\cdot\mathbf{q})\mathbf{q}], \tag{11b}$$

$$\mathbf{q}(c) = \frac{\mathbf{q}}{\sqrt{1+2c\mathbf{q}^2}}\,. \tag{11c}$$

For $c > 0$ this mapping is a singularity-free canonical transformation. For $c < 0$, on the other hand, the mapping is singular for $1 + 2c\mathbf{q}^2 = 0$, and *complex* whenever $1 + 2c\mathbf{q}^2 < 0$. There is no way to give such a mapping a global interpretation even if the singular points are omitted. Since, for $c > 0$, the map (11) satisfies the constraint $\mathbf{q}^2(c) < 1/2c$, it follows that a negative c map may be applied and treated as canonical only if it has been preceded by a positive c map of equal or greater strength (or its equivalent).

Entirely similar remarks apply to $f_3' = b'p^2q$ and $f_4' = c'\mathbf{p}^2(\mathbf{p}\cdot\mathbf{q})$. Other singular transformations exist, but we content ourselves with these few examples.

6.2 Waves

6.2.1 Configuration representation

It is familiar that the relation of geometrical optics to wave optics is the same as particle mechanics to quantum mechanics, and therefore it is no surprise that quantum mechanical methods applied to ray optics yield scalar wave optics. In the analogue of the Schrödinger representation, the wave field $\psi(\mathbf{q},z)$ replaces a ray $\mathbf{q}(z)$, or a collection of rays, and $\rho(\mathbf{q},z) = |\psi(\mathbf{q},z)|^2$ is interpreted as the relative probability to find a ray at $\mathbf{q}, z$ if a measurement were made. Here, again, we note that z assumes the role of time in these problems. The role of the Planck constant $\hbar$ is played by $\bar{\lambda} := \lambda/2\pi = 1/k$, the reduced wave length of the field. Thus ray optics becomes more appropriate for short wave lengths, i.e., for large wave numbers k. In the usual limit of many particles (rays) the relative probability is proportional to the optical field intensity, and so we may interpret $\psi(\mathbf{q},z)$ as the field amplitude and $|\psi(\mathbf{q},z)|^2$ as the field intensity at $(\mathbf{q},z)$ as well.

In the Schrödinger (configuration) representation the coordinates $\mathbf{Q}$ act as multiplication by $\mathbf{q}$, while the momenta $\mathbf{P}$ act as differential operators, $\mathbf{P} = ik^{-1}\partial/\partial\mathbf{q}$. Each real classical phase-space function $f(\mathbf{p},\mathbf{q})$ is transcribed into a Hermitian operator

$$F(\mathbf{P},\mathbf{Q}) := \mathsf{W}[f(\mathbf{P},\mathbf{Q})] \tag{12}$$

by some choice of ordering prescription chosen here as *Weyl ordering* and represented by the ordering symbol W. Apart from ordering ambiguities, Poisson brackets relate to quantum commutators according to the rule

$$ik^{-1}\{f,g\} \quad \leftrightarrow \quad [F,G]. \tag{13a}$$

Thus, again apart from ordering ambiguities, we are able to transcribe canonical maps into operator isomorphisms,

$$e^{:f:}g \quad \leftrightarrow \quad e^{-ikF}Ge^{ikF}. \tag{13b}$$

Such a view corresponds to the Heisenberg picture of quantum machanics, but we may adopt the Schrödinger picture as well, where one need only substitute ikF for $:f:$.

For example, free propagation for a distance ℓ in a medium of index n [*cf.* (1*a*)] arises for

$$\begin{aligned}\psi(\mathbf{q}, z+\ell) &= \exp(-i\frac{1}{2}\frac{k\ell}{n}\mathbf{P}^2)\psi(\mathbf{q}, z)\\ &= \frac{kn}{2\pi i\ell}\int d\mathbf{u}\, \exp(i\frac{1}{2}\frac{kn}{\ell}(\mathbf{q}-\mathbf{u})^2)\psi(\mathbf{u}, z).\end{aligned} \tag{14a}$$

The transformation induced by a thin axially symmetric lens [*cf.*, (1*b*)] is even simpler taking the form

$$\begin{aligned}\psi'(\mathbf{q}) &= \exp(i\tfrac{1}{2}\frac{k(\overline{n}-n)}{r}\mathbf{Q}^2)\,\psi(\mathbf{q})\\ &= \exp(i\tfrac{1}{2}\frac{k(\overline{n}-n)}{r}\mathbf{q}^2)\,\psi(\mathbf{q}).\end{aligned} \tag{14b}$$

More generally, corresponding to the classical generator $f(\mathbf{p}, \mathbf{q})$ we have the transformation

$$\psi_F(\mathbf{q}) = \exp\big(ikF(\mathbf{P}, \mathbf{Q})\big)\,\psi(\mathbf{q}), \tag{15}$$

where F is given by (12). If f is a homogeneous polynomial then so too is F. Since F is a Hermitian operator is it suggestive that $\exp(ikF)$ is a unique unitary transformation, but this is not always the case; it is only true if F is (essentially) self adjoint. Let us consider the quantum transformations induced by the two examples discussed at the end of the preceding section.

For $F_3 = \frac{1}{2}b(PQ^2 + Q^2P$, $P = P_x$, $Q = Q_x$ (P_y and Q_y are ignored) the singularity in the associated classical map translates into the fact that there are not one but many different unitary transformations $\exp(ikF)$ distinguished from one another by different boundary conditions at $|q| = \infty$. A natural choice of the appropriate unitary operator is not always evident, especially in more complicated examples of singular transformations.

For $F_4 = \frac{1}{2}c[(\mathbf{P}\cdot\mathbf{Q})\mathbf{Q}^2 + \mathbf{Q}^2(\mathbf{Q}\cdot\mathbf{P})]$ the nonglobal nature of the associated classical map for $c < 0$ translates into the assertion that $\exp(ikF_4)$ is *never* a unitary operator, as follows from the fact that as a differential equation $F_4\phi(\mathbf{q}) = -i\phi(\mathbf{q})$ has a square-integrable solution for any $c > 0$ but not if $c < 0$. Instead, $\exp(ikF_4)$ is *isometric* for $c > 0$ and *contractive* for $c < 0$, in the very same sense as for the operator $\exp(iG)$, $G = icd/du$, on the space of square-integrable functions with $0 < u < \infty$ [3].

Of course, remarks similar to the those above apply to the polynomials $F_3' = \frac{1}{2}b'(P^2Q + QP^2)$ and to $F_4' = \frac{1}{2}c'[\mathbf{P}^2(\mathbf{P}\cdot\mathbf{Q}) + (\mathbf{Q}\cdot\mathbf{P})\mathbf{P}^2]$, as well as to other polynomial operators.

6.2.2 Integral kernels

Corresponding to the standard classical representation (5) for an imaging element is a corresponding standard quantum representation of the same element given by

$$e^{ikF_2}e^{ikF_3}e^{ikF_4}\cdots. \tag{16}$$

In the configuration representation each of the component transformations has an integral kernel representation so that the field amplitude transforms as

$$\psi'(\mathbf{q}, z) = \int d\mathbf{q}_2 \int d\mathbf{q}_3 \cdots \int d\mathbf{q}'\, K_2(\mathbf{q}, \mathbf{q}_2)\, K_3(\mathbf{q}_2, \mathbf{q}_3)\, K_4(\mathbf{q}_3, \mathbf{q}_4) \cdots \psi(\mathbf{q}', z), \tag{17a}$$

where, in standard quantum notation,

$$K_n(\mathbf{q}_{n-1}, \mathbf{q}_n) := \langle \mathbf{q}_{n-1} \mid e^{ikF_n} \mid \mathbf{q}_n \rangle. \tag{17b}$$

Evaluation of such kernels and multiple integrals is generally not simple, and recourse to approximate evaluations, such as stationary phase approximations or sophisticated ray tracing approximations, are often useful. Greater care may be needed to adequately handle ray crossings or caustics where the integral kernel K or its approximation may diverge.

6.2.3 Coherent-state representation

Although the analog of the Schrödinger representation may be perfectly adequate to describe the problems of wave optics it may not be the most ideal representation in all circumstances. On one hand, we have already seen in the discussion of aberrations the symmetry between $\mathbf{p}$ and $\mathbf{q}$ that holds in the most general case. On the other hand, focussing on $\mathbf{q}$ alone, as in the Schrödinger representation, is part of the reason that approximations near caustics require extra care.

A natural representation of the field amplitude that addresses both aspects mentioned above is provided by the analog of the coherent-state representation. We emphasize that this representation is adopted here for convenience and for its special properties to be discussed, and its use should not be confused with its most famous application to yet another aspect of optics, namely quantum optics. That being understood we are certainly free to draw on the wealth of coherent-state techniques developed elsewhere, and we content ourselves here with a statement of the principal results needed [4].

The coherent-state representation of the field amplitude is denoted by $\psi(\mathbf{p}, \mathbf{q}, z)$, and to avoid confusion we shall hereafter generally refer to the configuration representation by $\psi(\mathbf{u}, z)$ with $\mathbf{u} = (x, y)$. The connection between the two representations is given by the relations

$$\psi(\mathbf{p}, \mathbf{q}, z) = \sqrt{\frac{k}{\pi}} \int d\mathbf{u}\, \exp(-k\tfrac{1}{2}\mathbf{u}^2 - ik\mathbf{p}\cdot\mathbf{u})\, \psi(\mathbf{u} + \mathbf{q}, z), \tag{18a}$$

$$\psi(\mathbf{u}, z) = \frac{\sqrt{k/\pi}}{\lambda^2} \int d\mathbf{p}\, d\mathbf{q}\, \exp(-k\tfrac{1}{2}(\mathbf{u}-\mathbf{q})^2 + ik\mathbf{p}\cdot(\mathbf{u}-\mathbf{q}))\, \psi(\mathbf{p}, \mathbf{q}, z), \tag{18b}$$

$$\psi(\mathbf{q}, z) = \frac{\sqrt{\pi/k}}{\lambda^2} \int d\mathbf{p}\, \psi(\mathbf{p}, \mathbf{q}, z). \tag{18c}$$

Note that (18b) and (18c) both lead to the configuration representation, with (18c) being clearly more economical.

In the bra-ket notation we set

$$|\,\mathbf{p}, \mathbf{q}\rangle := \exp(-i\mathbf{q}\cdot\mathbf{P}) \exp(i\mathbf{p}\cdot\mathbf{Q})|\,0\,\rangle, \tag{19a}$$

which are a collection of unit vectors with $|\,0\,\rangle$ the normalized solution of the conditions $(Q_1 + iP_1)|\,0\,\rangle = (Q_2 + iP_2)|\,0\,\rangle$. In this notation $\psi(\mathbf{p}, \mathbf{q}, z) = \langle \mathbf{p}, \mathbf{q} | \psi(z) \rangle$, which is a bounded, continuous function. The advertised symmetry between $\mathbf{p}$ and $\mathbf{q}$ is made clearly evident in the overlap

$$\langle \mathbf{p}_2, \mathbf{q}_2 \,|\, \mathbf{p}_1, \mathbf{q}_1 \rangle = \exp\{\tfrac{1}{2}k[i(\mathbf{p}_2 + \mathbf{p}_1)\cdot(\mathbf{q}_2 - \mathbf{q}_1) - \tfrac{1}{2}(\mathbf{p}_2 - \mathbf{p}_1)^2 - \tfrac{1}{2}(\mathbf{q}_2 - \mathbf{q}_1)^2]\} \tag{19b}$$

which, unlike the overlap of two Schrödinger bra-ket states $\langle \mathbf{y}_2 \,|\, \mathbf{y}_1 \rangle = \delta(\mathbf{y}_2 - \mathbf{y}_1)$, is bounded and continuous (and, up to overall factors, suitably analytic) in its arguments.

The function $\psi(\mathbf{p}, \mathbf{q}, z)$ is just the field amplitude averaged over a pencil of rays centered at position $\mathbf{q}$ and momentum $\mathbf{p}$. From the relations

$$\langle \mathbf{p}, \mathbf{q} \mid \mathbf{P} \mid \mathbf{p}, \mathbf{q} \rangle = \mathbf{p}, \tag{20a}$$

$$\langle \mathbf{p}, \mathbf{q} \mid \mathbf{Q} \mid \mathbf{p}, \mathbf{q} \rangle = \mathbf{q}, \tag{20b}$$

and

$$\langle \mathbf{p}, \mathbf{q} \mid (\mathbf{P} - \mathbf{p})^2 \mid \mathbf{p}, \mathbf{q} \rangle = k^{-1}, \tag{20c}$$

$$\langle \mathbf{p}, \mathbf{q} \mid (\mathbf{Q} - \mathbf{q})^2 \mid \mathbf{p}, \mathbf{q} \rangle = k^{-1}, \tag{20d}$$

we learn that the width of the pencil is roughly k^{-1} in both position and momentum. The integral

$$\lambda^{-2} \int_B d\mathbf{p}\, d\mathbf{q}\, |\psi(\mathbf{p}, \mathbf{q}, z)|^2 \tag{21a}$$

effectively measures the field intensity in the phase-space cell B, and in particular, if B is all of phase space then

$$\lambda^{-2} \int d\mathbf{p}\, d\mathbf{q}\, |\psi(\mathbf{p}, \mathbf{q}, z)|^2 = \int d\mathbf{u}\, |\psi(\mathbf{u}, z)|^2. \tag{21b}$$

The fundamental resolution of unity given by

$$1 = \int |\mathbf{p}, \mathbf{q}\rangle \frac{d\mathbf{p}\, d\mathbf{q}}{\lambda^2} \langle \mathbf{p}, \mathbf{q}| \tag{22}$$

readily permits the formulation of coherent-state-representation expressions [4]. For example, the analog of (17) now reads

$$\begin{aligned}\psi'(\mathbf{p}, \mathbf{q}, z) = \int \frac{d\mathbf{p}_2 d\mathbf{q}_2}{\lambda^2} \cdots \int \frac{d\mathbf{p}' d\mathbf{q}'}{\lambda^2} \langle \mathbf{p}, \mathbf{q} \mid e^{ikF_2} \mid \mathbf{p}_2, \mathbf{q}_2 \rangle \\ \times \langle \mathbf{p}_2, \mathbf{q}_2 \mid e^{ikF_3} \mid \mathbf{p}_3, \mathbf{q}_3 \rangle \langle \mathbf{p}_3, \mathbf{q}_3 \mid e^{ikF_4} \mid \mathbf{p}_4, \mathbf{q}_4 \rangle \cdots \psi(\mathbf{p}', \mathbf{q}', z).\end{aligned} \tag{23}$$

Note that each integral kernel involved here is a continuous function in all arguments which furthermore is uniformly bounded by one. In the context of Eq. (18c) it is worth emphasizing that peaks in the field amplitude, which exist typically at caustics or ray crossings, arise in the configuration representation through coherent superposition in the integral over $\mathbf{p}$. The reason for this phenomenon is physically plausible in a heuristic ray tracing picture: In the configuration representation of an imaging system, all rays that start at $\mathbf{u}_{\text{initial}}$ and end at $\mathbf{u}_{\text{final}}$ in a "time" z contribute; if many rays contribute the amplitude tends to peak. In the coherent-state representation, on the other hand, the *unique* ray (if any) that starts at $\mathbf{q}_{\text{initial}}$ with momentum ($\approx$ orientation) $\mathbf{p}_{\text{initial}}$ and ends at $\mathbf{q}_{\text{final}}$ with momentum $\mathbf{p}_{\text{final}}$ in a "time" z contributes, and so peaking in the coherent-state representation is not possible.

6.2.4 Coherent-state path-integral representation

For our final topic we turn to a special representation of the integral kernels in (23) that make up an element of the overall system. We have in mind a path-integral representation, one that is adapted to the coherent-state field representation. Indeed, the conventional configuration-space path integral is only well-suited to generators having a quadratic momentum dependence and thus is not appropriate in the present case. The conventional phase-space path integral can handle a wider class of generators, but, despite appearances, it does not respect the $\mathbf{p}, \mathbf{q}$ symmetry nor does it optimally deal with the simultaneous contribution of many rays. Only a coherent-state path integral overcomes these various problems [5].

Coherent-state path integrals, like other path integrals, may be defined as the limit of a lattice-space formulation as the lattice spacing goes to zero. Even for this kind of construction there are still several different lattice prescriptions that are possible. Instead we would prefer an alternatively regularized formulation that retains a close correspondence with the continuous "time" variable of the classical theory. Fortunately, a recently obtained precise formulation of coherent-state path integrals that is applicable for general polynomial generators appears ideally suited to our purposes [6]. Without offering any proofs we now present the basics of that formulation as transcribed into the present language of wave optics. Let $f(\mathbf{p},\mathbf{q})$ be a (real) classical polynomial generator of canonical transformations and $F(\mathbf{P},\mathbf{Q}) = \mathrm{W}[f(\mathbf{P},\mathbf{Q})]$ the hermitian polynomial operator determined by Weyl symmetrization. Associate to $f(\mathbf{p},\mathbf{q})$ the real phase-space polynomial

$$\tilde{f}(\mathbf{p},\mathbf{q}) = e^{-\Delta/(4k)} f(\mathbf{p},\mathbf{q}), \tag{24a}$$

$$\Delta := \frac{\partial^2}{\partial \mathbf{p}^2} + \frac{\partial^2}{\partial \mathbf{q}^2}; \tag{24b}$$

$\tilde{f}$ is sometimes called the *antinormal*-ordered symbol related to F. Then it may be shown [6] that

$$\langle \mathbf{p}'',\mathbf{q}'' \mid \exp(ikF(\mathbf{P},\mathbf{Q}))| \,\mathbf{p}',\mathbf{q}'\rangle = \lim_{D\to\infty} \lambda^2 e^D \int d\mu^D(\mathbf{p},\mathbf{q}) \exp\Big(ik\int_0^1 dt\,\{\mathbf{p}(t)\cdot\dot{\mathbf{q}}(t) + \tilde{f}[\mathbf{p}(t),\mathbf{q}(t)]\}\Big). \tag{25}$$

In this expression μ^D is a Wiener measure on independent phase-space paths $\mathbf{p}(t)$ and $\mathbf{q}(t), 0 \le t \le 1$, pinned so that $\mathbf{p}(0),\mathbf{q}(0) = \mathbf{p}',\mathbf{q}'$ and $\mathbf{p}(1),\mathbf{q}(1) = \mathbf{p}'',\mathbf{q}''$, with a total weight

$$\int d\mu^D(\mathbf{p},\mathbf{q}) = \frac{1}{2\pi D}\exp\Big(\frac{-(\mathbf{p}''-\mathbf{p}')^2-(\mathbf{q}''-\mathbf{q}')^2}{2D}\Big), \tag{26a}$$

and a normalized connected covariance given by $(0 \le t_< \le t_> \le 1)$

$$\begin{aligned}\langle p_\alpha(t_<)p_\beta(t_>)\rangle_c &= \langle q_\alpha(t_<)q_\beta(t_>)\rangle_c \\ &= \delta_{\alpha\beta}\, D\, t_<\,(1-t_>).\end{aligned} \tag{26b}$$

The integral $\int dt\,\mathbf{p}\cdot\dot{\mathbf{q}}$ should be interpreted as the stochastic integral $\int \mathbf{p}\cdot d\mathbf{q}$, in which case (25) is given as the limit of well-defined path integrals as the diffusion constant D diverges. One advantage of the formulation offered by (25) is the ability to make variable changes, *e.g.*, canonical transformations, in the path integral in a controlled way.

When F is essentially self adjoint on the space of finite linear combinations of coherent states then the integral kernel given by (25) determines a unique unitary operator. However, as noted earlier $F = F_3 = \frac{1}{2}b(PQ^2 + Q^2P)$ does not lead to a unique unitary operator. Because the path-integral representation implicitly contains more structure than the abstract operators, then for $\tilde{f} = \tilde{f}_3 = b[pq^2 - p/(2k)]$, appropriate to F_3, the right side of (25) automatically selects the natural choice for the associated unitary operator. This is an additional and desirable bonus offered by the coherent-state path integral representation. For $F = F_4 = \frac{1}{2}c[(\mathbf{P}\cdot\mathbf{Q})\mathbf{Q}^2 + \mathbf{Q}^2(\mathbf{Q}\cdot\mathbf{P})]$ the operator $\exp(icF_4)$ is isometric for $c > 0$ and contractive for $c < 0$, and these properties are fairly direct consequences of the coherent-state path integral as well, as must be the case. Whenever the transformation is isometric the total intensity (21b) is conserved in analogy with (8).

A straightforward extension of Eq. (25) permits concatenated transformations to be represented in nearly as concise a fashion. For example, it follows that

$$\langle \mathbf{p}'',\mathbf{q}'' \mid e^{ikF_2}e^{ikF_3}e^{ikF_4} \mid \mathbf{p}',\mathbf{q}'\rangle = \lim_{D\to\infty} \lambda^2 e^{3D} \int d\mu^D(\mathbf{p},\mathbf{q}) \exp\Big(ik\int_0^3 dt\,\{\mathbf{p}(t)\cdot\dot{\mathbf{q}}(t) + \tilde{f}[\mathbf{p}(t),\mathbf{q}(t),t]\}\Big), \tag{27}$$

where in the present case

$$\tilde{f}(\mathbf{p},\mathbf{q},t) := \tilde{f}_2(\mathbf{p},\mathbf{q}), \qquad 2 \le t \le 3, \tag{28a}$$
$$:= \tilde{f}_3(\mathbf{p},\mathbf{q}), \qquad 1 \le t < 2, \tag{28b}$$
$$:= \tilde{f}_4(\mathbf{p},\mathbf{q}), \qquad 0 \le t < 1, \tag{28c}$$

and μ^D is redefined accordingly.

We remark that such path integrals may be approximately evaluated by perturbation expansion in the aberration coefficients, or by an expansion in the nonquadratic terms in $\tilde{f}$ (which is not exactly the same as f since generally $\tilde{f} \neq f$). In addition a stationary-phase approximation may be used. Observe that a stationary-phase approximation for a coherent-state path integral differs notably from one for a configuration-space path integral. The appropriate version that results may be traced from work on the subject already in print [7].

It is clear that the coherent-state path integral —just as the configuration-space path integral [8]— can be applied to distributed systems, to inhomogeneous and nonisotropic propagation media, and to propagation media with fluctuations. These additional aspects will be dealt with in a separate article.

References

[1] A.J. Dragt, *J. Opt. Soc. Am.* **72**, 372 (1982); *ib.* ***Lectures on Nonlinear Orbit Dynamics***, AIP Conference Proceedings, Vol. 87 (1982).

[2] E. Forest, ***Lie Algebraic Methods for Charged Particle Beams and Light Optics***, University of Maryland, Department of Physics and Astrophysics, Ph.D. Thesis, 1984 (unpublished).

[3] See, *e.g.*, F. Riesz and B. Sz.-Nagy, ***Functional Analysis***, (Frederick Ungar, New York, 1955).

[4] See, *e.g.*, J.R. Klauder and E.C.G. Sudarshan, ***Fundamentals of Quantum Optics***, (Benjamin, New York, 1968), Chapter 7.

[5] L.S. Shulman, ***Techniques and Applications of Path Integration***, (Wiley-Interscience, New York, 1981).

[6] J.R. Klauder and I. Daubechies, ***Phys. Rev. Lett.*** **52**, 1161 (1984); I. Daubechies and J.R. Klauder, *J. Math. Phys.* **26**, 2239 (1985).

[7] J.R. Klauder, in ***Path Integrals***, Ed. by G.J. Papadopoulos and J.T. Devreese (Plenum, New York, 1978), p. 5; *ib. Phys. Rev.* **D19**, 2349 (1979).

[8] S.M. Flatté, D.R. Bernstein, and R. Dashen, *Phys. Fluids* **26**, 1701 (1983).

Chapter 7

Invariants and coherent states in fiber optics

by VLADIMIR I. MAN'KO

ABSTRACT: The aim of this chapter is to review recent results in the quantum mechanics of nonstationary systems, and to demonstrate how they are applied in paraxial fiber optics.

7.1 Introduction

The behaviour of light beams in media is described by the Maxwell equations. For *paraxial* beams, *i.e.*, those containing rays very close to parallel with the optical axis of the system, it was shown by Fock and Leontovich [1] that the Maxwell equations of the medium can be replaced with very good accuracy by a parabolic-type equation. This equation is formally, in fact, completely equivalent to the Schrödinger equation for the two-dimensional time-dependent potential-well quantum system. The statement of Leontovich and Fock [1] holds not only for light beams in optical media, but for any wave process, for example acoustic waves in water or solids, for elastic waves in the earth, etc.

The wave process of any kind is described by a Schrödinger-like equation if its propagation obeys the Helmholtz equation, and if the corresponding wave beam is paraxial. The Leontovich-Fock paraxial approximation used for electromagnetic waves in reference [1] was also applied for acoustic beams (see the book by Brechovskih [2]) and for elastic-wave geophysics [3]. During the last twenty years, the consideration of light beam propagation in nonhomogeneous media in the paraxial approximation was investigated in detail using the analogy with the quantum mechanical Schrödinger equation. First, the known quantum mechanical results were used for the stationary states in time-independent potential wells. The results are reviewed in the books by Marcuse [4] and Arnaud [5].

The main point of this approach is to use the results of quantum mechanics and to apply these results formally in the theory of light propagation in nonhomogeneous media. So, when we obtained new results in the quantum mechanics of systems with time-dependent Hamiltonians [6–11], we formulated the possibility to use these results in the problem of light propagation. These new results (which had not been used in references [4] and [5]) are connected with the coherent-state representation [12] for *time-dependent* quantum systems. New integrals of motion for nonstationary quantum systems were found explicitly for the many-dimensional forced quantum oscillator with time-dependent frequency in [12,14–16]. The $ISp(2n,\Re)$ inhomogeneous symplectic group was found to be the dynamical group [17] of this oscillator. Thus, in addition to the results of quantum mechanics used in the theory of optical waveguides in [4,5,18–22], in reference [12] we formulated the possibility of applying the Franck–Condon principle for the calculation of the coupling coefficients between two fibers. The known results on gaussian beam propagation in nonhomogeneous media (see the complex-ray theory in [4]) were interpreted in [12] as light beam coherent states propagating in these media.

The properties of the coherent-state representation (nonorthogonality and completeness) have been used in order to explain the completeness of complex rays. Using the coherent state properties, we formulated in [12] the possibility to use the matrix elements of the $ISp(2n,\Re)$ irreducible representations in the Fock basis, in order to reproduce the results obtained in [21]; in the latter it was shown that the mode coupling coefficients of two connected fibers are expressed in terms of Hermite polynomials of four variables. The new set of time-dependent invariants was an important point in our considerations of reference [12]. The approach, based on coherent states, the new invariants, the symplectic dynamical group $Isp(2n,\Re)$, the density matrix formalism, the analogue of the Franck–Condon principle for fibers, and the possibility of using the quantum invariants for the calculation of the correlation functions [12], were applied and developed in detail in references [23–29]. The analogues for light beams in heterolasers, of the quantum-mechanical Franck–Condon principle and of the Ramsauser effect, were formulated in references [30–32].

The Fock–Leontovich approach in the frame of group-theoretical considerations has also been investigated intensively by H. Bacry in [33,34]. The alternative Hamiltonian approach based on the Fermat principle and its quantized version (see [4,5,35]) has been studied in a series of articles by K.B. Wolf in [36–41], through the use of the symplectic group formalism. There, the aberrations of higher order have been calculated explicitly. Also in this Hamiltonian frame, in reference [42] we studied coherent states under spherical aberration, using the generating functional technique.

7.2 The parabolic equation

We consider a light beam of a fixed frequency ω and given by its field, or wavefunction, $\phi(x,y,z)$; then, for any projection of the magnetic or electric field, it obeys the Helmholtz equation

$$\frac{\partial^2\phi}{\partial x^2}+\frac{\partial^2\phi}{\partial y^2}+\frac{\partial^2\phi}{\partial z^2}+\frac{\omega^2}{c^2}n(x,y,z)^2\phi=0, \tag{1}$$

where n is the refraction index for a general, inhomogeneous medium. If we consider that the beam is almost parallel to the z axis and denote $n_0=n(0,0,z)$, then the field can be written

$$\phi(x,y,z)=\frac{1}{\sqrt{n_0}}\psi(x,y,z)\exp\left(i\kappa\int_0^z dz'\,n(0,0,z')\right); \tag{2}$$

the amplitude of ψ is assumed to vary slowly with z in comparison with the fast exponential variation factor. Substituting (2) into (1) and neglecting the second derivative $\partial^2\psi/\partial z^2$, which is small with respect to the term $i\kappa\partial\psi/\partial z$, this leads to the following equation:

$$\begin{aligned}\frac{i}{\kappa}\frac{\partial\psi}{\partial\tau}&=-\frac{1}{2\kappa^2}\left(\frac{\partial^2\psi}{\partial x^2}+\frac{\partial^2\psi}{\partial y^2}\right)+(n_0^2-n^2)\psi,\\ \tau&=-\int_0^z\frac{dz'}{n_0(z')}.\end{aligned} \tag{3}$$

We can see from (3) that the optical beam field in the paraxial approximation is described by a Schrödinger-like equation where the role of the Planck constant is played by the wavelenght $1/\kappa$, which we shall take as unity, and the role of *time* is played, roughly speaking, by the z–coordinate. In the case of media which are homogeneous under z–translation, the *time* coordinate τ is proportional to the coordinate z. The role of the potential well is played by the refraction index $n_0^2 - n(x,y,z)^2$, which depends on the two *space* coordinates x, y, and *time* $z(\tau)$. The problem of solving the Helmholtz equation in paraxial approximation reduces thus to the problem of solving the Schrödinger equation for a system with two degrees of freedom, with a time-dependent potential well. In the following section we shall discuss the integrals of the motion of quantum systems with time-dependent potentials.

7.3 Quantum invariants and propagator

In this section we follow references [10] and [12]. The evolution operator of the system, $U(z)$, is defined by the equality

$$U(z)\,\psi(x,y,0) = \psi(x,y,z). \tag{4}$$

If the refraction index n is a real function, this operator is unitary: $U^\dagger = U^{-1}$. In that case, we can introduce the two initial momentum projections $\mathbf{p}_0$ and the two initial coordinates $\mathbf{x}_0$, to form the *four integrals of the motion*:

$$\mathbf{p}_0 = U\mathbf{p}U^{-1}, \qquad \mathbf{x}_0 = U\mathbf{x}U^{-1}, \qquad \mathbf{x} = (x_1, x_2) := (x, y). \tag{5}$$

These integrals of the motion satisfy the conditions

$$\iint_{-\infty}^{\infty} dx\,dy\;\psi(x,y,z)^*\,\mathbf{p}_0\,\psi(x,y,z) = \iint_{-\infty}^{\infty} dx\,dy\;\psi(x,y,0)^*\,\mathbf{p}\,\psi(x,y,0), \tag{6}$$

$$\iint_{-\infty}^{\infty} dx\,dy\;\psi(x,y,z)^*\,\mathbf{x}_0\,\psi(x,y,z) = \iint_{-\infty}^{\infty} dx\,dy\;\psi(x,y,0)^*\,\mathbf{x}\,\psi(x,y,0). \tag{7}$$

Thus, the mean values $\langle\mathbf{p}_0\rangle$ and $\langle\mathbf{x}_0\rangle$ are conserved during the propagation of the beam along the z-axis. From formulae (5) and (6) it follows that the integrals of the motion $\mathbf{p}_0(z)$ and $\mathbf{x}_0(z)$, acting on the solution of the Schrödinger equation, produce a function which is also a solution to the same equation. The ***propagator*** or ***Green function*** $G(\mathbf{x},\mathbf{x}',z) := \langle\mathbf{x}\mid U\mid\mathbf{x}'\rangle$ then obeys the system of four equations

$$\mathbf{x}_0 G(\mathbf{x},\mathbf{x}',z) = \mathbf{x}' G(\mathbf{x},\mathbf{x}',z), \tag{8}$$

$$\mathbf{p}_0 G(\mathbf{x},\mathbf{x}',z) = i\frac{\partial}{\partial\mathbf{x}'} G(\mathbf{x},\mathbf{x}',z). \tag{9}$$

In (8) and (9), the operators $\mathbf{x}_0$ and $\mathbf{p}_0$ act on the variable $\mathbf{x}$ in the Green function $G(\mathbf{x},\mathbf{x}',z)$.

In media with gain or loss there is a complex refraction index; the evolution operator is not unitary; the operators $\mathbf{x}_0$ and $\mathbf{p}_0$ defined in (5) are *not* integrals of the motion. Nevertheless, they transform any solution of the Schrödinger equation into another such solution. The propagator G then obeys too the system of equations (8–9) with these operators. Formulae (6–7) are *not* correct for this case. Instead of the operators $\mathbf{x}_0$ and $\mathbf{p}_0$, in the case of nonunitary evolution operator $U(z)$, the operators $\mathbf{X}_{\rm inv}$ and $\mathbf{P}_{\rm inv}$, defined by

$$\mathbf{X}_{\rm inv} := \left(U^\dagger\right)^{-1}\mathbf{x}\,U^{-1}, \tag{10}$$

$$\mathbf{P}_{\rm inv} := \left(U^\dagger\right)^{-1}\mathbf{p}\,U^{-1}, \tag{11}$$

are integrals of the motion for media with gain or loss. They coincide with $\mathbf{x}_0$ and $\mathbf{p}_0$ for the unitary evolution operator case.

7.4 Quadratic media

Let us consider optical media in which the refraction index may be approximated by the following expression:

$$n(x,y,z)^2 = n_0^2 - \tfrac{1}{2}\Omega_x(z)^2\,x^2 - \tfrac{1}{2}\Omega_y(z)^2\,y^2 + \gamma(z)\,xy + f_1(z)\,x + f_2(z)\,y. \tag{12}$$

An optical medium with such refraction index will be called a ***quadratic*** medium. If $\gamma = f_1 = f_2 = 0$, and Ω_x and Ω_y are equal and z-independent, we have an inhomogeneous medium with a quadratically decreasing index; this models some fibers, as discussed in reference [3].

Let us now introduce the following notation: the column four-vector $\mathbf{Q}^\top = (p_1, p_2, x_1, x_2) = \{Q_\alpha\}_{\alpha=1,\ldots,4}$ (where $^\top$ means matrix or vector transposition), and the Hamiltonian

$$H = \tfrac{1}{2}\mathbf{Q}^\top\mathbf{B}(z)\mathbf{Q} + \mathbf{C}(z)^\top\mathbf{Q}, \tag{13}$$

where for example $\mathbf{C}(z)^\top\mathbf{Q} = \sum_{\alpha=1}^{4} C_\alpha(z)Q_\alpha$. The 4×4 matrix $\mathbf{B}(z)$ contains the coefficients of equation (3), which can be rewritten in the form

$$i\frac{\partial\psi}{\partial\tau} = [\tfrac{1}{2}\mathbf{Q}^\top\mathbf{B}(z)\mathbf{Q} + \mathbf{C}(z)^\top\mathbf{Q}]\,\psi. \tag{14}$$

The 4×4 matrix $\mathbf{B}(z)$ may be written as $\mathbf{B}(z) = \begin{pmatrix}\mathbf{b}_1 & \mathbf{b}_2\\ \mathbf{b}_3 & \mathbf{b}_4\end{pmatrix}$, where the 2×2 matrices $\mathbf{b}_i$, $i = 1,\ldots,4$ have the form

$$\mathbf{b}_1 = \begin{pmatrix}1 & 0\\ 0 & 1\end{pmatrix}, \quad \mathbf{b}_2 = \mathbf{b}_3 = \mathbf{0}, \quad \mathbf{b}_4 = \begin{pmatrix}\Omega_x^2 & -\gamma\\ -\gamma & \Omega_y^2\end{pmatrix}. \tag{15}$$

The four-vector $\mathbf{C}$ is equal to

$$\mathbf{C}^\top = \big(0, 0, -f_1(z), -f_2(z)\big). \tag{16}$$

7.5 Invariants for quadratic media

Let us study the invariants $\mathbf{p}_0$ and $\mathbf{x}_0$ for quadratic media without loss, in the form

$$\begin{pmatrix}\mathbf{p}_0\\ \mathbf{x}_0\end{pmatrix} = \mathbf{\Lambda}(z)\mathbf{Q} + \mathbf{\Delta}(z) = \begin{pmatrix}\boldsymbol{\lambda}_1 & \boldsymbol{\lambda}_2\\ \boldsymbol{\lambda}_3 & \boldsymbol{\lambda}_4\end{pmatrix}\begin{pmatrix}\mathbf{p}\\ \mathbf{x}\end{pmatrix} + \begin{pmatrix}\boldsymbol{\delta}_1\\ \boldsymbol{\delta}_2\end{pmatrix}, \tag{17}$$

where the 4×4 matrix $\mathbf{\Lambda}(z)$ consists of four 2×2 matrices $\boldsymbol{\lambda}_i$, $i = 1,\ldots,4$, and the four-vector $\mathbf{\Delta}$ consists of two two-vectors $\boldsymbol{\delta}_1$ and $\boldsymbol{\delta}_2$. The vectors $\mathbf{p}_0$ and $\mathbf{x}_0$ will be the quantum invariants if the matrix $\mathbf{\Lambda}$ and the vector $\mathbf{\Delta}$ obey the equations

$$\begin{aligned}\dot{\mathbf{\Lambda}} &= \mathbf{\Lambda\Sigma B},\\ \dot{\mathbf{\Delta}} &= \mathbf{\Lambda\Sigma C},\end{aligned} \qquad \mathbf{\Sigma} := \begin{pmatrix}\mathbf{0} & \mathbf{E}_2\\ -\mathbf{E}_2 & \mathbf{0}\end{pmatrix}, \quad \mathbf{E}_2 := \begin{pmatrix}1 & 0\\ 0 & 1\end{pmatrix}, \tag{18}$$

where the overdot indicates differentiation with respect to z. These equations may be put explicitly in terms of the matrices $\boldsymbol{\lambda}_i$ and vectors $\boldsymbol{\delta}_i$ thus:

$$\begin{aligned}
\dot{\boldsymbol{\lambda}}_1 &= \boldsymbol{\lambda}_1\mathbf{b}_3 - \boldsymbol{\lambda}_2\mathbf{b}_1, & \boldsymbol{\lambda}_1(0) &= \mathbf{E}_2,\\
\dot{\boldsymbol{\lambda}}_2 &= \boldsymbol{\lambda}_1\mathbf{b}_4 - \boldsymbol{\lambda}_2\mathbf{b}_4, & \boldsymbol{\lambda}_2(0) &= \mathbf{0},\\
\dot{\boldsymbol{\lambda}}_3 &= \boldsymbol{\lambda}_3\mathbf{b}_3 - \boldsymbol{\lambda}_4\mathbf{b}_1, & \boldsymbol{\lambda}_3(0) &= \mathbf{0},\\
\dot{\boldsymbol{\lambda}}_4 &= \boldsymbol{\lambda}_3\mathbf{b}_4 - \boldsymbol{\lambda}_4\mathbf{b}_2, & \boldsymbol{\lambda}_4(0) &= \mathbf{E}_2,\\
\dot{\boldsymbol{\delta}}_1 &= \boldsymbol{\lambda}_1\mathbf{c}_2 - \boldsymbol{\lambda}_2\mathbf{c}_1, & \boldsymbol{\delta}_1(0) &= \mathbf{0},\\
\dot{\boldsymbol{\delta}}_2 &= \boldsymbol{\lambda}_3\mathbf{c}_2 - \boldsymbol{\lambda}_4\mathbf{c}_1, & \boldsymbol{\delta}_2(0) &= \mathbf{0}.
\end{aligned} \tag{19}$$

Here we use the general matrix $\mathbf{B}(z)$ and vector $\mathbf{C}(z)$.

It follows from (18) and (19) that

$$\mathbf{\Lambda}\Sigma\mathbf{\Lambda}^\top = \Sigma. \tag{20}$$

Thus the matrix $\mathbf{\Lambda}$ is a real symplectic matrix belonging to the group *Sp(4,$\Re$)*. The linear transformation (17) is an element of the inhomogeneous symplectic group *ISp(4,$\Re$)*. In terms of the matrices λ_i we can obtain from (20) that

$$\begin{array}{lll} \lambda_2\lambda_1^\top = \lambda_1\lambda_2^\top, & \lambda_3\lambda_4^\top = \lambda_4\lambda_3^\top, & \lambda_4\lambda_1^\top - \lambda_3\lambda_2^\top = \mathbf{E}_2, \\ \lambda_4^\top\lambda_2 = \lambda_2^\top\lambda_4, & \lambda_1^\top\lambda_3 = \lambda_3^\top\lambda_1, & \lambda_4^\top\lambda_1 - \lambda_2^\top\lambda_3 = \mathbf{E}_2, \\ \det\mathbf{\Lambda} = 1, & \mathbf{\Lambda}^{-1} = -\Sigma\mathbf{\Lambda}^\top\Sigma, & \mathbf{\Lambda}^{-1} = \begin{pmatrix} \lambda_4^\top & -\lambda_2^\top \\ -\lambda_3^\top & \lambda_1^\top \end{pmatrix}. \end{array} \tag{21}$$

The linear quantum invariants for quadratic media (17) are operators in the Schrödinger representation. They depend on z and are expressed in terms of the 4×4 matrix $\mathbf{\Lambda}(z)$ and the four-vector $\mathbf{\Delta}(z)$. If we can solve explicitly the linear equation (18) for $\mathbf{\Lambda}$ and $\mathbf{\Delta}$, we shall know explicitly the linear invariants $\mathbf{p}_0$ and $\mathbf{x}_0$. Since any function of quantum invariants is an invariant also, we can take any function of the integrals of the motion $\mathbf{p}_0$ and $\mathbf{x}_0$, for example the second-order polynomials. They are quantum invariants. All the quadratic forms of the operators $\mathbf{p}_0$ and $\mathbf{x}_0$, together with the linear form and the identity operator, form a basis for the representation of the *isp(4,$\Re$)* algebra of the *Isp(4,$\Re$)* group. This algebra may be considered as a *spectrum-generating algebra* for light beams in quadratic media. The generators of this representation are the quantum integrals of the motion.

7.6 The propagator in quadratic media

The propagator in quadratic media can be found using equations (8) and (9), and the explicit form of the integrals of the motion (17). The coordinate dependence of the propagator $G(\mathbf{x},\mathbf{x}',z)$ may be fixed through solving the system (8–9); the z-dependent phase factor can be obtained when the propagator is put into the Schrödinger equation with initial conditions

$$\left(i\frac{\partial}{\partial\tau} - H\right)G = i\delta(\mathbf{x}-\mathbf{x}')\,\delta(\tau), \qquad G(\mathbf{x},\mathbf{x}',0) = \delta(\mathbf{x}-\mathbf{x}'). \tag{22}$$

The system of equations (8) and (9) for quadratic media is a linear system of first order differential equations, with coefficients which are linear with respect to the coordinates. The solution to this system is the quadratic exponential [16]

$$\begin{aligned} G(\mathbf{x},\mathbf{x}',z) = [-2\pi i]^{-1}[\det\lambda_3]^{-1/2}\exp\Big(&-\frac{1}{2}i[\mathbf{x}^\top\lambda_3^{-1}\lambda_4\mathbf{x} - 2\mathbf{x}^\top\lambda_3^{-1}\mathbf{x}' + \mathbf{x}'^\top\lambda_1\lambda_3^{-1}\mathbf{x}' + 2\mathbf{x}^\top\lambda_3^{-1}\boldsymbol{\delta}_2 \\ &+ 2\mathbf{x}'^\top(\boldsymbol{\delta}_1 - \lambda_1\lambda_3^{-1}\boldsymbol{\delta}_2) + \boldsymbol{\delta}_2^\top\lambda_1\lambda_3^{-1}\boldsymbol{\delta}_2 - 2\int_0^\tau d\tau'\,\dot{\boldsymbol{\delta}}_1\boldsymbol{\delta}_2]\Big). \end{aligned} \tag{23}$$

In momentum representation, the propagator $G(\mathbf{p},\mathbf{p}',z)$ has the form

$$\begin{aligned} G(\mathbf{p},\mathbf{p}',z) = [2\pi i]^{-1}[\det\lambda_2]^{-1/2}\exp\Big(&\frac{1}{2}i[\mathbf{p}^\top\lambda_2^{-1}\lambda_1\mathbf{p} - 2\mathbf{p}^\top\lambda_2^{-1}\mathbf{p}' + \mathbf{p}'^\top\lambda_4\lambda_2^{-1}\mathbf{p}' + 2\mathbf{p}^\top\lambda_2^{-1}\boldsymbol{\delta}_1 \\ &+2\mathbf{p}'^\top(\boldsymbol{\delta}_2 - \lambda_4\lambda_2^{-1}\boldsymbol{\delta}_1) + \boldsymbol{\delta}_1^\top\lambda_4\lambda_2^{-1}\boldsymbol{\delta}_1 - 2\int_0^\tau d\tau'\,\dot{\boldsymbol{\delta}}_2\boldsymbol{\delta}_1]\Big). \end{aligned} \tag{24}$$

The initial conditions are: $G(\mathbf{p},\mathbf{p}',0) = \delta(\mathbf{p}-\mathbf{p}')$.

7.7 Coherent states, gaussian beams

Let us now consider gaussian beams in quadratic media; these can be described as the ***coherent states*** of references [12,25–29]. To this end we construct the two complex integrals of the motion given by the components of the vector $\mathbf{A}(z)$ defined by

$$\begin{aligned} \mathbf{A}(z) &:= \frac{1}{\sqrt{2}}(\mathbf{x}_0(z) + i\mathbf{p}_0(z)) := \frac{1}{\sqrt{2}}(\boldsymbol{\lambda}_p \mathbf{p} + \boldsymbol{\lambda}_x \mathbf{x} + \boldsymbol{\delta}), \\ \boldsymbol{\lambda}_p &= \boldsymbol{\lambda}_3 + i\boldsymbol{\lambda}_1, \qquad \boldsymbol{\lambda}_x = \boldsymbol{\lambda}_4 + i\boldsymbol{\lambda}_2, \qquad \boldsymbol{\delta} = \boldsymbol{\delta}_2 + i\boldsymbol{\delta}_1 . \end{aligned} \tag{25}$$

These integrals of the motion obey the commutation relations of the boson creation and annihilation operators

$$[A_j, A_k^\dagger] = \delta_{jk}, \qquad [A_j, A_k] = 0, \qquad [A_j^\dagger, A_k^\dagger] = 0, \quad j,k = 1,2. \tag{26}$$

If we now look for the eigenvectors of the operator $\mathbf{A}(z)$,

$$\mathbf{A}\Psi_{\boldsymbol{\alpha}}(\mathbf{x}, z) = \boldsymbol{\alpha}\,\Psi_{\boldsymbol{\alpha}}(\mathbf{x}, z), \tag{27}$$

where $\boldsymbol{\alpha}$ is a complex two-vector. The function $\Psi_{\boldsymbol{\alpha}}(\mathbf{x}, z)$ can be easily found in the form of a Gaussian through solving the equation

$$\begin{aligned} \Psi_{\boldsymbol{\alpha}}(\mathbf{x}, z) = (\det \boldsymbol{\lambda}_p)^{-1/2} \exp\Big(&-\frac{1}{2} i\mathbf{x}^\top \boldsymbol{\lambda}_p^{-1}\boldsymbol{\lambda}_x \mathbf{x} - i\mathbf{x}^\top \boldsymbol{\lambda}_p^{-1}(\boldsymbol{\delta} - \sqrt{2}\,\boldsymbol{\alpha}) + \frac{1}{2}\boldsymbol{\alpha}^\top \boldsymbol{\lambda}_p^* \boldsymbol{\lambda}_p^{-1}\boldsymbol{\alpha} \\ &+ \frac{1}{\sqrt{2}}\boldsymbol{\alpha}^\top(\boldsymbol{\delta}^* - \boldsymbol{\lambda}_p^*\boldsymbol{\lambda}_p^{-1}\boldsymbol{\delta}) + \frac{1}{4}\boldsymbol{\delta}^\top \boldsymbol{\lambda}_p^*\boldsymbol{\lambda}_p^{-1}\boldsymbol{\delta} - \frac{1}{4}|\boldsymbol{\delta}|^2 - \frac{1}{2} i \int_0^T d\tau' \, \mathrm{Im}(\boldsymbol{\delta}^\top \dot{\boldsymbol{\delta}}^*) - \frac{1}{2}|\boldsymbol{\alpha}|^2 \Big), \end{aligned} \tag{28}$$

where * means complex conjugation.

The properties of the coherent state gaussian beams are the following: they are normalized,

$$\iint_{-\infty}^{\infty} d\mathbf{x}\, \Psi_{\boldsymbol{\alpha}}(\mathbf{x}, z)^* \, \Psi_{\boldsymbol{\alpha}}(\mathbf{x}, z) = 1, \tag{29}$$

but are not orthogonal:

$$\iint_{-\infty}^{\infty} d\mathbf{x}\, \Psi_{\boldsymbol{\alpha}}(\mathbf{x}, z)^* \, \Psi_{\boldsymbol{\beta}}(\mathbf{x}, z) = \exp(-\tfrac{1}{2}|\boldsymbol{\alpha}|^2 - \tfrac{1}{2}|\boldsymbol{\beta}|^2 + \boldsymbol{\alpha}^{\top *}\boldsymbol{\beta}). \tag{30}$$

The coherent gaussian beams are a complete (overcomplete) set of functions:

$$\begin{aligned} &\iint_{C^2} \frac{d^2\boldsymbol{\alpha}}{\pi^2}\, \Psi_{\boldsymbol{\alpha}}(\mathbf{x}, z)^* \, \Psi_{\boldsymbol{\alpha}}(\mathbf{x}', z) = \delta(\mathbf{x} - \mathbf{x}') \\ &d^2\boldsymbol{\alpha} := d\mathrm{Re}\alpha_1 \, d\mathrm{Re}\alpha_2 \, d\mathrm{Im}\alpha_1 \, d\mathrm{Im}\alpha_2 . \end{aligned} \tag{31}$$

The integration is performed on the two complex planes of the projections of the complex vector $\boldsymbol{\alpha}$.

Any field distribution can be considered as the superposition of coherent state Gaussian beams. In fact, the form of the coherent state gaussian beam coincides with the complex ray gaussian beam in reference [5], which was observed in [12]. The importance of this observation is that the formal theory of coherent states [44] is elaborated, and afterwards we obtain the identification of complex ray gaussian beams with coherent state gaussian beams, with properties (29–31). See references [12, 24–29].

The initial mean values of coordinates and momenta of coherent Gaussian beams are described by the complex vector $\boldsymbol{\alpha}$,

$$\langle \mathbf{x}_0(0)\rangle = \sqrt{2}\,\mathrm{Re}\,\boldsymbol{\alpha}, \qquad \langle \mathbf{p}_0(0)\rangle = \sqrt{2}\,\mathrm{Im}\,\boldsymbol{\alpha}. \tag{32}$$

The initial widths of the coherent state $\Psi_{\boldsymbol{\alpha}}(\mathbf{x},0)$ obey, in dimensionless variables, the equality

$$\Delta x_i = \Delta p_i = \tfrac{1}{2}, \qquad i = 1,2. \tag{33}$$

The center of the gaussian beam (28) propagates in the medium as follows:

$$\begin{pmatrix} \langle \mathbf{p}\rangle(z) \\ \langle \mathbf{x}\rangle(z) \end{pmatrix} = \begin{pmatrix} \boldsymbol{\lambda}_4^\top & -\boldsymbol{\lambda}_2^\top \\ -\boldsymbol{\lambda}_3^\top & \boldsymbol{\lambda}_1^\top \end{pmatrix} \begin{pmatrix} \sqrt{2}\,\mathrm{Im}\boldsymbol{\alpha} - \boldsymbol{\delta}_1(z) \\ \sqrt{2}\,\mathrm{Re}\boldsymbol{\alpha} - \boldsymbol{\delta}_2(z) \end{pmatrix}. \tag{34}$$

The widths of the beam (17) in the process of propagation can be expressed in terms of the symplectic matrix $\boldsymbol{\Lambda}(z)$. For the dispersion matrix

$$\sigma_{\alpha\beta}(z) := \langle Q_\alpha Q_\beta\rangle - \langle Q_\alpha\rangle\langle Q_\beta\rangle, \tag{35}$$

we thus have the expression

$$\sigma_{\alpha\beta}(z) = \Lambda^{-1}_{\alpha\gamma}(z)\,\Lambda^{-1}_{\beta\mu}(z)\,\sigma_{\gamma\mu}(0), \tag{36}$$

where $\sigma_{\gamma\mu}(0)$ is the dispersion matrix for the initial gaussian distribution function in the plane $z = 0$, and summation over repeated indices is implied. The matrix $\boldsymbol{\Lambda}$ is given by equations (21). All the higher momenta for the field distribution function in coherent state beams, are expressed in terms of the mean values (34) and the dispersion matrix $\boldsymbol{\sigma}$ in (35).

7.8 Modes in quadratic media

In the previous section we constructed the coherent state gaussian beams which are eigenfunctions of the nonhermitian integral of the motion, $\mathbf{A}(z)$ in (25). We shall now consider the eigenfunctions of the hermitian integral of the motion $\mathbf{A}^\dagger(z)\mathbf{A}(z)$; it is a two-vector with components $A_1^\dagger A_1$, $A_2^\dagger A_2$. We may solve directly the eigenvalue problem for the mode solution $\Psi_{\mathbf{n}}(\mathbf{x},z)$,

$$\mathbf{A}^\dagger\mathbf{A}\Psi_{\mathbf{n}}(\mathbf{x},z) = \mathbf{n}\Psi_{\mathbf{n}}(\mathbf{x},z), \qquad \mathbf{n}^\top := (n_1, n_2). \tag{37}$$

But there is also the property of the coherent state $\Psi_{\boldsymbol{\alpha}}(\mathbf{x},z)$ to be the generating function for the mode solutions [12,13],

$$\Psi_{\boldsymbol{\alpha}}(\mathbf{x},z) = e^{-|\boldsymbol{\alpha}|^2/2} \sum_{n_1=0}^{\infty}\sum_{n_2=0}^{\infty} \frac{\alpha_1^{n_1}\alpha_2^{n_2}}{\sqrt{n_1!\,n_2!}}\Psi_{n_1 n_2}(\mathbf{x},z), \tag{38}$$

where the mode solutions obey the conditions of orthogonality

$$\iint_{-\infty}^{\infty} d\mathbf{x}\,\Psi_{\mathbf{n}}(\mathbf{x},z)^*\,\Psi_{\mathbf{m}}(\mathbf{x},z) = \delta_{\mathbf{n},\mathbf{m}}\,, \tag{39}$$

and completeness

$$\sum_{n_1=0}^{\infty}\sum_{n_2=0}^{\infty} \Psi_{\mathbf{n}}(\mathbf{x},z)^*\,\Psi_{\mathbf{n}}(\mathbf{x}',z) = \delta(\mathbf{x}-\mathbf{x}'). \tag{40}$$

So, expanding the function $\exp(\frac{1}{2}|\boldsymbol{\alpha}|^2)\,\Psi_{\boldsymbol{\alpha}}(\mathbf{x},z)$ in series with respect to the complex parameters α_1, α_2, and comparing this function with the known generating function for the Hermite polynomials of two variables, we obtain the mode solution $\Psi_{\mathbf{n}}(\mathbf{x},z)$ in the following form:

$$\begin{aligned}\Psi_{\mathbf{n}}(\mathbf{x},z) = \frac{1}{\sqrt{\pi\, n_1!\, n_2!\, \det\boldsymbol{\lambda}_p}} \exp\Big(&-\frac{1}{2} i\mathbf{x}^\top\boldsymbol{\lambda}_p^{-1}\boldsymbol{\lambda}_x\mathbf{x} + i\mathbf{x}^\top\boldsymbol{\lambda}_p^{-1}\boldsymbol{\delta} + \frac{1}{4}\boldsymbol{\delta}^\top\boldsymbol{\lambda}_p^{*}\boldsymbol{\lambda}_p^{-1}\boldsymbol{\delta} \\ &- \frac{1}{4}|\boldsymbol{\delta}|^2 - \frac{1}{2} i\int_0^\tau d\tau'\,\mathrm{Im}(\boldsymbol{\delta}^\top\dot{\boldsymbol{\delta}}^*)\Big) \\ &\times H_{n_1n_2}\big(\tfrac{1}{\sqrt{2}}(\boldsymbol{\delta} - \boldsymbol{\lambda}_p\boldsymbol{\lambda}_p^{*-1}\boldsymbol{\delta}^*) - i\sqrt{2}\,\boldsymbol{\lambda}_p^{\dagger-1}\mathbf{x}\big).\end{aligned} \tag{41}$$

Here the function $H_{n_1n_2}(\xi)$ is the Hermite polynomial of two variables determined by the matrix $\mathbf{D} := -\boldsymbol{\lambda}_p^{*}\boldsymbol{\lambda}_p^{-1}$. Its generating function is [16]:

$$\exp(-\tfrac{1}{2}\mathbf{z}^\top\mathbf{D}\mathbf{z} + \mathbf{x}^\top\mathbf{D}\mathbf{z}) = \sum_{n_1=0}^{\infty}\sum_{n_2=0}^{\infty}\frac{z_1^{n_1}z_2^{n_2}H_{n_1n_2}(\mathbf{x})}{n_1!\,n_2!}. \tag{42}$$

7.9 Mode coupling coefficients and the Franck–Condon principle

Let us now consider the situation when the fiber with nonquadratic refraction index consists of two different pieces, one with $n_1(\mathbf{x})$ and the second with $n_2(\mathbf{x})$. The light beam energy concentrated in a mode $\Psi_{\mathbf{n}}(\mathbf{x},1)$ in the first fiber, at the place of the connection between the two fibers ($z=0$) will be distributed among the modes $\Psi_{\mathbf{m}}(\mathbf{x},z)$ of the second fiber. The problem is how to calculate the mode coupling coefficients. These are, in fact, determined by the overlap integrals of the modes of both fibers,

$$C_{\mathbf{n}}^{\mathbf{m}} := \iint_{-\infty}^{\infty} d\mathbf{x}\,\Psi_{\mathbf{n}}(\mathbf{x},1)^*\,\Psi_{\mathbf{m}}(\mathbf{x},z)\Big|_{z=0}. \tag{43}$$

This overlap integral was first calculated by Arnaud [21] for quadratic media, and expressed in terms of Hermite polynomials of four variables. The overlap integral for N–dimensional oscillators was calculated in terms of Hermite polynomials of $2N$ variables in reference [6]. For nonquadratic media, the mode coupling coefficients $C_{\mathbf{n}}^{\mathbf{m}}$ cannot be evaluated explicitly for general coordinate dependence of the refraction index. But in [12] it was observed that the same overlap integral (43) arises when one calculates the transition probability between the vibronic states of triatomic molecules in the process of electron transition. The vibronic structure of the electronic lines is determined by precisely the same integrals. In the quantum mechanics of molecular spectra, the Franck–Condon principle was formulated to evaluate the maximal probability $|C_{\mathbf{n}}^{\mathbf{m}}|^2$ from the vibronic level $\mathbf{n}$ of the initial electronic state to the level $\mathbf{m}$ of the final electronic state. This principle can be applied in order to predict which mode in the second fiber will be excited with maximal probability, when the field energy was contained in a given mode of the first fiber.

This evaluation can be done following the prescription contained in the Figure next page, which we present in the one-dimensional case of plane waveguides. In the figure, the two "potential wells" $n^2 - n_0^2$ for the first and the second fibers are shown. If the mode with index $\mathbf{n}$ is excited in the first fiber, the mode with index $\mathbf{m}$ maximally excited in the second fiber will be obtained on the intersection of the perpendicular from the $\mathbf{n}^{\text{th}}$ level of the first potential, with the potential well curve of the second fiber. This principle applies for any coordinate dependence of the refraction index, of arbitrarily complicated form, in both fibers.

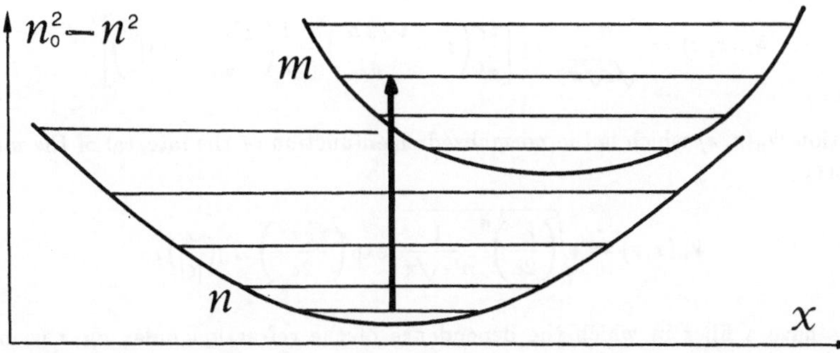

Figure. Transition between two vibronic levels following the Franck–Condon principle.

For connections between two quadratic fibers, extra information can be extracted from the Franck–Condon principle. It is known [43] that for molecular systems, the maximal probabilities $|C_{\mathbf{n}}^{\mathbf{m}}|^2$ on the two-variable plane of $\mathbf{n}$ and $\mathbf{m}$, lie on the *Condon parabola* and an extra direct line. This means that if one mode is excited in the first fiber, with maximal probability there will be *three* excited modes in the second fiber. One of the latter modes is given by the Franck–Condon principle depicted in the figure above and, with smaller but large enough probability, there will be two extra excited modes corresponding to the two extra points on the Condon parabola and direct line. If the refraction index is described by a function with a maximum on the optical axis, the parabolic profile may be considered as a good approximation for the lowest modes. The above prediction about the excitation of three modes in the second fiber by a single mode in the first is, therefore, applicable for the lowest modes in nonquadratic media as well.

7.10 One-dimensional quadratic media and the analogue of the Ramsauer effect

We now consider one-dimensional quadratic media with the Hamiltonian

$$H = \tfrac{1}{2}p_x^2 + \tfrac{1}{2}\omega(z)^2\, x^2. \tag{44}$$

This is the particular case of the general Hamiltonian of the form (13) with

$$\mathbf{C} = \mathbf{0}, \qquad \mathbf{B}(z) = \begin{pmatrix} 1 & 0 \\ 0 & \omega(z)^2 \end{pmatrix}. \tag{45}$$

There is one complex integral of the motion which has the form

$$A = \frac{i}{\sqrt{2}}\big(\epsilon(\tau)\, p - \dot{\epsilon}(\tau)\, x\big), \qquad [A, A^\dagger] = 1, \tag{46}$$

where the complex function ϵ obeys the classical equation of motion for the oscillator with τ–dependent frequency

$$\ddot{\epsilon} + \omega^2 \epsilon = 0. \tag{47}$$

The normalized coherent state gaussian beam $\Psi_\alpha(x,z)$ for the medium is expressed explicitly in terms of the function $\epsilon(\tau)$ thus:

$$\Psi_\alpha(x,z) = \frac{1}{\sqrt{\epsilon\sqrt{\pi}}}\exp\left[\frac{i\dot{\epsilon}}{2\epsilon}\left(x - \frac{i\sqrt{2}\,\alpha}{\dot{\epsilon}}\right)^2 - \frac{1}{2}\left(\frac{\dot{\epsilon}^*\alpha^2}{\dot{\epsilon}} + |\alpha|^2\right)\right]. \tag{48}$$

The mode solution $\Psi_n(x,z)$ which is the normalized eigenfunction of the integral of the motion $A^\dagger A$ in (46), has the form

$$\Psi_n(x,z) = \sqrt{\left(\frac{\dot{\epsilon}^*}{2\epsilon}\right)^n \frac{1}{n!\,\epsilon\sqrt{\pi}}}\exp\left(\frac{i\dot{\epsilon}\,x^2}{2\epsilon}\right)H_n\left(\frac{x}{|\epsilon|}\right). \tag{49}$$

Let us have a fiber in which the dependence of the refraction index on z is modeled by a function $\omega(z)$ which for large negative z is a constant ω_i, and for large positive z is another constant ω_f. This dependence describes also the connection between two fibers where the refraction index changes abruptly from ω_i to ω_f. Let us now introduce $\xi(\tau)$ and $\eta(\tau)$, linear combinations of ϵ and $\dot{\epsilon}$, given by

$$\begin{aligned}\xi &:= \tfrac{1}{2}e^{-i\omega_f\tau}\left(\epsilon\sqrt{\omega_f} - i\dot{\epsilon}/\sqrt{\omega_f}\right),\\ \eta &:= \tfrac{1}{2}e^{-i\omega_f\tau}\left(\epsilon\sqrt{\omega_f} + i\dot{\epsilon}/\sqrt{\omega_f}\right).\end{aligned} \tag{50}$$

The commutation relations of the integrals of the motion A, $A^\dagger$ in (46), yield the constraint

$$|\xi|^2 - |\eta|^2 = 1. \tag{51}$$

The propagator $G(\gamma,\alpha,z)$ in the coherent state gaussian beam basis, with overlap integral given by (48), has the form

$$G(\gamma,\alpha,z) = \frac{1}{\sqrt{\xi}}\exp\left(\alpha^2\frac{\eta^*}{2\xi} + \frac{\alpha\gamma^*}{\xi} - (\gamma^*)^2\frac{\eta}{2\xi} - \frac{1}{2}|\alpha|^2 - \frac{1}{2}|\gamma|^2\right). \tag{52}$$

Expanding the analytic function $\exp\frac{1}{2}(|\alpha|^2 + |\gamma|^2)\,C(\gamma,\alpha,z)$ in series with respect to degrees of α and γ^*, we obtain the propagator $G(m,n,z)$ in the discrete mode basis. In fact, this propagator can be expressed in terms of Hermite polynomials of two variables, in which the argument is equal to zero. In that special case, the Hermite polynomial of two variables is expressed in terms of Legendre functions $P_\tau^\bullet(x)$ [10]. Thus we have

$$\begin{gathered}G(m,n) = \sqrt{\frac{n!}{m!\,\xi}}\exp\left[i\tfrac{1}{2}(m-n)\phi_\eta - i\tfrac{1}{2}(m+n)\phi_\xi\right]P_{(m+n)/2}^{(m-n)/2}(\cos\tfrac{1}{2}\theta),\\ \phi_\eta := \arg\eta, \qquad \phi_\xi := \arg\xi, \qquad \cos\theta := 1 - 2|\eta/\xi|^2.\end{gathered} \tag{53}$$

The mode coupling coefficients $|G(m,n)|^2$ between the nth mode in the first fiber with the mth mode of the second fiber, is expressed for $m > n$ as follows:

$$|G(m,n)|^2 = \frac{n!}{m!\,|\xi|}\left|P_{(m+n)/2}^{(m-n)/2}(\cos\tfrac{1}{2}\theta)\right|^2. \tag{54}$$

Thus, if we do not want to have excitation of modes other than the nth in the second fiber, we must have the equality $|G(m,n)|^2 = \delta_{mn}$. This can be in the case when the argument of the Legendre function is equal to unity. From this observation follows the interpretation of the equality. In fact, we introduce the parameter

$$R := \sin^2 \tfrac{1}{2}\theta = |\eta/\xi|^2, \tag{55}$$

which may be interpreted as the *reflection coefficient* of a fictitious quantum particle with energy $\frac{1}{2}\omega_{\text{in}}^2 + U(-\infty)$, scattered by the potential well $U(z)$ determined by the equation

$$\omega(z)^2 = \omega_{\text{in}}^2 + 2\big(U(-\infty) - U(z)\big). \tag{56}$$

This follows from the formal correspondence between the classical equation of motion for ϵ in (47), and the stationary Schrödinger equation in one-dimensional quantum mechanics, as well as from equations (50).

In quantum mechanics, moreover, there is the well-known *Ramsauer effect*: the potential is transparent for scattered particles if the energy of the particle is chosen in an appropriate manner. In the quantum mechanics of atoms and molecules, the potential is fixed, and one has the possibility to vary the energy (the *de Broglie wavelength*) of the particle to be scattered. For appropriate energies we have the Ramsauer effect which makes a given potential well transparent. In fiber optics we have the fixed parameters of the two connected fibers; this is to have a fixed energy for the fictitious particle. We may vary, however, the "potential well" in the place of connection of the two fibers. We can make the z-dependence of the refraction index at the place of connection in such a manner so as to have the analogue of the Ramsauer effect in fiber optics, *i.e.* a transparent connection. The analogy to the Ramsauer effect can be made also for the two-dimensional case.[1]

7.11 Concluding remarks

We have considered the use of quantum 'time'-dependent invariants for systems with nonstationary potentials in fiber optics, in the frame of the Fock–Leontovich paraxial approximation. The considerations were based on the formal analogy between the equation for paraxial light beam propagation in fibers, with the Schrödinger equation for two-dimensional systems which model fibers, and also one-dimensional systems which model plane waveguides. By means of the invariants which exist for arbitrary coordinate dependence of the refraction index, we constructed light beams as coherent states. For quadratic media, we provided explicit forms for the gaussian beams.

The analogy between the quantum mechanical Schrödinger equation and the parabolic light beam propagation equation in optics, gave us the possibility to formulate for fiber optics the analogues of well-known quantum mechanical effects. The Franck–Condon and Ramsauer effects have been used for the theory of dissimilar fiber connections.

For quadratic media, the Hamiltonian (13) belongs to the Lie algebra of the symplectic group $ISp(4,\Re)$ in fibers, and to $ISp(2,\Re)$ in plane waveguides. Thus, the propagators represent finite transformations of these groups. The matrix elements of these transformations coincide with the propagators in coordinate representation (23), coherent state representation (52), or discrete mode representation (53). This correspondence between the propagators of quadratic-profile fiber optics, and the symplectic group representation matrix elements, gives us the possibility to use the known properties of the latter in order to obtain some unexpected summation rules for the mode coupling coefficients in the former, which have been obtained earlier in the theory of vibronic structure in triatomic molecular spectra [43].

[1] It was Professor Louis Michel who pointed out to the author (private communication) that the effect of transparency at the connection of two fibers with appropriate z-dependent refraction index is in complete analogy with the Ramsauer effect in quantum mechanics.

It should be emphasized that the formalism of symplectic groups, of coherent state representations, and of Hermite polynomials in several variables that have been used in the theory of the vibronic spectra of polyatomic molecules, can be used almost without change in fiber optics. This has been shown in references [25–29]. The correlation functions and generating functionals can be calculated in the frame of Fock–Leontovich fiber optics, in the same manner as was done for the quantized version of the optical Hamiltonian in reference [42] which was based on the Fermat principle. Professor John Klauder introduced the continuous representations in quantum mechanics, as a particular case of those for arbitrary Lie groups [44]. These continuous representations for Lie groups were later rederived, and are sometimes called *coherent states for Lie groups*. The continuous representations of Klauder (coherent states for Lie groups) can be used also in the fiber optics of *non*quadratic media, in those cases when the Hamiltonians may be considered as the linear form of the Lie group representation with z-dependent coefficients.

Acknowledgement

I would like to thank Bernardo Wolf for the hospitality extended at IIMAS–UNAM, and for helpful discussions. Also, I would like to thank Profs. L. Michel, I.A. Arnaud, J.R. Klauder, and H. Bacry, for illuminating discussions on the material of this article.

References

[1] M. Leontovich and V. Fock, Solution of the problem of electromagnetic waves along the earth suface by the method of parabolic equations, *Sov. J. Phys.* **10**, 13 (1946).

[2] L.M. Brechovskih, *Waves in Slab Media* (in Russian), (Nauka, Moscow, 1973).

[3] J.A. Hudson, A parabolic approximation for elastic waves, *Wave Motion* **2**, 207 (1980).

[4] D. Marcuse, *Light Transmission Optics*, (Van Nostrand, New York, 1972).

[5] J.A. Arnaud, *Beam and Fiber Optics*, (Academic Press, New York, 1976).

[6] I.A. Malkin and V.I. Man'ko, Coherent states and excitation of n-dimensional non-stationary forced oscillator, *Phys. Lett.* **32A**, 243 (1970).

[7] I.A. Malkin, V.I. Man'ko, and D.A. Trifonov, Linear adiabatic invariants and coherent states, *J. Math. Phys.* **14**, 576 (1973).

[8] V.V. Dodonov, I.A. Malkin, and V.I. Man'ko, Integrals of the motion, Green function and coherent states of dynamical systems, *Intern. J. Theor. Phys.* **14**, 37 (1975).

[9] V.V. Dodonov and V.I. Man'ko, Coherent states and the resonance of a quantum damped oscillator, *Phys. Rev.* **A20**, 550 (1979).

[10] I.A. Malkin and V.I. Man'ko, *Dynamical Symmetries and Coherent States of Quantum Mechanics* (in Russian), (Nauka, Moscow, 1979).

[11] V.I. Man'ko, Invariants and symmetries of dynamical systems, In *Proceedings of the Second International Colloquium on Group Theoretical Methods in Physics*, Vol. 1, (University of Nijmegen, The Netherlands, June 25–29, 1973); pp. A107 *et seq.*

[12] V.I. Man'ko, Possible applications of the integrals of the motion and coherent state methods for dynamical systems. In *Quantum Electrodynamics with Outer Fields* (in Russian), (Tomsk State University and Tomsk Pedagogical Institute, publishers; Tomsk, USSR, 1977); pp. 101–120.

[13] R. Glauber, Coherent states of the quantum oscillator, *Phys. Rev. Lett.* **10**, 84 (1963).

[14] I.A. Malkin, V.I. Man'ko, and D.A. Trifonov, Coherent states and excitation charge moving in a time-dependent electromagnetic field, *Phys. Rev.* **2D**, 1370 (1970).

[15] I.A. Malkin and V.I. Man'ko, Coherent states of the charged oscillator in a nonstationary magnetic field, *Zhurn. Exper. i Teor. Fiz.* **59**, 1746 (1970).

[16] V.V. Dodonov and V.I. Man'ko, Integrals of the motion and dynamics of general nonstationary quadratic Fermi-Bose systems (in Russian), *P.N. Lebedev Inst. Trudy* **152**, 145–193 (1983).

[17] I.A. Malkin, V.I. Man'ko, and D.A. Trifonov, Dynamical symmetries of nonstationary systems, *Nuovo Cimento* **4A**, 773 (1971).

[18] J.A. Arnaud, Hamilton theory of beam mode propagation. In *Progress in Optics*, Vol. XI (North Holland, 1973); pp. 249–302.

[19] J.A. Arnaud, Nonorthogonal optical waveguides and resonators, *BSTJ* **49**, 2311 (1970).

[20] H. Kogelnik, On the propagation of gaussian beams of light through lenslike media incuding those with a loss or gain variation, *Applied Optics* **4**, 1562–1569 (1965).

[21] J.A. Arnaud, Mode coupling in first-order optics, *J. Opt. Soc. Amer.* **61**, 751 (1971).

[22] E.E. Bergmann, Optical resonators with paraxial modes, *Applied Optics* **11**, 113 (1972).

[23] S.G. Krivoshlykov, N.I. Petrov, and I.N. Sissakian, Coherent properties of optical fields in nonhomogeneous media. In *Proceedings of the Second International Seminar on Group Theoretical Methods in Physics*, Vol. II (Zvenigorod, USSR, 1980); pp. 235 *et seq.*

[24] S.G. Krivoshlykov, E.V. Kurmyshev, and I.N. Sissakian, Correlated coherent states in the problems of wave propagations in nonhomogeneous quadratic media. In *Proceedings of the Second International Seminar on Group Theoretical Methods in Physics*, Vol. II (Zvenigorod, USSR, 1980); pp. 226 *et seq.*

[25] S.G. Krivoshlykov and I.N. Sissakian, Quantum mechanical theory of optical beam propagation in inhomogeneous media, P.N. Lebedev Institute preprint # 117 (1979).

[26] S.G. Krivoshlykov and I.N. Sissakian, Mode coupling between two connected multimode parabolic-index optical waveguides, *Opt. Quant. Electr.* **11**, 393–405 (1979).

[27] S.G. Krivoshlykov and I.N. Sissakian, Coherent states and light propagation in inhomogeneous media (in Russian), *Kvantovaîa Elektronika* **7**, 553–564 (1980).

[28] S.G. Krivoshlykov, N.I. Petrov, and I.N. Sissakian, Spatial coherence of the light source with $\cos^m \vartheta$ diagram in continuous-inhomogeneous media with quadratic refraction index profile (in Russian), *Pis'ma Zh. Teor. Fiz.* **9**, 1489–1494 (1980).

[29] S.G. Krivoshlykov, N.I. Petrov, and I.N. Sissakian, Density matrix formalism in problems of propagation of optical fields in weakly inhomogeneous media (in Russian), Institute of General Physics preprint # 10 (1985), to be published in *Optical and Quantum Electronics.*

[30] M. A. Man'ko, Analogue of the Franck–Condon principle for heterolasers (in Russian), *Kvantovaîa Elektronika* (1985, in press).

[31] M. A. Man'ko, Analogue of the Ramsauer effect for heterolasers (in Russian), *Kvantovaîa Elektronika* (1985, in press).

[32] M. A. Man'ko, Modes in heterolasers, *P.N. Lebedev Inst. Trudy* **166** (1986, in press).

[33] H. Bacry and M. Cadilhac, Metaplectic group and Fourier optics, *Phys. Rev.* **A23**, 2533–2536 (1981).

[34] H. Bacry, Group theory and paraxial optics. In *Proceedings of the XIII$^{\rm th}$ International Colloquium on Group Theoretical Methods in Physics (College Park, Maryland, 21–25 May, 1984)*, W.W. Zachary, Ed. (World Scientific Publ., Singapore, 1984); pp. 215–224.

[35] A. Dragt, Lie algebraic theory of geometrical optics and optical aberrations, *J. Opt. Soc. Amer.* **72**, 372–379 (1982).

[36] K.B. Wolf, A group-theoretic treatment of gaussian optics and third-order aberrations. In *Proceedings of the XII*th *International Colloquium on Group Theoretical Methods in Physics (Trieste, Italy, 5–10 Sept. 1983).* Lecture Notes in Physics # 201, (Springer Verlag, 1984); pp. 133–136.

[37] M. Navarro-Saad and K.B. Wolf, Factorization of the phase-space transformation produced by an arbitrary refracting surface. Preprint CINVESTAV, Mexico (April 1984); to appear in *J. Opt. Soc. America.*

[38] K.B. Wolf, Group theoretical methods in aberrating systems. In *Proceedings of the XIII*th *International Colloquium on Group Theoretical Methods in Physics (College Park, Maryland, 21–25 May, 1984),* W.W. Zachary, Ed. (World Scientific Publ., Singapore, 1984); pp. 111–114.

[39] M. Navarro-Saad and K.B. Wolf, Application of a factorization theorem for ninth-order aberration optics. Comunicaciones Técnicas IIMAS preprint (Desarrollo) # 41, (February 1985). To appear in *Journal of Symbolic Computation.*

[40] K.B. Wolf, Symmetry in Lie optics. Reporte de Investigación, Depto. de Matemáticas, Universidad Autónoma Metropolitana, **4**, # 1 (1985); submitted for publication.

[41] M. Navarro-Saad and K.B. Wolf, The group theoretical treatment of aberrating systems. I. Aligned lens systems in third aberration order. Comunicaciones Técnicas IIMAS preprint # 363 (1984); K.B. Wolf, *ibid.* II. Axis-symmetric inhomogeneous systems and fiber optics in third aberration order. Comunicaciones Técnicas IIMAS preprint # 366 (1984). To appear in *J. Math. Phys.*

[42] V.I. Man'ko and K.B. Wolf, The influence of aberrations in the optics of gaussian beam propagation. Reporte de Investigación del Depto. de Matemáticas, Universidad Autónoma Metropolitana, **4**, # 3 (1985).

[43] E.V. Doktorov, I.A. Malkin, and V.I. Man'ko, Dynamical symmetry and the Franck–Condon principle for polydimensional molecules (in Russian), *P.N. Lebedev Inst. Trudy* **152**, 202 (1983).

[44] J.R. Klauder, Continuous representation theory. II. Generalized relation between quantum and classical dynamics, *J. Math. Phys.* **5**, 177–1187 (1964).

Chapter 8

The influence of spherical aberration on gaussian beam propagation

by VLADIMIR I. MAN'KO and KURT BERNARDO WOLF

ABSTRACT: Gaussian beams include a number of field or wavefunctions which have a clear quantum mechanical analogue: coherent states, correlated coherent states and discrete modes for quantum oscillators. These are used to model optical fibers and to describe the output of laser devices. When these beams leave their source and travel freely through space, they loose their coherence and *aberrate*. The source of this aberration is purely geometrical, and is termed *spherical* aberration. We describe this process in the framework of the Fermat–Hamilton formulation of optics, studying the behaviour of the center of the beam, its width, and the way in which the initial uncorrelation of position and momentum is lost.

8.1 Introduction

There are several interrelated approaches to the study of light beam propagation. The approach we will develop in this chapter is based on the Fermat principle and proceeds through the formalism of Hamilton [1]. It considers light beams as governed by the minimal action principle. In this approach, the ray evolution is governed by the *optical Hamiltonian*[1]

$$H = -\sqrt{n^2 - p^2}, \tag{1.1}$$

which generates a ray ***path,i.e.***a one-dimensional group of canonical transformations of the points of ***optical phase space***[2] When the canonical transformation has a nonlinear part, we identify this nonlinearity as the effect of ***aberrations***.

From here we follow the ***ansatz*** that, to extrapolate from geometrical optics to *wave* optics, we must consider the canonical transformation applying to a pair of canonically conjugate operators satisfying the Heisenberg commutation relations [2]. This is the usual line of ascent of quantum mechanics

[1]For an introduction to this subject and the derivation of equation (1.1), see Chapter 4 in this volume. (Editor's note.)

[2]Optical phase space has coordinates of position and momentum; the former correspond to the intersection of the ray with a $z = 0$ reference plane, and the latter describe the ray *direction* with respect to the normal. Ray propagation is referred to the system's ***optical axis***, taken to be in the z–direction.

in its Hamilton–Schrödinger formulation, applied to systems of nonrelativistic point particles. The optical-mechanical connection has been used mainly in the *paraxial* approximation [3,4], *i.e.*, for *linear* canonical transformations produced by Hamiltonians *quadratic* in the coordinates of phase space. The free propagation Hamiltonian (1.1) aberrates out of sheer geometry, however. It is quadratic only in the paraxial approximation [see ahead, Eq. (3.1)]. The terms beyond second order produce a particular type of aberration, termed generically ***spherical*** aberration.

Nevertheless, models of the quantum harmonic oscillator —which is a quadratic Hamiltonian— have been used successfully to model *coherent states* in optical fibers [5,6]. *Correlated* coherent states [7] and *discrete* modes [8] have been investigated also; these belong to a class of beams for which we use the term ***gaussian beams*** in this chapter. These are of fundamental interest in the description of the output of lasers, parabolic-profile optical fiber models, and other optical devices of technological relevance.

Another approach that should be mentioned so as to avoid possible confusion, is that associated with the names of Fock and Leontovich [9]. This uses also quantum mechanical methods, but is based on the approximate solution of the wave-Helmholtz equation. In the paraxial regime, light propagation in nonhomogeneous media formally coincides with the Schrödinger equation for a time-dependent potential [8,10]. Due to this, many results of quantum mechanics apply in the physics of paraxial beam propagation, in particular the Franck–Condon principle and the Ramsauer effect.[3] What must be emphasized is that, in the Fock–Leontovich approximation, the momentum-dependent term in the Hamiltonian is of the usual quadratic form of kinetic energy, $p^2/2n$. In the frame of the Fermat–Hamilton model that we study in *this* chapter, the Hamiltonian is (1.1) and the momentum-dependent terms responsible for aberrations appear as an infinite series beyond the quadratic term.

As will be seen below, there is the question of convergence of the momentum series in the Hamiltonian. This problem stems from the *global* properties of the momentum variable in optics: ray direction is described by the geometry of a *circle*, and does *not* range over the full real line, as it *does* in nonrelativistic quantum mechanics. Nevertheless, our approach here follows both the well-established treatment for spherical aberration in optics, as well as the usual gaussian beam characterization.

The plan of this chapter is as follows: Section 2 defines and describes gaussian beams, coherent states, correlated coherent states, and discrete correlated modes. Section 3 presents the optical Hamiltonian of free propagation and its straightforward ***wavization*** leading to the Green function in momentum space. The unfolding of gaussian beams under the action of this Hamiltonian is studied in Section 4, in terms of their center, width, and correlation functions. The generating functional method yields the evolution of the quantities of interest. The concluding Section 5 offers some further comments on the scope of these results.

8.2 Gaussian beams

In the next four subsections we present the gaussian wavefunctions: the fundamental mode, coherent states, correlated coherent states, and discrete correlated modes.

8.2.1 The fundamental gaussian mode

The fundamental gaussian mode is a scalar field or wavefunction which at the initial $z = 0$ plane is given by the gaussian exponential function

$$\psi(\mathbf{q}, 0) = N\ \exp(Aq_x^2 - Bq_y^2 + Cq_xq_y + Dq_x + Eq_y), \tag{2.1}$$

---

[3]See the preceding chapter in this volume. (Editor's note.)

where the coefficients A, B, ..., E may be complex but must satisfy certain conditions (see below), and N is a normalization constant. If the medium is homogeneous, we may choose a coordinate system whose origin and orientation are such that the linear and cross terms in (2.1) vanish. In what follows, we need thus only consider

$$\psi(\mathbf{q},0) = \frac{1}{\sqrt{\omega_x \omega_y \pi}} \exp\left(-\frac{1}{2\omega_x^2} q_x^2\right) \exp\left(-\frac{1}{2\omega_y^2} q_y^2\right). \tag{2.2}$$

Here, ω_x and ω_y are defined as the ***widths***[4] of the beam in the x– and y–directions. The Gaussian in (2.2) is normalized in $\mathcal{L}^2(\Re^2)$, so that $\iint_{-\infty}^{\infty} d\mathbf{q}\,|\psi(\mathbf{q},0)|^2 = 1$.

In momentum space (which here has the meaning of frequency at directions across the optical axis), the wavefunction is also a fundamental gaussian exponential,

$$\tilde{\psi}(\mathbf{p},0) = \sqrt{\frac{\omega_x \omega_y}{\pi}} \exp\left(-\tfrac{1}{2}\omega_x^2 p_x^2\right) \exp\left(-\tfrac{1}{2}\omega_y^2 p_y^2\right). \tag{2.3}$$

It has widths ω_x^{-1} and ω_y^{-1}, and is of course also normalized.

The basic properties of the fundamental gaussian mode (considered as a quantum-mechanical wavefunction, as the notation implies) are the following.

i. The ***average*** of position and of direction are zero: $\langle\hat{\mathbf{q}}\rangle = 0$, $\langle\hat{\mathbf{p}}\rangle = 0$. This is simply due to parity.

ii. The ***width*** of the mode in position and momentum space are: $\langle\hat{q}_i^2\rangle = \frac{1}{2}\omega_i^2$, $\langle\hat{p}_i^2\rangle = \frac{1}{2}\omega_i^{-2}$, $i = x, y$.

iii. The fundamental mode satisfies with *equality* the Heisenberg uncertainty relation, $\Delta q\,\Delta p \geq \frac{1}{2}$, in each of the two components. The ***dispersion*** of an observable x is defined as usual, *i.e.*,

$$\Delta x := \sqrt{\langle(\hat{x}^2 - \langle\hat{x}\rangle^2)\rangle}. \tag{2.4}$$

iv. Position and momentum are (initially) ***uncorrelated***: $\langle(\hat{q}_i\hat{p}_i + \hat{p}_i\hat{q}_i)\rangle = 0$ ($i = x, y$, no sum).

8.2.2 Coherent states

The fulfillment of equality in the Heisenberg uncertainty relation is a defining characteristic of all ***coherent states*** [5,6]. These are gaussian fundamental modes of *unit* width, so that the width in configuration and momentum spaces are *equal*,[5] and of dispersion $1/\sqrt{2}$ in each direction. They are *translated* in both configuration and momentum space, so that they represent gaussian beams that cross the reference ($z = 0$) plane at any point and at any angle. This phase-space translation is described by a complex parameter $\boldsymbol{\alpha}$, such that its averages are $\langle\hat{\mathbf{q}}\rangle_\alpha = \sqrt{2}\,\mathrm{Re}\,\boldsymbol{\alpha}$ and $\langle\hat{\mathbf{p}}\rangle_\alpha = \sqrt{2}\,\mathrm{Im}\,\boldsymbol{\alpha}$. The wavefunctions of these beams are customarily built as the $\mathcal{L}^2(\Re^2)$–normalized eigenfunctions of the annihilation operator $\hat{\mathbf{a}} = (\hat{\mathbf{q}} + i\hat{\mathbf{p}})/\sqrt{2}$, with eigenvalue $\boldsymbol{\alpha}$. They have the form

$$\psi(0, 1/\sqrt{2}; \boldsymbol{\alpha}; \mathbf{q}) = \frac{1}{\sqrt{\pi}} \exp\left(-\tfrac{1}{2}q^2 + \sqrt{2}\,\boldsymbol{\alpha}\cdot\mathbf{q} - \tfrac{1}{2}(\alpha^2 + |\alpha|^2)\right). \tag{2.5}$$

---

[4]Note that ω^{-2} appears here in the exponent, while ω^{-1} is used in Ref. [12]. The conversion from one definition to the other is immediate, but should be done with care. (Editor's note.)

[5]Equivalently, q will be measured in units of ω, and p in units of ω^{-1}.

As for the fundamental gaussian mode, position and momentum are uncorrelated. We shall refer to (2.5) alternatively as *Glauber* coherent states [6].

The above wavefunctions and all other quantities of interest are separable in x–y cartesian coordinates. We shall henceforth work with each of the one-dimensional position and momentum coordinates, for notational convenience and without loss of generality. The factor $1/\sqrt{\pi}$ will be thus replaced by $\pi^{-1/4}$.

8.2.3 Correlated coherent states

In reference [7], Dodonov, Kurmyshev, and Man'ko introduced the ***correlated*** coherent states. Their construction may be given in terms of *complex canonical transforms* [11; 12, Sect. 9.2], recalling that the Glauber coherent states (2.5) can be defined as the *Bargmann* transform of a Dirac $\delta(\mathbf{q}-\boldsymbol{\alpha})$ [12, Subsect. 9.3.5].[6] We first define the appropriate creation and annihilation operators as a canonical transform $\mathbb{C}_\mathbf{M}$ of the canonical pair $\hat{\mathbf{q}}$, $\hat{\mathbf{p}}$. This transform is the product of the Bargmann transform $\mathbb{C}_\mathbf{B}$ and a general real *geometric* transform $\mathbb{C}_\mathbf{G}$,

$$\mathbb{C}_\mathbf{M} = \mathbb{C}_\mathbf{GB} = \mathbb{C}_\mathbf{G}\mathbb{C}_\mathbf{B}, \tag{2.6a}$$

$$\mathbf{G} = \begin{pmatrix} \omega & 0 \\ \kappa & \omega^{-1} \end{pmatrix}, \qquad \mathbf{B} = \frac{1}{\sqrt{2}}\begin{pmatrix} 1 & -i \\ -i & 1 \end{pmatrix}, \qquad \mathbf{M} = \frac{1}{\sqrt{2}}\begin{pmatrix} \omega & -i\omega \\ \kappa - i\omega^{-1} & \omega^{-1} - i\kappa \end{pmatrix}. \tag{2.6b}$$

This transform yields

$$\begin{pmatrix} \hat{a} \\ i\hat{a}^\dagger \end{pmatrix} = \mathbb{C}_\mathbf{M}\begin{pmatrix} \hat{q} \\ \hat{p} \end{pmatrix}\mathbb{C}_{\mathbf{M}^{-1}} = \mathbf{M}^{-1}\begin{pmatrix} \hat{q} \\ \hat{p} \end{pmatrix} = \begin{pmatrix} [(\omega^{-1} - i\kappa)\hat{q} + i\omega\hat{p}]/\sqrt{2} \\ [(\omega^{-1} + i\kappa)\hat{q} - i\omega\hat{p}]/\sqrt{2} \end{pmatrix}. \tag{2.7}$$

This transformation is quantum-canonical in the sense that $[\hat{a}, \hat{a}^\dagger] = 1$. The transformation $\mathbb{C}_\mathbf{G}$ is called *geometric* since, $\mathbf{G}$ being lower-triangular, its action on $\mathcal{L}^2(\mathfrak{R})$ is a *Lie* transformation [12, Subsect. 9.1.5]:

$$(\mathbb{C}_\mathbf{G} f)(q) = \frac{1}{\sqrt{\omega}} e^{i\kappa q^2/2\omega} f\left(\frac{q}{\omega}\right). \tag{2.8}$$

It may only change the scale (by ω) and multiply the function by an imaginary-width (oscillating) gaussian exponential.

Correlated coherent states are now defined as $\mathcal{L}^2(\mathfrak{R})$–normalized eigenfunctions of $\hat{a}(\omega,\kappa)$ with eigenvalue α in the complex plane,

$$\hat{a}\psi(\alpha; q) = \alpha\psi(\alpha; q), \tag{2.9a}$$

The creation and annihilation operators, $\hat{a}$ and $\hat{a}^\dagger$, and the eigenvalue α, should bear the parameters ω and κ as labels, but we shall omit them. Equivalently, the last equation may be written as a canonical transform kernel [12, Subsect. 9.1.2]:

$$\psi(\alpha; q) = (\mathbb{C}_\mathbf{M}\delta_\alpha)(q) = C_\mathbf{M}(q, \alpha). \tag{2.9b}$$

This may be seen to be, for ω and κ fixed,

$$\psi(\alpha; q) = \sqrt{\omega\sqrt{\pi}}\, \exp\left(-\frac{q^2}{2\omega^2}(1 - i\omega\kappa) + \sqrt{2}\frac{\alpha q}{\omega} - \frac{1}{2}(\alpha^2 + |\alpha|^2) \right). \tag{2.9c}$$

[6]Note that all factors of $(2\pi)^{-1/4}$ in Eqs. (9.90–91) there, should read $(2\pi)^{1/4}$.

The notation in [7] makes use of the correlation coefficient r defined in [13, Sect. 10.3],

$$\begin{aligned} r &:= \frac{\langle \frac{1}{2}[(\hat{p}-\langle\hat{p}\rangle)(\hat{q}-\langle\hat{q}\rangle)+(\hat{q}-\langle\hat{q}\rangle)(\hat{p}-\langle\hat{p}\rangle)]\rangle}{\sqrt{\langle(\hat{p}-\langle\hat{p}\rangle)^2\rangle\langle(\hat{q}-\langle\hat{q}\rangle)^2\rangle}} \\ &= \frac{\omega\kappa}{\sqrt{1+\omega^2\kappa^2}}. \end{aligned} \tag{2.10}$$

We thus introduce

$$\eta := \frac{\omega}{\sqrt{2}}, \qquad \kappa = \frac{r/\omega}{\sqrt{1-r^2}}, \qquad 0 \le r < 1, \tag{2.11}$$

to write hereafter

$$\hat{a} = \frac{R}{2\eta}\hat{q} + i\eta\hat{p}, \qquad R := 1 - \frac{ir}{\sqrt{1-r^2}}, \tag{2.12a}$$

$$\hat{a}^\dagger = \frac{R^*}{2\eta}\hat{q} - i\eta\hat{p}, \qquad R^* = 1 + \frac{ir}{\sqrt{1-r^2}}, \tag{2.12b}$$

$$\psi(r,\eta;\alpha;q) = \sqrt{\eta\sqrt{2\pi}}\,\exp\left(-\frac{R}{4\eta^2}q^2 + \frac{\alpha}{\eta}q - \frac{1}{2}(\alpha^2 + |\alpha|^2)\right). \tag{2.13}$$

The Fourier transform of the correlated state (2.9)–(2.13) may be expressed in terms of the canonical transform kernel [12] of $e^{i\pi/4}(\mathbb{C}_{\mathbf{FGB}}\delta_\alpha)(q) = e^{i\pi/4}C_{\mathbf{FM}}(q,\alpha)$, namely

$$\tilde{\psi}(r,\eta;\alpha;p) = \sqrt{\frac{\eta}{R}\sqrt{\frac{2}{\pi}}}\,\exp\left[-\frac{\eta^2}{R}\left(p + i\frac{\alpha}{\eta}\right)^2 - \frac{1}{2}(\alpha^2 + |\alpha|^2)\right]. \tag{2.14}$$

In order to evaluate the moments of the correlated state $\psi_\alpha := \psi(r,\eta;\alpha;q) = \mathbb{C}_{\mathbf{G}}\psi_\alpha^0$ in terms of those of the Glauber state ψ_α^0 given in **2.2**, we note that

$$\begin{aligned} \begin{pmatrix} \langle\hat{q}\rangle_\alpha \\ \langle\hat{p}\rangle_\alpha \end{pmatrix} &= \begin{pmatrix} \langle \mathbb{C}_{\mathbf{G}}\psi_\alpha^0|\hat{q}|\mathbb{C}_{\mathbf{G}}\psi_\alpha^0\rangle \\ \langle \mathbb{C}_{\mathbf{G}}\psi_\alpha^0|\hat{p}|\mathbb{C}_{\mathbf{G}}\psi_\alpha^0\rangle \end{pmatrix} = \left\langle \psi_\alpha^0 \middle| \begin{pmatrix} \mathbb{C}_{\mathbf{G}}^{-1}\,\hat{q}\mathbb{C}_{\mathbf{G}} \\ \mathbb{C}_{\mathbf{G}}^{-1}\,\hat{p}\mathbb{C}_{\mathbf{G}} \end{pmatrix} \middle| \psi_\alpha^0 \right\rangle \\ &= \left\langle \psi_\alpha^0 \middle| \mathbf{G}\begin{pmatrix} \hat{q} \\ \hat{p} \end{pmatrix} \middle| \psi_\alpha^0 \right\rangle = \mathbf{G}\begin{pmatrix} \langle\hat{q}\rangle_\alpha^0 \\ \langle\hat{p}\rangle_\alpha^0 \end{pmatrix} = \sqrt{2}\begin{pmatrix} \omega & 0 \\ \kappa & \omega^{-1} \end{pmatrix}\begin{pmatrix} \operatorname{Re}\alpha \\ \operatorname{Im}\alpha \end{pmatrix}, \end{aligned} \tag{2.15a}$$

that is,

$$\langle\hat{q}\rangle_\alpha = 2\eta\operatorname{Re}\alpha, \qquad \langle\hat{p}\rangle_\alpha = \frac{r\operatorname{Re}\alpha}{\eta\sqrt{1-r^2}} + \frac{\operatorname{Im}\alpha}{\eta}. \tag{2.15b}$$

The coordinate average is thus multiplied by the factor $\omega = \eta\sqrt{2}$, and (2.8) tells us that the quadratic factor q^2 in the exponential becomes q^2/ω^2, replacing the original width by a factor ω. The *dispersion* of a correlated coherent state is thus ω times the dispersion of the Glauber coherent state, which is $1/\sqrt{2}$. For *correlated* states, therefore, we have

$$\Delta_{\psi_\alpha} = \eta, \qquad \Delta_{\tilde{\psi}_\alpha} = \frac{1}{2\eta}\frac{1}{\sqrt{1-r^2}}. \tag{2.16}$$

The result on the dispersion of $\tilde{\psi}_\alpha$ may be read off from the width of the function appearing in the integral $\langle\hat{p}\rangle_\alpha$. It is given by $1/\tilde{\omega}_e^2 = 1/\tilde{\omega}^2 + 1/\tilde{\omega}^{*2}$, where $\tilde{\omega}^2 = \frac{1}{2}R$. *Correlated* coherent states thus no longer abide the minimum in the Heisenberg uncertainty relation, but $\Delta_{\psi_\alpha}\Delta_{\tilde{\psi}_\alpha} = 1/\sqrt{1-r^2} \geq \frac{1}{2}$. This may be interpreted as an ***enlargement*** of the Heisenberg constant $\hbar$ to $\hbar/\sqrt{1-r^2} \geq \hbar$. Thus $r=0$ brings us back to the uncorrelated states, and $r \to 1$ means *complete* correlation.

8.2.4 Discrete corrrelated modes

As is the case for the uncorrelated Glauber states, the correlated coherent states, $\psi(r,\eta;\alpha;q)$ in (2.13–14), are the generating functions, in powers of the parameter α, of the *discrete correlated modes* $\psi_n(r,\eta;q)$ [17]:

$$\psi(r,\eta;\alpha;q) = e^{-|\alpha|^2/2} \sum_{n=0}^{\infty} \frac{\alpha^n}{\sqrt{n!}}\, \psi_n(r,\eta;q), \tag{2.17}$$

and similarly for $\tilde{\psi}(r,\eta;\alpha;p)$. Out of the generating function for Hermite polynomials, we obtain

$$\psi_n(r,\eta;q) = \frac{1}{\sqrt{2^n\, n!\, \eta\sqrt{2\pi}}} \exp\left(-\frac{Rq^2}{4\eta^2}\right) H_n\left(\frac{q}{\eta\sqrt{2}}\right), \tag{2.18}$$

$$\tilde{\psi}_n(r,\eta;p) = \frac{1}{\sqrt{\left(\frac{2}{1-2/R}\right)^n \frac{R\, n!}{\eta}\sqrt{\frac{\pi}{2}}}} \exp(-R\eta^2p^2)\, H_n\left(-ip\frac{\eta}{2R}[1-\frac{2}{R}]\right). \tag{2.19}$$

The uncorrelated discrete modes are recovered for $r=0$ (so $R=1$) and $\eta = 1/\sqrt{2}$.

The discrete correlated modes are orthonormal and complete in $\mathcal{L}^2(\Re)$:

$$\langle\psi_n \mid \psi_m\rangle = \delta_{n,m}, \qquad \sum_{n=0}^{\infty} \psi_n(q)^*\psi_n(q') = \delta(q-q'), \tag{2.20}$$

and similarly for $\tilde{\psi}_n$. The discrete correlated modes may be built also as the normalized eigenstates of $\hat{a}^\dagger\hat{a}$, defined in (2.12), with eigenvalue n, *i.e.*, as harmonic oscillator-type wavefunctions. A light beam ψ_n will have an intensity distribution $|\psi_n|^2$ and will present a (one-dimensional) pattern of n dark spots corresponding to the n zeros of the Hermite polynomial, with a cosecant-type intensity maximum envolvent, and a gaussian exponential decrease thereafter. One may calculate the center and width of these modes, as was done for the simpler cases in the previous subsections.

8.3 The free propagation Hamiltonian

In this section we present the dynamics of freely propagating light beams governed by the optical Hamiltonian (1.1) of a homogeneous system.

8.3.1 The Hamiltonian as a series

As was indicated in the first section, the free propagation Hamiltonian [1] is

$$\begin{aligned} H &= -\sqrt{n^2-p^2} & |p| \leq n \\ &= n\left[-1+\frac{1}{2}\left(\frac{p^2}{n}\right)+\frac{1}{8}\left(\frac{p}{n}\right)^4+\frac{1}{16}\left(\frac{p}{n}\right)^6+\frac{5}{128}\left(\frac{p}{n}\right)^8+\frac{7}{256}\left(\frac{p}{n}\right)^{10}+\cdots\right] \\ &= n\sum_{m=0}^{\infty}\frac{(2m-3)!!}{(2m)!!}\left(\frac{p}{n}\right)^m, & |p| < n \end{aligned} \tag{3.1}$$

where we recall that $1!! = 1$, $0!! = 1$, $(-1)!! = 1$, and $(-3)!! = -1$.

As we also remarked earlier, the formalism of gaussian beams has been developed with attention to their properties under propagation in parabolic-profile fiber models, or free space in the *paraxial* approximation. There, $H \approx -n + p^2/2n$, and the mathematics is that of the linear symplectic group of transformations $Sp(2,\Re)$. *Aberration* refers to the rest of the series expansion of the optical Hamiltonian (3.1). We remarked that this is *spherical* aberration, which is thus unavoidable in any free-propagaation system, because it is of purely *geometric* origin.

In order to study the behaviour of the most general type of Hamiltonian of the free-propagation kind, we propose the form

$$H(p,z) = \sum_{m=0}^{\infty} h_m(z)\frac{(p^2)^m}{m!}, \tag{3.2}$$

where the coefficients h_m may depend, at most, on the distance z along the optical axis, representing a stratified, inhomogeneous medium. In the actual optical case (3.1), we have

$$h_m = m!\,\frac{(2m-3)!!}{(2m)!!}\,n^{-2m+1}. \tag{3.3}$$

We shall examine the situation mainly where the medium is homogeneous, $n =$ constant, and only later indicate the replacements needed for stratified media $n = n(z)$.

8.3.2 The equations of motion

The standard hamiltonian formalism [14] shows that the time evolution of an observable $f(q,p;z)$ under a classical Hamiltonian $H(z)$ is given by

$$\frac{d}{dz}f(q,p;z) = \{H,f\} + \frac{\partial}{\partial z}f(q,p;z), \tag{3.4}$$

where $\{\cdot,\cdot\}$ is the Poisson bracket $\{f,g\} := \dfrac{\partial f}{\partial q}\dfrac{\partial g}{\partial p} - \dfrac{\partial f}{\partial p}\dfrac{\partial g}{\partial q}$. For the system (3.2), this leads to the equations of motion for the canonically conjugate pair (q,p) given by

$$\begin{aligned} \dot{q} &= \frac{dH}{dp} = 2\sum_{m=1}^{\infty} h_m \frac{p^{2m-1}}{(m-1)!}, \\ \dot{p} &= -\frac{dH}{dq} = 0, \end{aligned} \tag{3.5}$$

where overdots indicate z–derivatives. These equations may be integrated for z from z_1 to z_2. If we denote by unprimed q and p their values at $z_1 = 0$ and by primed q' and p' their values at $z = z_2$,

$$q'(q,p;z) = q + \int_0^z dz' \frac{dH(z')}{dp} = q + 2\sum_{m=1}^{\infty}\left(\int_0^z dz'\, h_m(z')\right)\frac{p^{2m-1}}{(m-1)!}, \qquad p'(p,z) = p. \tag{3.6a}$$

In the actual free-propagation case (3.3),

$$q'(q,p;z) = q + z\frac{p}{\sqrt{n^2-p^2}}, \qquad p' = p. \tag{3.6b}$$

8.3.3 *Wavization*

We follow the process of *wavization* of the Hamiltonian (3.2) through the familiar Schrödinger replacement

$$q \mapsto \hat{q} = q\cdot, \qquad p \mapsto \hat{p} = -i\lambda\!\!\!^{-}_0 \frac{d}{dq}. \tag{3.7}$$

The fundamental unit of length is $\lambda\!\!\!^{-}_0 = \lambda/2\pi$, which we set equal to unity in all our spatial measurements of q. The conjugate momentum p will have *no* units.

Due to the Stone–von Neumann theorem [15], we cannot have here a quantum canonically conjugate pair of self-adjoint position and momentum operators in $\mathcal{L}^2(\Re)$ fulfilling a Heisenberg–Weyl algebra, $[\hat{q},\hat{p}] = i\hat{1}$, since the *range* of classical p is bounded by n. Nevertheless, we shall give operational meaning to the quantities involved in terms of the matrix elements between phase functions $\psi(q)$ which are ***forward-concentrated***, *i.e.*, whose momentum distribution $\tilde{\psi}(p)$ is such that $|\tilde{\psi}(p)|^2$ is of negligible integral for values of p beyond a ***system width*** $W \ll n$.

8.3.4 The evolution operator

The evolution operator of the system will be denoted $\hat{U}(z)$. Its matrix elements may be obtained through the Van Vleck formula [16] from the classical hamiltonian generating function. It can be also defined equivalently as the solution to the Schrödinger equation with identity initial conditions:

$$i\frac{d\hat{U}(z)}{dz} = \hat{H}\hat{U}(z), \qquad \hat{U}(0) = \hat{1}, \tag{3.8a}$$

whose formal solution is

$$\hat{U}(z) = \exp\left(-i\int_0^z dz'\,\hat{H}(z)\right) = \exp\left(-i\sum_{m=0}^{\infty}[\int_0^z dz'\,h_m(z)]\frac{(\hat{p}^2)^m}{m!}\right). \tag{3.8b}$$

This formula may be upheld for matrix elements between forward-concentrated states. The unitarity of this representation is violated as one moves beyond the set of forward-concentrated functions; but *within* this set, the equality of the matrix elements

$$\hat{U}(z)\hat{p} = \hat{p}\hat{U}(z), \qquad \hat{U}(z)\hat{q} = \hat{q}'\hat{U}(z), \tag{3.9}$$

holds exactly for the first equation, and approximately for the second.

8.3.5 The Green function in momentum space

The Green function $G(x_1, x_2; z)$ can be found as the matrix element of $\hat{U}(z)$ in any representation basis $\{x\}$; in the cases of interest these bases are continuous and $G(x_1, x_2; z)$ is an integral kernel. For our description, we need the ***momentum*** representation, which is *diagonal*:

$$\begin{aligned} \tilde{G}(p_1, p_2; z) &= \langle p_1|\hat{U}(z)|p_2\rangle \\ &= \delta(p_1 - p_2)\, \exp\Big(-i \sum_{m=0}^{\infty} [\int_0^z dz'\, h_m(z)] \frac{p_1^{2m}}{m!}\Big) \\ &=: \delta(p_1 - p_2)\, \tilde{G}(p_1; z). \end{aligned} \tag{3.10}$$

In the case of the homogeneous-medium optical Hamiltonian (3.1), the last factor is given by

$$\tilde{G}(p; z) = \exp\big(iz\sqrt{n^2 - p^2}\big). \tag{3.11}$$

For any initial distribution in momentum space $\tilde{\psi}_{\text{in}}(p)$, the propagation over distance z will yield the distribution function

$$\tilde{\psi}(p; z) = \tilde{G}(p; z)\, \tilde{\psi}_{\text{in}}(p). \tag{3.12}$$

We noted that within the set of forward-concentrated phase functions, $|\tilde{\psi}_{\text{in}}(p)|$ should be negligible beyond the system's width $W \ll n$, and so $\tilde{G}(p; z)$ may be replaced by zero for $p > W$. The exponential increasing (or decreasing) behaviour of (3.11) for $p > n$ is thus of no consequence for $\tilde{\psi}(p; z)$, even though $\tilde{G}(n; z) = 1$, since it will be also forward-concentrated. Analogous arguments apply to the more general case in (3.10).

A wavefunction $\psi_{\text{in}}(q)$ at $z = 0$ will evolve through a distance z to a wavefunction $\psi(q; z)$, obtained as the Fourier transform of (3.12). The integration over p may be restricted to the interval $(-W, W)$. When it is extended beyond this interval for the purpose of analytic solution, into regions where the hamiltonian series diverges, it is to be expected that any phase divergence of $\tilde{G}(p; z)$ will be recognized and the integral regularized. We may evaluate *explictly* the *propagator* $G(q_1, q_2; z)$, *i.e.*, the Fourier transform of $\tilde{G}(p_1, p_2; z)$, only when working in the *linear* approximation (where $h_0 = -n$, $h_1 = 1/n$, and $h_m = 0$ for $m \geq 2$). This $\tilde{G}(p_1, p_2; z)$ is the Green function of the Schrödinger free particle [3] times e^{inz}.

8.4 Characteristics of gaussian beams under free propagation

We shall now describe the characteristic properties of gaussian free propagation, avoiding the use of the propagator $G(q_1, q_2; z)$. For that, we need the generating functional and the generating function for the system (3.2).

8.4.1 The correlation functions

The two-point correlation functions [13] for a system in a state $\psi(q; z)$, at z_1 and z_2, are

$$K_q^{(2)}(z_1, z_2) := \langle \hat{q}(z_1)\hat{q}(z_2)\rangle - \langle \hat{q}(z_1)\rangle\langle \hat{q}(z_2)\rangle, \tag{4.1a}$$

$$K_p^{(2)}(z_1, z_2) := \langle \hat{p}(z_1)\hat{p}(z_2)\rangle - \langle \hat{p}(z_1)\rangle\langle \hat{p}(z_2)\rangle, \tag{4.1b}$$

$$K_{qp}^{(2)}(z_1, z_2) := \langle \hat{q}(z_1)\hat{p}(z_2)\rangle - \langle \hat{q}(z_1)\rangle\langle \hat{p}(z_2)\rangle, \tag{4.1c}$$

$$K_{pq}^{(2)}(z_1, z_2) := \langle \hat{p}(z_1)\hat{q}(z_2)\rangle - \langle \hat{p}(z_1)\rangle\langle \hat{q}(z_2)\rangle. \tag{4.1d}$$

For $z_1 = z = z_2$, we have the coordinate and momentum ***dispersions***

$$\sigma_q^2(z) := K_q^{(2)}(z,z), \qquad \sigma_p^2(z) := K_p^{(2)}(z,z). \tag{4.2a}$$

Although they will not be used in the sequel, we may also define the *mixed* dispersion

$$\sigma_{pq}^2(z) := \tfrac{1}{2}\big(K_{qp}^{(2)}(z,z) + K_{pq}^{(2)}(z,z)\big). \tag{4.2b}$$

In terms of this, the correlation ***coefficient*** is

$$\Gamma(z) = \sigma_{pq}(z)/\sigma_p(z)\,\sigma_q(z). \tag{4.3}$$

The generalization of (4.1*a*) to n points is

$$K_q^{(n)}(z_1, z_2, \ldots, z_n) := \frac{1}{n!}\sum_{\text{perm } z_i} \langle \hat{q}(z_1)\,\hat{q}(z_2)\cdots\hat{q}(z_n)\rangle, \tag{4.4}$$

where the sum is performed over all permutations of the indices of $z_1, z_2, \ldots, z_n$. In terms of this, the ***average*** is clearly $\langle\hat{q}\rangle = K_q^{(1)}(z)$. We also have $\langle\hat{p}\rangle = K_p^{(1)}(z)$. The ***two-point correlation coefficient*** [13] is

$$\begin{aligned} R(z_1,z_2) :&= \frac{\frac{1}{2}\langle[\hat{q}(z_1)\hat{q}(z_2) + \hat{q}(z_2)\hat{q}(z_1)]\rangle - \langle\hat{q}(z_1)\rangle\langle\hat{q}(z_2)\rangle}{\sigma_q(z_1)\,\sigma_q(z_2)} \\ &= \frac{K_q^{(2)}(z_1,z_2) - K_q^{(1)}(z_1)K_q^{(1)}(z_2)}{\sqrt{[K_q^{(2)}(z_1,z_1) - (K_q^{(1)}(z_1))^2][K_q^{(2)}(z_2,z_2) - (K_q^{(1)}(z_2))^2]}}. \end{aligned} \tag{4.5}$$

8.4.2 The generating functional

When the system is described by a wavefunction $\psi(z)$, its ***density matrix*** will be denoted

$$\hat{\rho}(z) := |\psi(z)\rangle\langle\psi(z)|. \tag{4.6}$$

We now define the ***functional*** $\phi(f)$, dependent on an arbitrary function $f(z)$, as

$$\begin{aligned} \phi(f) :&= \left\langle \exp\Big(\int dz\,\hat{q}(z)\,f(z)\Big)\right\rangle \\ &= \mathrm{Tr}\left\{\hat{\rho}(0)\,\exp\Big(\int dz\,\hat{q}(z)\,f(z)\Big)\right\}, \end{aligned} \tag{4.7}$$

where the integral is indefinite. The usefulness of this functional is that, through Taylor expansion of the exponential and n-fold functional differentiation we obtain the n-point corrrelation function (4.4):

$$K^{(n)}(z_1, z_2, \ldots, z_n) = \left.\frac{\delta^n \phi(f)}{\delta f(z_1)\,\delta f(z_2)\cdots\delta f(z_n)}\right|_{f=0} \tag{4.8}$$

We may bring the generating functional to a form better suited for evaluation through using (3.6) for $\hat{q}(z)$ and the introduction of an *effective* Hamiltonian h_e, leading the system from an *effective* time $\tau = 0$ to $\tau = i$ (the imaginary unit),

$$\phi(f) = \mathrm{Tr}\left\{ \hat{\rho}(0) \exp(-i\tau \hat{h}_e)\Big|_{\tau=i} \right\}, \tag{4.9}$$

$$\begin{aligned} \hat{h}_e &= \int dz \left(\hat{q}(0) + \int_0^z dz' \frac{\partial \hat{H}(z')}{\partial p} \right) f(z) \\ &= \mu \hat{q}(0) + \int dz\, f(z) \int_0^z dz' \frac{\partial \hat{H}(z')}{\partial p} \\ &=: \mu \hat{q}(0) + F(f, \hat{p}). \end{aligned} \tag{4.10a}$$

In the last two lines, we have used the functional

$$\mu := \int dz\, f(z). \tag{4.10b}$$

When the medium is homogeneous in z and the Hamiltonian is (3.1), we may write explicitly

$$\hat{h}_e = \mu \hat{q} + \nu \frac{\hat{p}}{\sqrt{n^2 - \hat{p}^2}}, \tag{4.11}$$

with the functional

$$\nu := \int dz\, z f(z). \tag{4.12}$$

The evaluation of $\hat{U}_e(\tau) := \exp(-i\tau \hat{h}_e)$ for *general* $\hat{h}_e$ obeying the Schrödinger equation (3.8a) with $\hat{h}_e$ in place of $\hat{H}$, is aided by noting that the form of $\hat{h}_e$ allows us to write

$$\hat{U}_e(\tau) = \exp(-i\tau \hat{h}_e) = \exp(-i\tau\mu\hat{q})\, \hat{W}(f, \tau). \tag{4.13}$$

Because of (3.8a), the Schrödinger equation satisfied by the *reduced* evolution operator $\hat{W}(f, \tau)$ is

$$\begin{aligned} i\frac{\partial \hat{W}(f,\tau)}{\partial \tau} &= e^{i\tau\mu\hat{q}} F(f, \hat{p}) e^{-i\tau\mu\hat{q}} \hat{W}(f, \tau) \\ &= F(f, \hat{p} - \tau\mu\hat{1})\, \hat{W}(f, \tau). \end{aligned} \tag{4.14}$$

The Green function in momentum space of the reduced system transforming through $\hat{W}(f, \tau)$, is evaluated as in (3.10):

$$G_W(p_1, p_2; \tau) = \delta(p_1 - p_2) \exp\left(-i \int_0^\tau d\tau' F(f, p_1 - \tau\mu) \right). \tag{4.15}$$

We may now return to the evaluation of $\phi(f)$ in (4.9) through the evaluation of the trace in momentum space,

$$\begin{aligned} \mathrm{Tr}\left\{ \hat{\rho}(0) \exp(-i\tau \hat{h}_e) \right\} &= \int dp_1 \int dp_2 \int dp_3\, \tilde{\psi}(p_1, 0)\, \tilde{\psi}(p_2, 0)^* \langle p_2 | e^{-i\mu\tau\hat{q}} | p_3 \rangle\, G_W(p_3, p_1; \tau) \\ &= \int dp\, \tilde{\psi}(p, 0)^*\, \tilde{\psi}(p + \mu\tau, 0) \exp\left(-i \int_0^\tau d\tau' F(f, p + \mu[\tau - \tau']) \right), \end{aligned} \tag{4.16}$$

valuated at $\tau = i$ and in company with forward-concentrated states $\tilde{\psi}(p,0)$. We recall in this expression that $F(f,p)$ as defined in (4.10a) is, for the Hamiltonian (3.3) expanded in series,

$$\begin{aligned} F(f,p) &= \int dz\, f(z) \int_0^z dz' \, \frac{\partial H(z')}{\partial p} \\ &= 2\sum_{m=1}^{\infty} \frac{p^{2m-1}}{(m-1)!} \int dz\, f(z) \int_0^z dz'\, h_m(z') \\ &= 2\nu \sum_{m=1}^{\infty} \frac{h_m}{(m-1)!} p^{2m-1} \qquad \text{[when } h_m \text{ constant]} \\ &= \frac{\nu p}{\sqrt{n^2 - p^2}} \qquad \text{[for the optical Hamiltonian (3.1)].} \end{aligned} \tag{4.17}$$

In the case when the Hamiltonian of the system is *independent* of z, the integral on τ' in (4.16) may be performed, yielding

$$\phi(f) = \int dp\, \tilde{\psi}(p,0)^* \, \tilde{\psi}(p + i\mu) \exp\!\left(-i\frac{\nu}{\mu}[H(p) - H(p + i\mu)]\right). \tag{4.18}$$

8.4.3 The generating function for moments

For a system described by a density matrix (4.6), the ***generating function*** for its moments is:

$$\chi(\lambda; z) := \sum_{m=0}^{\infty} \frac{\lambda^m}{m!} \langle (\hat{q}(z))^m \rangle = \langle e^{\lambda \hat{q}(z)} \rangle = \mathrm{Tr}\left\{ \hat{\rho}(0)\, e^{\lambda \hat{q}(z)} \right\}. \tag{4.19}$$

This particularizes the generating functional (4.7) for $f = \lambda\delta$ (the Dirac δ and λ constant), whereby we set $\mu \mapsto \lambda$ in (4.10*b*) and, when the system is z-dependent, $\nu \mapsto \lambda z$ in (4.12). The previous derivation to (4.16) and (4.17) for $F(\lambda\delta, p)$ then yields, for the general z-dependent case,

$$\chi(\lambda; z) = \int dp\, \tilde{\psi}(p,0)^* \tilde{\psi}(p + i\lambda, 0) \exp\!\left(-i \int_0^z dz' \, [H(p + i\lambda, z') - H(p, z')]\right). \tag{4.20}$$

The nth moment of the foward-concentrated beam is then found as

$$\langle (\hat{q}(z))^n \rangle = \left. \frac{\partial^n \chi(\lambda; z)}{\partial \lambda^n} \right|_{\lambda = 0}. \tag{4.21}$$

Since in our model H depends only on p (and may depend on z), we need not inquire into the time evolution of the *momentum* correlation functions, nor the moments of a beam. They are conserved.

In what follows, we shall need the expectation value, width, and the higher moments of gaussian exponential functions with *shape parameters* γ and δ, which were found in reference [17]. We denote

$$\tilde{\psi}_{\gamma\delta}(p) := N \exp[-\gamma(p - \delta)^2]. \tag{4.22}$$

Then,

$$\langle\hat{p}\rangle_{\gamma\delta} = \frac{\gamma\delta + \gamma^*\delta^*}{\gamma + \gamma^*} = \frac{\operatorname{Re}\gamma\delta}{\operatorname{Re}\gamma}, \tag{4.23}$$

$$\sigma_p^2 = \langle(\hat{p}^2 - \langle\hat{p}\rangle_{\gamma\delta}^2)\rangle_{\gamma\delta} = \frac{1}{2(\gamma+\gamma^*)} = \frac{1}{4\operatorname{Re}\gamma}, \tag{4.24}$$

$$\langle\hat{p}^m\rangle_{\gamma\delta} = s^m H_m(\rho), \tag{4.25}$$

where

$$s^2 := -\tfrac{1}{2}\sigma_p^2, \qquad \rho := \tfrac{1}{2}\langle\hat{p}\rangle_{\gamma\delta}/s. \tag{4.26}$$

When the averaging function is a correlated coherent state $\psi(r,\eta;\alpha,q)$ given by (2.13–14), then

$$\gamma = \frac{\eta^2}{R}, \qquad \delta = -i\frac{\alpha}{\eta}, \tag{4.27}$$

while $\langle\hat{p}\rangle_\alpha$ is given by (2.15b) and

$$\sigma_p^2 = \frac{1}{4\eta(1-r^2)}, \tag{4.28}$$

which is independent of α.

8.4.4 Aberrations in the moments of correlated coherent states

The generating function $\chi(\lambda;z)$ in (4.20) can be written in the following form, which reveals the structure of the dependence on the parameters of a gaussian beam (4.22):

$$\chi(\lambda;z) = \int dp\, W_G(p)\exp[S_0(\lambda,p;z) + S_\psi(\lambda,p)]. \tag{4.29}$$

Here $W_G(p)$ is the intensity distribution of the initial gaussian beam at $z=0$, namely

$$W_G(p) := \tilde{\psi}_{\gamma\delta}(p;0)^*\tilde{\psi}_{\gamma\delta}(p;0) = \frac{1}{\sqrt{2\pi\sigma_p^2}}\exp\!\left(-\frac{(p-\langle\hat{p}\rangle_{\gamma\delta})^2}{2\sigma_p^2}\right), \tag{4.30}$$

S_0 is independent of the parameters of the incoming beam,

$$S_0(\lambda,p;z) := -i\int_0^z dz'\,[H(p+i\lambda,z') - H(p,z')], \tag{4.31}$$

while S_ψ is what is left after extracting $W_G(p)$ from $\tilde{\psi}_{\gamma\delta}(p;0)^*\tilde{\psi}_{\gamma\delta}(p+i\lambda;0)$, namely

$$S_\psi(\lambda,p) := \gamma\lambda^2 - 2i\gamma\lambda(p-\delta), \tag{4.32}$$

and does not depend on z.

We may now find the first two moments using (4.21). The influence on aberrations on the center of the light beam is thus given by

$$\begin{aligned}\langle\hat{q}(z)\rangle_{\gamma\delta} &= \frac{\partial}{\partial\lambda}\chi(\lambda)\Big|_{\lambda=0} = \langle(S_0' + S_\psi')|_{\lambda=0}\rangle_{\gamma\delta} \\ &= 2i\gamma\delta - 2i\gamma\langle\hat{p}\rangle_{\gamma\delta} + \int_0^z dz'\left\langle\frac{\partial\hat{H}(p,z')}{\partial p}\right\rangle_{\gamma\delta} \\ &= \langle\hat{q}(0)\rangle_{\gamma\delta} + \int_0^z dz'\left\langle\frac{\partial\hat{H}(p,z')}{\partial p}\right\rangle_{\gamma\delta},\end{aligned} \tag{4.33}$$

$$\begin{aligned}\langle\hat{q}^2(z)\rangle_{\gamma\delta} &= \frac{\partial^2}{\partial\lambda^2}\chi(\lambda)\Big|_{\lambda=0} = \langle((S_0')^2 + S_0'' + 2S_\psi' S_0' + S_\psi'' + (S_\psi')^2)|_{\lambda=0}\rangle_{\gamma\delta} \\ &= \langle\hat{q}^2(0)\rangle_{\gamma\delta} - 2i(\gamma-\gamma^*)\int_0^z dz'\left\langle\hat{p}\frac{\partial\hat{H}(p,z')}{\partial p}\right\rangle_{\gamma\delta} \\ &\quad + 2i(\gamma\delta-\gamma^*\delta^*)\int_0^z dz'\left\langle\frac{\partial\hat{H}(p,z')}{\partial p}\right\rangle_{\gamma\delta} \\ &\quad + \int_0^z dz'\int_0^z dz''\left\langle\frac{\partial\hat{H}(p,z')}{\partial p}\frac{\partial\hat{H}(p,z'')}{\partial p}\right\rangle_{\gamma\delta},\end{aligned} \tag{4.34}$$

$$\begin{aligned}\sigma_q^2(z) &= \sigma_q^2(0) - 2i(\gamma-\gamma^*)\int_0^z dz'\left\langle\hat{p}\frac{\partial\hat{H}(p,z')}{\partial p}\right\rangle_{\gamma\delta} \\ &\quad + 2i(\gamma\delta-\gamma^*\delta^* - i\langle\hat{q}(0)\rangle_{\gamma\delta})\int_0^z dz'\left\langle\frac{\partial\hat{H}(p,z')}{\partial p}\right\rangle_{\gamma\delta} \\ &\quad + \int_0^z dz'\int_0^z dz''\left\langle\frac{\partial\hat{H}(p,z')}{\partial p}\frac{\partial\hat{H}(p,z'')}{\partial p}\right\rangle_{\gamma\delta} \\ &\quad - \left(\int_0^z dz'\left\langle\frac{\partial\hat{H}(p,z')}{\partial p}\right\rangle_{\gamma\delta}\right)^2.\end{aligned} \tag{4.35}$$

When the Hamiltonian H is independent of z, we see that single integrals $\int_0^z dz'\langle\cdots\rangle_{\gamma\delta}$ lead to terms $z\langle\cdots\rangle_{\gamma\delta}$ *linear* in z, while $\int_0^z dz'\int_0^z dz''\langle\cdots\rangle_{\gamma\delta}$ and $\left(\int_0^z dz'\langle\cdots\rangle_{\gamma\delta}\right)^2$ lead to terms *quadratic* in z. *No other z-dependence is possible.*

It is at this point that we can introduce a truncated series expansion of the Hamiltonian by aberration order. This will translate into a truncated series expansion in $\langle\hat{p}^{2m}\rangle_{\gamma\delta}$, which in turn may be evaluated through (4.25) in a truncated series expansion in $\langle\hat{p}\rangle_{\gamma\delta}^{2k}$, with k ranging up to m. This was exemplified to seventh aberration order in Ref. [18], but will not be repeated here.

It is worthwhile to detail the structure of the mean value and dispersion for the case of the optical homogeneous-medium Hamiltonian (3.1) in a correlated coherent state $\psi(r,\eta;\alpha;q)$, (2.13–14), with the values (4.27) for the parameters γ and δ, and certain constants J_{mn}. It is

$$\langle\hat{q}(z)\rangle_\alpha = 2\eta\,\mathrm{Re}\,\alpha + zJ_{11}, \tag{4.36}$$

$$\sigma_q^2(z) = \eta^2 + 4iz[\tfrac{1}{4}n^2 J_{03} + \gamma(J_{11}\langle\hat{p}\rangle_\alpha - J_{21})] + z^2[J_{22} - (J_{11})^2], \tag{4.37}$$

$$J_{mm'} := \langle\hat{p}^m(n^2-\hat{p}^2)^{-m'/2}\rangle_\alpha. \tag{4.38}$$

On (4.33) and (4.36) we should remark that this reproduces the usual classical formula $q(z) = q(0)+z\dot{q}(0)$ with the quantum averaging $\dot{q}(0) \mapsto \langle\hat{\dot{q}}(0)\rangle_\alpha$, with $\hat{\dot{q}} = \partial\hat{H}/\partial p$. The z^2-term coefficient in the dispersion (4.37) is in fact the velocity dispersion at the initial point, $\sigma_{\dot{q}}^2(0) = \langle(d\hat{H}/dp)^2\rangle_\alpha - (\langle d\hat{H}/dp\rangle_\alpha)^2$. We should note carefully that the *center* of the gaussian beam follows the classical trajectory predicted by geometrical optics. In this connection we recall that the Ehrenfest theorem in quantum mechanics implies that the averaged position and momentum follow the classical trajectory only for *quadratic* Hamiltonians. The optical Hamiltonian (beyond the gaussian approximation) is outside this set, but *does* mantain the correspondence.

Finally, we would like to remark that the integrals in (4.38), and others which will appear below, are understood in the *regularized* sense for $m' > 0$, since the integrand is troublesome at $p = \pm n$. This means that we can reduce them to convergent integrals through integration by parts, since they appear in company with a decreasing gaussian exponential. Thus, for example,

$$\left\langle\frac{d^2\hat{H}}{dp^2}\right\rangle = \frac{1}{\sigma_p^2}\left(\left\langle\hat{p}\frac{d\hat{H}}{dp}\right\rangle - \langle\hat{p}\rangle\left\langle\frac{d\hat{H}}{dp}\right\rangle\right), \tag{4.39a}$$

$$\left\langle\frac{d\hat{H}}{dp}\right\rangle = \frac{1}{\sigma_p^2}(\langle\hat{p}\hat{H}\rangle - \langle\hat{p}\rangle\langle\hat{H}\rangle), \tag{4.39b}$$

$$\left\langle\hat{p}\frac{d\hat{H}}{dp}\right\rangle = -\langle\hat{H}\rangle + \frac{1}{\sigma_p^2}(\langle\hat{p}^2\hat{H}\rangle - \langle\hat{p}\rangle\langle\hat{p}\hat{H}\rangle). \tag{4.39c}$$

In this way all integrals $J_{mm'}$ are seen to be convergent.

8.4.5 Aberration of the two-point correlation functions

We shall now apply the procedure of last subsection to the generating functional of **4.2**, $\phi(f)$ in equations (4.18) for the z-independent Hamiltonian case, in order to find the two-point correlation functions of the gaussian beam (4.22) through its functional derivatives (4.8). The procedure for the z-dependent case follows closely that of last section.

We again write $\phi(f)$ in the form (4.29),

$$\phi(f) = \int dp\, W_G(p) \exp[S_0(\mu,\nu) + S_\psi(\mu)], \tag{4.40}$$

where $W_G(p)$ is given by (4.30), but now

$$S_0(\mu,\nu) := -i\frac{\nu}{\mu}[H(p) - H(p+i\mu)], \tag{4.41}$$

$$S_\psi(\mu) := \gamma\mu^2 - 2i\gamma\mu(p-\delta), \tag{4.42}$$

γ and δ being the beam parameters in (4.22). We note again that S_ψ is independent of ν. As in the first of the three members of (4.33) and (4.34),

$$K^{(1)}(z) = \left.\frac{\delta\phi(f)}{\delta f(z)}\right|_{f=0} = \left\langle\frac{\delta(S_0+S_\psi)}{\delta f(z)}\exp(S_0+S_\psi)\Big|_{f=0}\right\rangle, \tag{4.43}$$

$$\begin{aligned} K^{(2)}(z_1,z_2) &= \left.\frac{\delta^2\phi(f)}{\delta f(z_1)\,\delta f(z_2)}\right|_{f=0} \\ &= \left\langle\left(\frac{\delta^2(S_0+S_\psi)}{\delta f(z_1)\,\delta f(z_2)} + \frac{\delta(S_0+S_\psi)}{\delta f(z_1)}\frac{\delta(S_0+S_\psi)}{\delta f(z_2)}\right)\exp(S_0+S_\psi)\Big|_{f=0}\right\rangle. \end{aligned} \tag{4.44}$$

Now, $\delta\mu/\delta f(z) = 1$, $\delta\nu/\delta f(z) = z$, $\mu|_{f=0} = 0$, $\nu|_{f=0} = 0$, $\delta S_0/\delta\mu|_{f=0} = 0$, and $\delta S_0/\delta\nu|_{f=0} = -\partial H/\partial p$. Chain-ruling functional derivatives yields

$$\left.\frac{\delta S_0}{\delta f(z)}\right|_{f=0} = -z\frac{\partial H(p)}{\partial p}, \qquad \left.\frac{\delta S_\psi}{\delta f(z)}\right|_{f=0} = -2i\gamma(p-\delta), \tag{4.45}$$

$$\left.\frac{\delta^2 S_0}{\delta f(z_1)\,\delta f(z_2)}\right|_{f=0} = i(z_1+z_2)\frac{\partial^2 H(p)}{\partial p^2}, \qquad \left.\frac{\delta^2 S_\psi}{\delta f(z_1)\,\delta f(z_2)}\right|_{f=0} = -2\gamma. \tag{4.46}$$

Applied to (4.43), equations (4.45) yield again the result (4.33) for $\langle\hat{q}(z)\rangle$. In (4.44), equations (4.45) and (4.46) give

$$\begin{aligned} K^{(2)}(z_1,z_2) &= \eta^2[1+4(\operatorname{Re}\alpha)^2] \\ &\quad + i(z_1+z_2)\left[\left\langle\frac{\partial^2\hat{H}}{\partial p^2}\right\rangle - 2\gamma\left(\delta\left\langle\frac{\partial\hat{H}}{\partial p}\right\rangle - \left\langle\hat{p}\frac{\partial\hat{H}}{\partial p}\right\rangle\right)\right] \\ &\quad + z_1 z_2\left\langle\left(\frac{\partial\hat{H}}{\partial p}\right)^2\right\rangle, \end{aligned} \tag{4.47}$$

where the expectation values are $\langle\cdots\rangle_{\gamma\delta}$. For the optical Hamiltonian (3.1), we may use the constants $J_{mm'}$ defined in (4.8) to write

$$\begin{aligned} K^{(2)}(z_1,z_2) = \eta^2&[1-4(\operatorname{Re}\alpha)^2] \\ &+ i(z_1+z_2)[\tfrac{1}{2}n^2 J_{03} + 2\gamma(\delta J_{11} - J_{21})] \\ &+ z_1 z_2 J_{22}. \end{aligned} \tag{4.48}$$

This result may be compared with (4.37) through (4.2).

8.4.6 Aberration for discrete correlated modes

The results obtained in the last two sections for correlated coherent states, ψ_α in (2.13–14), may be extended to the discrete correlated modes, ψ_n in (2.18–19), using the fact that the former are the generating functions in α of the latter, as seen in equation (2.17). From the Taylor expansion we obtain the discrete correlated states as the nth derivatives of ψ_α with respect to α, valuated at $\alpha = 0$.

If we denote by ϕ_α and χ_α the generating functional and function [(4.7) and (4.19)] for the former, and ϕ_n and χ_n for the latter states, we may relate them through

$$\phi_n(f) = \left.\frac{1}{n!}\frac{\partial^n}{\partial\alpha^n}\frac{\partial^n}{\partial\alpha^{*n}}e^{|\alpha|^2}\phi_\alpha(f)\right|_{\alpha=0}, \tag{4.49}$$

$$\chi_n(\lambda) = \left.\frac{1}{n!}\frac{\partial^n}{\partial\alpha^n}\frac{\partial^n}{\partial\alpha^{*n}}e^{|\alpha|^2}\chi_\alpha(\lambda)\right|_{\alpha=0}. \tag{4.50}$$

Due to this connection between generating functionals, we have a similar connection between all their functional derivatives and, consequently, of all correlation functions and moments described by gaussian initial field distributions and discrete modes. For example, the mean value of the coordinate and the two-point correlation function are

$$\langle\psi_n|\hat{q}(z)^k|\psi_n\rangle = \left.\frac{1}{n!}\frac{\partial^n}{\partial\alpha^n}\frac{\partial^n}{\partial\alpha^{*n}}e^{|\alpha|^2}\langle\psi_\alpha|\hat{q}(z)^k|\psi_\alpha\rangle\right|_{\alpha=0}, \tag{4.51}$$

$$K_n^{(2)}(z_1,z_2) = \left.\frac{1}{n!}\frac{\partial^n}{\partial\alpha^n}\frac{\partial^n}{\partial\alpha^{*n}}e^{|\alpha|^2}K_\alpha^{(2)}(z_1,z_2)\right|_{\alpha=0}. \tag{4.52}$$

When applying this to (4.36) and (4.48), we should note that the complex number α is contained only in δ, $\langle\hat{p}\rangle_\alpha$, $\langle\hat{q}^2(0)\rangle_\alpha$ and in the expressions for $J_{mm'}$. The first two are given in (4.27) and (2.15b), the third is $\eta^2[1+4(\operatorname{Re}\alpha)^2]$. Thus,

$$\frac{\partial\langle\hat{p}\rangle_\alpha}{\partial\alpha} = \frac{-iR^*}{2\eta}, \qquad \frac{\partial\langle\hat{p}\rangle_\alpha}{\partial\alpha^*} = \frac{iR}{2\eta}, \tag{4.53}$$

$$\frac{\partial\langle\hat{q}(0)^2\rangle_\alpha}{\partial\alpha} = 4\eta^2 \operatorname{Re}\alpha = \frac{\partial\langle\hat{q}(0)^2\rangle_\alpha}{\partial\alpha^*}. \tag{4.54}$$

We see from (4.28) that σ_p^2 is independent of α. In order to calculate the α-derivative of $J_{mm'}$ in (4.38), we observe that for any integral of the form

$$J_{\varsigma\alpha} := \frac{1}{\sqrt{2\pi\sigma_p^2}} \int dp\, \varsigma(p) \exp\left(-\frac{(p-\langle\hat{p}\rangle_\alpha)^2}{2\sigma_p^2}\right), \tag{4.55}$$

we have

$$\frac{\partial}{\partial\alpha}\frac{\partial}{\partial\alpha^*} e^{|\alpha|^2} J_{\varsigma\alpha}\Big|_{\alpha=0} = \frac{1}{\sigma_p^2}\langle 0|\hat{p}^2\,\varsigma(\hat{p})|0\rangle, \tag{4.56}$$

$$\frac{\partial}{\partial\alpha}\frac{\partial}{\partial\alpha^*} e^{|\alpha|^2} J_{mm'}\Big|_{\alpha=0} = \frac{1}{\sigma_p^2} J_{m+2,m'}\Big|_{\alpha=0}, \tag{4.57}$$

and that $J_{mm'}|_{\alpha=0} = 0$ when m is odd, since the average is taken with an even distribution function.

For the $n=1$ discrete mode, for example, from (4.37) we obtain

$$\sigma_{q1}^2(z) = \left\{3\eta^2 + iz\left[\frac{n^2}{\sigma_p^2}J_{23} + 4\gamma\left(\frac{2\gamma^*}{\eta^2}J_{21} - \frac{1}{\sigma_p^2}J_{41}\right)\right] + z^2\frac{1}{\sigma_p^2}J_{42}\right\}\Big|_{\alpha=0}. \tag{4.58}$$

A similar process will derive all other quantities of interest.

8.5 Concluding remarks

We have worked with the classical optical Hamiltonian which follows from the Fermat principle, containing terms which are responsible for spherical aberration. We have *wavized* this aberration along the lines of the Schrödinger representation of the position and momentum operators on the Hilbert space of square-integrable functions on the full real line. This has permitted us to describe the properties of the propagating light beams keeping the Heisenberg uncertainty relation in its usual form. In this frame, the model turned out to be explicitly solvable for general gaussian beams. The technique for obtaining these results was the use of the generating function and functional. There are three aspects to be discussed which are connected with the physical applicability of the Hamilonian (1.1) and with its wavized version.

The first point concerns the geometrical optics limit for media without discontinuities in the refraction index, which should yield the ray path for every original wavelength. The possible generalization of the optical Hamiltonian would be to Hamiltonians with the structure

$$H = \sum_{n=0}^{\infty} H_n \lambda^n. \tag{5.1}$$

This would entail correction terms to the eikonal equation connected with the Fermat–Maupertuis principle.

The second point concerns the propagation of beams in media with sharp jumps in the refraction index, such as lenses. The sharpness of the jump should have a bearing on the way in which ***wavization*** is performed, since it is to be expected that *reflection* phenomena arise. In order to take the reflected beam into account, a spin–$\frac{1}{2}$ formalism as in quantum mechanics has been used in reference [19]. The analogy of the pair-creation mechanism in quantum electrodynamics may be even more fruitful, especially since the reflected beam is the particle moving in the opposite *time* direction, as an *antiparticle*.

The last point which must be brought to discussion is the recognition of the fact that position and momentum in optics do not have quite the same meaning as position and momentum in quantum mechanics, since ray direction ranges over a *circle*, and **not** over the real line. The latter is the approximation we have assumed and followed here. The formalism of the exact situation appears to be described within a Heisenberg–Weyl approach using the formalism developed in [2, Sect. 6] for quantum mechanics on a compact space. The main implication is that configuration space must be taken as an infinite set of equidistant, discrete points [21].[7] This may be reasonable for wave phenomena in view of the Whittaker–Shannon sampling theorem [22], of which the paraxial approximation is a well-defined limiting case. In reference [21] we define a fundamental gaussian beam as the evolution over imaginary time (*i.e.* $t = i\omega$, ω being the width of the Gaussian) of a Dirac δ. The real time evolution of this Gaussian, as well as the propagator are found in closed form.

This last model must be compared in turn with the present approach and with the Fock–Leontovich paraxial approximation of the Helmholtz equation mentioned in the introduction. It would be interesting —and ultimately necessary— to compare the results on the influence of higher-order aberrations on propagating gaussian beams for the three approaches. We want to emphasize that the predictions and region of applicability of the approach followed in this chapter must be further clarified in comparison with experiment.

References

[1] D. Marcuse, *Light Transmission Optics*, (Van Nostrand, New York, 1972).

[2] K.B. Wolf, The Heisenberg–Weyl ring in quantum mechanics. In *Group Theory and its Applications*, Vol. III, E.M. Loebl, Ed. (Academic Press, New York, 1975).

[3] M. Nazarathy and J. Shamir, Fourier optics described by operator algebra, *J. Opt. Soc. America* **70** 150–158 (1980); *ib.* First-order optics —a canonical operator representation: lossless systems, *J. Opt. Soc. America* **72**, 356–364 (1982).

[4] H. Bacry and M. Cadilhac, Metaplectic group and Fourier optics, *Phys. Rev.* **A23**, 2533–2536 (1981).

[5] J.R. Klauder, Continuous representation theory. II. Generalized relation between quantum and classical dynamics, *J. Math. Phys.* **5**, 177–1187 (1964).

[6] R. Glauber, Coherent states of the quantum oscillator, *Phys. Rev. Lett.* **10**, 84 (1963).

[7] V.V. Dodonov, E.V. Kurmyshev, and V.I. Man'ko, Generalized uncertainty relation in correlated coherent states, *Phys. Lett.* **79A**, 150–152 (1980).

[8] J.A. Arnaud, *Beam and Fiber Optics*, (Academic Press, New York, 1976).

[9] M. Leontovich and V. Fock, Solution of the problem of electromagnetic waves along the earth suface by the method of parabolic equations, *Sov. J. Phys.* **10**, 13 (1946).

[7] See Chapter 5, Sect. 5.3, in this volume. (Editor's note.)

[10] V.I. Man'ko, Possible applications of the integrals of the motion and coherent state methods for dynamical systems. In *Quantum Electrodynamics with Outer Fields* (in Russian), (Tomsk State University and Tomsk Pedagogical Institute, Tomsk, USSR, 1977); pp. 101–120.

[11] K.B. Wolf, Canonical transforms. I. Complex linear transforms, *J. Math. Phys.* **15**, 1295–1301 (1974).

[12] K.B. Wolf, ***Integral Transforms in Science and Engineering***, (Plenum, New York, 1979).

[13] M. Born and E. Wolf, ***Principles of Optics***, (Pergamon Press, Oxford, 1959).

[14] H. Goldstein, ***Classical Mechanics***, (Addison-Wesley, Reading, Mass. 1950).

[15] M.H. Stone, Linear transformations in Hilbert space. I. Geometrical aspects, *Proc. Nat. Acad. Sci. U.S.* **15**, 198–200 (1929); II. Analytical aspects; *ibid.* 423–425; III. Operator methods and group theory; *ibid.* **16**, 172–175 (1930); J. von Neumann, Die Eindeutigkeit der Schrödingerschen Operatoren, *Math. Ann.* **104**, 570–578 (1931).

[16] J.H. Van Vleck, The correspondence principle in the statistical interpretation of quantum mechanics, *Proc. Nat. Acad. Sci. USA*, **14**, 178–188 (1982).

[17] V.V. Dodonov, V.I. Man'ko, and V.V. Semionov, The density matrix of the canonical transformed Hamiltonian in the Fock basis, P.N. Lebedev Institute of Physics preprint N54, Moscow, 1983.

[18] V.I. Man'ko and K.B. Wolf, The influence of aberrations in the optics of gaussian beam propagation. Reporte de Investigación del Depto. de Matemáticas, Universidad Autónoma Metropolitana, preprint **4**, # 3 (1985).

[19] G. Eichmann, Quasi-geometric optics of media with inhomogeneous index of refraction, *J. Opt. Soc. America* **61**, 161–168 (1971).

[20] M. García-Bullé, W. Lassner, and K.B. Wolf, The metaplectic group within the Heisenberg–Weyl ring. Reporte de Investigación del Depto. de Matemáticas, Universidad Autónoma Metropolitana, preprint **4**, # 1 (1985); to appear in *J. Math. Phys.*

[21] K.B. Wolf, A euclidean algebra of hamiltonian observables in Lie optics, *Kinam* **6**, 141–156 (1985).

[22] J.W. Goodman, ***Introduction to Fourier Optics***, (Mc Graw-Hill, New York, 1968), Section 2-3.

Appendix A

The symplectic groups, their parametrization and cover

ABSTRACT: In this appendix we summarize the parametrization and some of the properties of connectivity and covering of the symplectic groups. This material was developed by Valentin Bargmann in his early work on the three-dimensional Lorentz group and on Hilbert spaces of analytic functions, and has been shown to be particularly relevant for Lie optics. Wave optics —at least in its paraxial approximation— seems to work with the *double* cover of the symplectic group of geometric first-order optics. This is strongly reminiscent of the double cover which the spin group affords over the classical rotation group, and makes necessary a closer acquaintance with the *metaplectic* groups.

A.1 Rank one: *SL(2,ℜ)*, *Sp(2,ℜ)*, *SU(1,1)*, *SO(2,1)*, and $\overline{Sp(2,\Re)}$

Lie groups of rank one present several accidental homomorphisms. Among the compact groups, the three-dimensional rotation group *SO(3)* and the two-dimensional special unitary group *SU(2)* are probably the most famous pair of homomorphic groups, the latter covering the former twice and allowing the description of phenomena such as spin. Non-compact groups of rank one present a fourfold such homomorphism: the group of 2×2 real matrices *SL(2,ℜ)*, is isomorphic to the two-dimensional symplectic group *Sp(2,ℜ)* and to the two-dimensional pseudo-unitary group *SU(1,1)*, and covers twice the three-dimensional pseudo-orthogonal group *SO(2,1)*. These groups are themselves infinitely connected, and possess a common universal cover $\overline{Sp(2,\Re)}$. A particularly relevant group for Lie optics is the *metaplectic* group, *Mp(2,ℜ)*; it covers *Sp(2,ℜ)* twice. We start with *SL(2,ℜ)* and relate to it all other homomorphic groups.

A.1.1 *SL(2,ℜ)*

We denote by *SL(2,ℜ)* the set of 2×2 real, unimodular matrices

$$\mathbf{g} := \begin{pmatrix} a & b \\ c & d \end{pmatrix}, \qquad \det \mathbf{g} = ad - bc = 1, \tag{1}$$

with the group product defined as ordinary matrix multiplication. It is a three-parameter noncompact semisimple group, connected and infinitely-connected. The latter facts are not obvious, and will be further elaborated in Section A.2, below.

A.1.2 *Sp(2,ℜ)*

The set of 2×2 matrices (1) is, at the same time, a set of ***symplectic***[1] matrices, *i.e.*, they satisfy

$$\mathbf{g}\mathbf{M}_{Sp(2)}\mathbf{g}^{\top} = \mathbf{M}_{Sp(2)}, \qquad \mathbf{M}_{Sp(2)} := \begin{pmatrix} 0 & +1 \\ -1 & 0 \end{pmatrix}, \tag{2}$$

where $\mathbf{g}^{\top}$ is the transpose of $\mathbf{g}$ and $\mathbf{M}_{Sp(2)}$ is the symplectic ***metric*** matrix. The unimodularity condition in (1) implies the validity of (2), as may be verified through elementary algebra. The set of matrices satisfying (2) constitute the *group*, denoted *Sp(2,ℜ)*, of two-dimensional real symplectic matrices.

A.1.3 *SU(1,1)*

The above two groups are isomorphic to a third one, the group of complex *"1+1"* unimodular pseudo-unitary matrices, denoted *SU(1,1)*, whose elements $\mathbf{u}$ satisfy

$$\mathbf{u}\mathbf{M}_{E(1,1)}\mathbf{u}^{\dagger} = \mathbf{M}_{E(1,1)}, \qquad \mathbf{M}_{E(1,1)} := \begin{pmatrix} +1 & 0 \\ 0 & -1 \end{pmatrix}, \tag{3}$$

where $\mathbf{u}^{\dagger} := \mathbf{u}^{\top *}$ is the ***adjoint*** (transpose, complex conjugate) of $\mathbf{u}$, and $\mathbf{M}_{E(1,1)}$ the pseudo-euclidean metric matrix.

It is easy to show that the most general matrix $\mathbf{u} \in SU(1,1)$ has the form

$$\mathbf{u} = \begin{pmatrix} \alpha & \beta \\ \beta^{*} & \alpha^{*} \end{pmatrix}, \qquad |\alpha|^2 - |\beta|^2 = 1. \tag{4}$$

The elements of the *SL(2,ℜ)* = *Sp(2,ℜ)* and *SU(1,1)* groups realized as 2×2 matrices, are related through a *similarity transformation*, an outer isomorphism by the complex unitary matrix

$$\mathbf{W} := \frac{1}{\sqrt{2}}\begin{pmatrix} \omega_4^{-1} & \omega_4^{-1} \\ -\omega_4 & \omega_4 \end{pmatrix} = \mathbf{W}^{\dagger -1}, \qquad \omega_4 := e^{i\pi/4} = \frac{1}{\sqrt{2}}(1+i). \tag{5}$$

We display this isomorphism explicitly in terms of the group parameters as

$$\begin{pmatrix} a & b \\ c & d \end{pmatrix} = \mathbf{g}(\mathbf{u}) = \mathbf{W}\mathbf{u}\mathbf{W}^{-1} = \begin{pmatrix} \mathrm{Re}\,(\alpha+\beta) & -\mathrm{Im}\,(\alpha-\beta) \\ \mathrm{Im}\,(\alpha+\beta) & \mathrm{Re}\,(\alpha-\beta) \end{pmatrix}, \tag{6a}$$

$$\begin{pmatrix} \alpha & \beta \\ \beta^{*} & \alpha^{*} \end{pmatrix} = \mathbf{u}(\mathbf{g}) = \mathbf{W}^{-1}\mathbf{g}\mathbf{W} = \frac{1}{2}\begin{pmatrix} (a+d) - i(b-c) & (a-d) + i(b+c) \\ (a-d) - i(b+c) & (a+d) + i(b-c) \end{pmatrix}. \tag{6b}$$

Since $\mathbf{W}^{-1}\mathbf{M}_{Sp(2)}\mathbf{W} = -i\,\mathbf{M}_{E(1,1)}$, (2) for $\mathbf{g}$ is equivalent to (3) for $\mathbf{u}$.

Of course, any other matrix $\mathbf{W}' = \mathbf{g}_0\mathbf{W}$ or $\mathbf{W}\mathbf{u}_0$, for fixed $\mathbf{g}_0 \in Sp(2,\Re)$ or $\mathbf{u}_0 \in SU(1,1)$, may be used to define equivalent isomorphisms between the two groups. What makes (5) particularly convenient for us is that it establishes the appropriate link between the *standard* realizations of the groups.

---

[1]SYMPLECTIC (simplektik), ***adjective and substantive***, first appearance: 1839. [*Adaptation from Greek συμπλεκτικόs, formed on σύν* SYM- +πλέκειν, TO TWINE, PLAIT, WEAVE.] **A.** adjective Epithet of a bone of the suspensorium in the skull of fishes, between the hyomandibular and the quadrate bones. **B.** substantive The symplectic bone. —*The Oxford Universal Dictionary on Historical Principles*, third ed., 1955. The use of this name for the Cartan C–family of semisimple groups is due to Hermann Weyl, in *The theory of groups and quantum mechanics*, 2nd ed. (Dover, New York, 1930).

A.1.4 *SO(2,1)*

For every real 2×2 matrix $\mathbf{g}$ in (1), we construct the real 3×3 matrix

$$\mathbf{\Gamma}(\mathbf{g}) := \begin{pmatrix} \frac{1}{2}(a^2+b^2+c^2+d^2) & \frac{1}{2}(a^2-b^2+c^2-d^2) & -cd-ab \\ \frac{1}{2}(a^2+b^2-c^2-d^2) & \frac{1}{2}(a^2-b^2-c^2+d^2) & cd-ab \\ -bd-ac & bd-ac & ad+bc \end{pmatrix}. \tag{7}$$

It has the properties

$$\mathbf{\Gamma}\mathbf{M}_{E(1,2)}\mathbf{\Gamma}^\top = \mathbf{M}_{E(1,2)}, \qquad \mathbf{M}_{E(1,2)} := \begin{pmatrix} +1 & 0 & 0 \\ 0 & -1 & 0 \\ 0 & 0 & -1 \end{pmatrix}, \tag{8a}$$

$$\det \mathbf{\Gamma} = 1, \tag{8b}$$

$$\mathbf{\Gamma}(\mathbf{g}_1)\mathbf{\Gamma}(\mathbf{g}_2) = \mathbf{\Gamma}(\mathbf{g}_1\mathbf{g}_2), \qquad \mathbf{\Gamma}(\mathbf{1}) = \mathbf{1}, \quad \mathbf{\Gamma}(\mathbf{g}^{-1}) = \mathbf{\Gamma}(\mathbf{g})^{-1}. \tag{8c}$$

The matrices $\mathbf{\Gamma}$ satisfying (8*a*) are called *"1+2"* pseudo-orthogonal matrices, and the statement of unimodularity in (8*b*) reduces the matrices under consideration to a set connected to the identity. They form a group denoted *SO(2,1)*. The rows and columns numbering the elements $\Gamma_{\mu,\nu}$ range here over μ, $\nu = 0$, 1, 2; the value 0 corresponds to the first row, *i.e.*, the time-like coordinate.

Since $\mathbf{\Gamma}(\mathbf{g})$ is quadratic in the parameters of $\mathbf{g}$, and $\mathbf{\Gamma}(\mathbf{g}) = \mathbf{\Gamma}(-\mathbf{g})$, (7) defines a 2:1 homomorphism between *Sp(2,ℜ)* and *SO(2,1)*, with kernel $\{-\mathbf{1}, \mathbf{1}\} = \mathfrak{Z}_2 \subset$ *Sp(2,ℜ)*. The *SO(2,1)* matrix corresponding to any *pair of* given *SU(1,1)* matrices may be obtained through (6) and (7). It is

$$\mathbf{\Gamma}(\mathbf{u}) = \begin{pmatrix} |\alpha|^2+|\beta|^2 & 2\operatorname{Re}\alpha\beta^* & 2\operatorname{Im}\alpha\beta^* \\ 2\operatorname{Re}\alpha\beta & \operatorname{Re}(\alpha^2+\beta^2) & \operatorname{Im}(\alpha^2-\beta^2) \\ -2\operatorname{Im}\alpha\beta & -\operatorname{Im}(\alpha^2+\beta^2) & \operatorname{Re}(\alpha^2-\beta^2) \end{pmatrix}, \tag{9a}$$

and conversely,

$$\alpha = \pm\sqrt{\tfrac{1}{2}(\Gamma_{11}+\Gamma_{12}) + \tfrac{1}{2}i(\Gamma_{12}-\Gamma_{21})}, \qquad \beta = \frac{1}{2\alpha}(\Gamma_{10} - i\Gamma_{20}). \tag{9b}$$

A.2 Connectivity

The connectivity of a three-dimensional manifold is, when multiple, a challenge to the mind. If we are to picture this intuitively, we must build up a proper representation of the manifold and follow one-parameter lines. If these lines turn out to be circles, then the covering of these by the real line becomes plausible within the group. As is so often the case, the complex plane is useful and *SU(1,1)* is, among the groups of last section, the easiest to analyze.

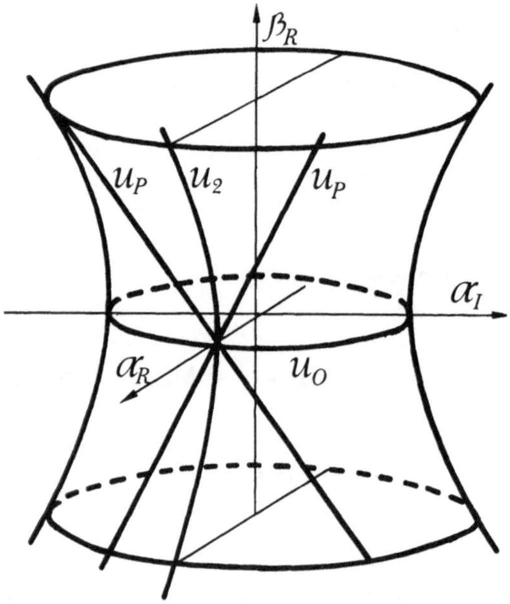

Figure. The $SU(1,1) \simeq Sp(2,\Re)$ group manifold is three-dimensional, connected and infinitely connected. It is $\Re^3$ pierced by a one-sheeted equilateral hyperboloid. Three representatives of one-parameter subgroups are drawn: $\mathbf{u}_0(\tau)$ (circle), $\mathbf{u}_2(\tau)$ (one branch of a hyperbola), and $\mathbf{u}_{P\pm}(\tau)$ (straight lines), lying on the $\beta_I = 0$ hyperboloid.

A.2.1 The connectivity of *SU(1,1)*

The connectivity properties of *Sp(2,$\Re$)* and its isomorphic groups is seen best in terms of the *SU(1,1)* parameters (4). We write the real and imaginary parts of α and β as $\alpha = \alpha_R + i\alpha_I$ and $\beta = \beta_R + i\beta_I$, so the unimodularity condition reads $\alpha_R^2 + \alpha_I^2 - \beta_R^2 = 1 + \beta_I^2 \geq 1$. For fixed β_I, the remaining three parameters are constrained to a one-sheeted revolution hyperboloid with its circular waist in the α-plane. As we let β_I range over $\Re$, we fill twice a region of $\Re^3$-space which is bounded by the $\beta_I = 0$ equilateral revolution hyperboloid. See the figure above. Topologically, the group manifold of *Sp(2,$\Re$)* is $\mathfrak{S}_1 \times \Re^2$, the circle times the cartesian plane.

The points $(\alpha_R, \alpha_I, \beta_R)$ in $\Re^3$ which do *not* belong to the group manifold as described above, form a simply connected tubular region. Any plane for which this region has an elliptic section is an infinitely connected space (the plane minus the unit disk), and hence so is a foliation of $\Re^3$ by such planes. One-parameter subgroups of *SU(1,1)* are lines in $\Re^3$ which must pass through the group identity: the point $(\alpha_R = 1, \alpha_I = 0, \beta_R = 0)$ on the circular waist.

The class representatives of the three nonequivalent subgroup are:

Elliptic: $\mathbf{u}_0(\tau)$, $\tau \in \Re \pmod{4\pi}$ given by $$\begin{cases} \alpha_R = \cos\frac{1}{2}\tau, \\ \alpha_I = \sin\frac{1}{2}\tau, \\ \beta_R = 0. \end{cases}$$

Hyperbolic: $\mathbf{u}_2(\tau)$, $\tau \in \Re$ given by $$\begin{cases} \alpha_R = \cosh\frac{1}{2}\tau, \\ \alpha_I = 0, \\ \beta_R = -\sinh\frac{1}{2}\tau. \end{cases}$$

Parabolic: $\mathbf{u}_{P\pm}(\tau)$, $\tau \in \Re$ given by $$\begin{cases} \alpha_R = 1, \\ \alpha_I = \frac{1}{2}\tau, \\ \beta_R = \pm\frac{1}{2}\tau. \end{cases}$$

The elliptic subgroup is the circular waist in the figure, the hyperbolic subgroup is one of the two branches of the equilateral hyperbola on the $\alpha_I = 0$ plane, while the parabolic ones are the straight lines contained in the hyperboloid which pass through the identity parallel to the α_I–β_R plane. The only subgroup class with a nontrivial covering is thus the elliptic one.

A.2.2 The covering group of $SU(1,1)$

Following Bargmann's pivotal work in reference [1], we now present the simply-connected universal covering group of $SU(1,1)$, making use of the (maximal) compact subgroup $U(1) = SO(2)$. This corresponds to the Iwasawa decomposition which factorizes globally a noncompact semisimple group into its maximal compact subgroup, times a solvable subgroup.

We write $\mathbf{u} \in SU(1,1)$ in (4), in the form

$$\begin{pmatrix} \alpha & \beta \\ \beta^* & \alpha^* \end{pmatrix} = \begin{pmatrix} e^{i\omega} & 0 \\ 0 & e^{-i\omega} \end{pmatrix} \begin{pmatrix} \lambda & \mu \\ \mu^* & \lambda \end{pmatrix}, \qquad \omega \in \mathfrak{S}_1,\ \lambda > 0,\ \mu \in \mathfrak{C} \tag{10a}$$

i.e., ω and λ are the argument and absolute value of α:

$$\begin{aligned} \omega &:= \arg \alpha = \tfrac{1}{2} i \ln(\alpha^* \alpha^{-1}), \\ \lambda &:= |\alpha| > 0, \\ \mu &:= e^{-i\omega}\beta = \sqrt{\frac{\alpha^*}{\alpha}}\, \beta, \end{aligned} \tag{10b}$$

with all multivalued functions evaluated on the *principal sheet*. Conversely, of course, $\alpha = e^{i\omega}\lambda$ and $\beta = e^{i\omega}\mu$. We note that $|\alpha|^2 - |\beta|^2 = \lambda^2 - |\mu|^2 = 1$, so $|\mu| < \lambda$.

This parametrization will be generalized to $Sp(2N,\Re)$ following Bargmann [2] in Section A.4. Actually, in the first treatment of $SU(1,1)$, Bargmann [1] used in 1947 an equivalent set of parameters

$$\begin{gathered} (\omega, \gamma), \qquad \omega \in \mathfrak{S}_1, \quad \gamma \in \mathfrak{C}, \quad |\gamma| < 1 \\ \gamma := \frac{\mu}{\lambda} = \frac{\beta}{\alpha}, \qquad \lambda = \frac{1}{\sqrt{1-|\gamma|^2}}, \quad \mu = \frac{\gamma}{\sqrt{1-|\gamma|^2}}. \end{gathered} \tag{10c}$$

In some respects this parametrization is more convenient, but it does not generalize easily to N dimensions. We shall here prefer the former and call $\{\omega, \lambda, \mu\}$ the *Bargmann parameters* of $SU(1,1)$, writing $\mathbf{u}\{\omega, \lambda, \mu\}$ when they are used.[2]

The $Sp(2,\Re) = SL(2,\Re)$ parameters (1) are expressed in terms of the Bargmann parameters through

$$\begin{pmatrix} a & b \\ c & d \end{pmatrix} = \begin{pmatrix} \cos\omega & -\sin\omega \\ \sin\omega & \cos\omega \end{pmatrix} \begin{pmatrix} \lambda + \operatorname{Re}\mu & \operatorname{Im}\mu \\ \operatorname{Im}\mu & \lambda - \operatorname{Re}\mu \end{pmatrix}. \tag{11a}$$

This displays the global decomposition of any nonsingular matrix into the product of an orthogonal and a positive definite symmetric matrix. Conversely,

$$\omega = \arg[(a+d) - i(b-c)], \qquad \mu = e^{-i\omega}[(a-d) + i(b+c)]. \tag{11b}$$

[2] Although λ is a redundant parameter (since $\lambda = \sqrt{1+|\mu|^2}$), it will be kept for the sake of easy comparison with the N-dimensional case.

A.2.3 The covering group $\overline{Sp(2,\Re)}$

Through matrix multiplication $\mathbf{u}\{\omega,\lambda,\mu\} = \mathbf{u}\{\omega_1,\lambda_1,\mu_1\}\mathbf{u}\{\omega_2,\lambda_2,\mu_2\}$ of either (10) or (11) we obtain

$$\omega = \omega_1 + \omega_2 + \arg\nu, \tag{12a}$$
$$\lambda = \lambda_1\,|\nu|\,\lambda_2, \tag{12b}$$
$$\mu = e^{-i\arg\nu}\,[\lambda_1\mu_2 + e^{-2i\omega_2}\mu_1\lambda_2], \tag{12c}$$

where

$$\nu := 1 + e^{-2i\omega_2}\lambda_1^{-1}\,\mu_1\,\mu_2^*\,\lambda_2^{-1}, \quad |\nu - 1| < 1. \tag{12d}$$

The last inequality stems from $|\mu_i/\lambda_i| < 1$, and implies that ν is within a circle of radius less than one, centered at $\nu = 1$; hence $\nu \neq 0$.

The group unit is $\mathbf{1} = \mathbf{u}\{\omega = 0, \lambda = 1, \mu = 0\}$ and the inverse is given by $\mathbf{u}\{\omega,\lambda,\mu\}^{-1} = \mathbf{u}\{-\omega,\lambda,-e^{2i\omega}\mu\}$. Note that the subset of $\mathbf{u}$'s given by $\mathbf{u}\{\omega = 0,\lambda,\mu\}$ are naturally coset representatives of $U(1)\backslash SU(1,1)$, but do *not* constitute a group.

From (10) and (11) it is clear that $SU(1,1)$ and $Sp(2,\Re) = SL(2,\Re)$ are described when ω is counted modulo 2π, that is $\omega \in \mathfrak{S}_1$, *i.e.* $\omega \equiv \omega (\mathrm{mod}\ 2\pi)$. If we drop this identification and consider $\omega \in \Re$ *with no modular condition*, defining the composition law through (12), we describe a *covering* of $SU(1,1)$ whose elements we denote by $\overline{u}\{\omega,\lambda,\mu\}$ ($\omega \in \Re$, $\mu \in \mathfrak{C}$). The manifold of this group is $\Re \times \mathfrak{C} = \Re^3$, and this is $\underline{\text{simply connected.}}$ The composition rule (12) for $\overline{u}\{\omega,\lambda,\mu\}$ thus describes the ***universal covering*** group $\overline{Sp(2,\Re)} = \overline{SL(2,\Re)} = \overline{SU(1,1)}$ of $Sp(2,\Re) = SL(2,\Re) = SU(1,1)$.

A.2.4 The metaplectic group $Mp(2,\Re)$

The center of $\overline{Sp(2,\Re)}$ is the set of elements $\mathfrak{Z}_\infty = \{\overline{u}\{n\pi,1,0\},\ n \in \mathfrak{Z}\}$, so that the pseudo-orthogonal group is $SO(2,1) = \overline{Sp(2,\Re)}/\{\overline{u}\{n\pi,1,0\},\ n \in \mathfrak{Z}\}$, and the symplectic group is $Sp(2,\Re) = \overline{Sp(2,\Re)}/\{\overline{u}\{2n\pi,1,0\},\ n \in \mathfrak{Z}\}$. Various M-fold coverings of $Sp(2,\Re)$ may be obtained from the universal cover $\overline{Sp(2,\Re)}$, modulo $\{\overline{u}\{2Mn\pi,1,0\},\ n \in \mathfrak{Z}\}$. In particular, we are interested in the *two-fold cover* of $Sp(2,\Re)$, the ***metaplectic*** group

$$Mp(2,\Re) = \overline{Sp(2,\Re)}/\{\overline{u}\{n\pi,1,0\},\ n \in \mathfrak{Z}\}. \tag{13}$$

Its elements will be written $\tilde{g}(\omega,\lambda,\mu)$, with $\mu \in \mathfrak{C}$, $\omega \equiv \omega (\mathrm{mod}\ 4\pi)$. The 2:1 mapping from $Mp(2,\Re)$ to $Sp(2,\Re)$ is given by (11) [(10) on $SU(1,1)$], and assigns the same image to $\tilde{g}(\omega,\lambda,\mu)$ and $\tilde{g}(\omega + 2\pi,\lambda,\mu)$.

Neither the metaplectic group nor its covers have a *matrix* representation (by *finite*-dimensional matrices, that is). They *do* have representations which are infinite-dimensional, as by integral kernels. This fact accounts for some of the difficulty we encounter when working with covers of $Sp(2,\Re)$.

Single-valued functions on covering groups may give rise to multivalued functions on the original group. The phase of the canonical transform[3] kernel $\theta_{\mathbf{g}}$ is a prime example of a single-valued function $\theta_{\tilde{g}}$ on $Mp(2,\Re)$, yielding a two-valued function on $Sp(2,\Re)$.

---

[3]See, for example, K.B. Wolf, *Integral Transforms in Science and Engineering*, (Plenum Publ. Corp., New York, 1979), part 4.

A.3 Subgroups

We list below some useful one-parameter subgroups $\mathbf{g}(\tau)$ of $Sp(2,\Re) = SL(2,\Re)$, together with their counterparts $\mathbf{u} \in SU(1,1)$, $\mathbf{\Gamma} \in SO(2,1)$, and one $\overline{u} \in \overline{Sp(2,\Re)}$:

A.3.1 Elliptic subgroup

$$\mathbf{g}_0(\tau) = \begin{pmatrix} \cos\frac{1}{2}\tau & -\sin\frac{1}{2}\tau \\ \sin\frac{1}{2}\tau & \cos\frac{1}{2}\tau \end{pmatrix}, \qquad \mathbf{u}_0(\tau) = \begin{pmatrix} e^{i\tau/2} & 0 \\ 0 & e^{-i\tau/2} \end{pmatrix},$$
$$\mathbf{\Gamma}_0(\tau) = \begin{pmatrix} 1 & 0 & 0 \\ 0 & \cos\tau & \sin\tau \\ 0 & -\sin\tau & \cos\tau \end{pmatrix}, \qquad \overline{u}_0\{\tfrac{1}{2}\tau, 1, 0\}. \tag{14a}$$

A.3.2 Hyperbolic subgroups

$$\mathbf{g}_1(\tau) = \begin{pmatrix} \cosh\frac{1}{2}\tau & -\sinh\frac{1}{2}\tau \\ -\sinh\frac{1}{2}\tau & \cosh\frac{1}{2}\tau \end{pmatrix}, \qquad \mathbf{u}_1(\tau) = \begin{pmatrix} \cosh\frac{1}{2}\tau & -i\sinh\frac{1}{2}\tau \\ i\sinh\frac{1}{2}\tau & \cosh\frac{1}{2}\tau \end{pmatrix},$$
$$\mathbf{\Gamma}_1(\tau) = \begin{pmatrix} \cosh\tau & 0 & \sinh\tau \\ 0 & 1 & 0 \\ \sinh\tau & 0 & \cosh\tau \end{pmatrix}, \qquad \overline{u}_1\{0, \cosh\tfrac{1}{2}\tau, -i\sinh\tfrac{1}{2}\tau\}; \tag{14b}$$

$$\mathbf{g}_2(\tau) = \begin{pmatrix} e^{-\tau/2} & 0 \\ 0 & e^{\tau/2} \end{pmatrix}, \qquad \mathbf{u}_2(\tau) = \begin{pmatrix} \cosh\frac{1}{2}\tau & -\sinh\frac{1}{2}\tau \\ -\sinh\frac{1}{2}\tau & \cosh\frac{1}{2}\tau \end{pmatrix},$$
$$\mathbf{\Gamma}_2(\tau) = \begin{pmatrix} \cosh\tau & -\sinh\tau & 0 \\ -\sinh\tau & \cosh\tau & 0 \\ 0 & 0 & 1 \end{pmatrix}, \qquad \overline{u}_2\{0, \cosh\tfrac{1}{2}\tau, -\sinh\tfrac{1}{2}\tau\}. \tag{14c}$$

A.3.3 Parabolic subgroups

$$\mathbf{g}_+(\tau) = \begin{pmatrix} 1 & -\tau \\ 0 & 1 \end{pmatrix}, \qquad \mathbf{u}_+(\tau) = \begin{pmatrix} 1 + i\frac{1}{2}\tau & -i\frac{1}{2}\tau \\ i\frac{1}{2}\tau & 1 - i\frac{1}{2}\tau \end{pmatrix},$$
$$\mathbf{\Gamma}_+(\tau) = \begin{pmatrix} 1 + \frac{1}{2}\tau^2 & -\frac{1}{2}\tau^2 & \tau \\ \frac{1}{2}\tau^2 & 1 - \frac{1}{2}\tau^2 & \tau \\ \tau & -\tau & 1 \end{pmatrix}, \qquad \overline{u}_+\{\arg[1 + i\tfrac{1}{2}\tau], |1 + i\tfrac{1}{2}\tau|, e^{-i(\omega+\pi/2)}\tfrac{1}{2}\tau\}; \tag{14d}$$

$$\mathbf{g}_-(\tau) = \begin{pmatrix} 1 & 0 \\ \tau & 1 \end{pmatrix}, \qquad \mathbf{u}_-(\tau) = \begin{pmatrix} 1 + i\frac{1}{2}\tau & i\frac{1}{2}\tau \\ -i\frac{1}{2}\tau & 1 - i\frac{1}{2}\tau \end{pmatrix},$$
$$\mathbf{\Gamma}_-(\tau) = \begin{pmatrix} 1 + \frac{1}{2}\tau^2 & \frac{1}{2}\tau^2 & -\tau \\ -\frac{1}{2}\tau^2 & 1 - \frac{1}{2}\tau^2 & \tau \\ -\tau & -\tau & 1 \end{pmatrix}, \qquad \overline{u}_-\{\arg[1 + i\tfrac{1}{2}\tau], |1 + i\tfrac{1}{2}\tau|, e^{-i(\omega-\pi/2)}\tfrac{1}{2}\tau\}. \tag{14e}$$

In all but the first (the elliptic subgroup), the correspondence between $Sp(2,\Re)$ and $SO(2,1)$ is one-to-one.

A.3.4 Conjugation and trace

All one-parameter subgroups can be obtained through similarity conjugation out of $\mathbf{g}_0(\tau)$, $\mathbf{g}_2(\tau)$, and $\mathbf{g}_+(\tau)$. In the above list,

$$\mathbf{g}_0(\tfrac{1}{2}\pi)\,\mathbf{g}_2(\tau)\,\mathbf{g}_0(\tfrac{1}{2}\pi)^{-1} = \mathbf{g}_1(\tau) \quad \text{and} \quad \mathbf{g}_0(\pi)\,\mathbf{g}_+(\tau)\,\mathbf{g}_0(\pi)^{-1} = \mathbf{g}_-(\tau).$$

The two parabolic subgroups listed at the end of **A.2.1** and displayed in the figure, are related to those of **A.2.3** through

$$\mathbf{g}_0(\pm\tfrac{1}{2}\pi)\,\mathbf{g}_+(\tau)\,\mathbf{g}_0(\pm\tfrac{1}{2}\pi)^{-1} = \mathbf{g}_{P\pm}(\tau).$$

Under similarity conjugation, the 2×2 *trace* of the matrices, $T := \operatorname{tr}\mathbf{u} = 2\operatorname{Re}\alpha = a + d$, is left invariant. For the three subgroup cases, we have $T_0(\tau) = 2\cos\frac{1}{2}\tau \in [-2, 2]$, $T_2(\tau) = 2\cosh\frac{1}{2}\tau \in [2, \infty)$, and $T_P(\tau) = 2$. [Note $T(\tau = 0) = 2$ in all cases.] If these subgroups are drawn as lines in the group manifold of the figure, the elliptic subgroups will be represented by plane ellipses —in any plane containing the α_R axis— passing through the identity $\mathbf{1}$ ($\alpha_R = 1$, $\alpha_I = 0$, $\beta_R = 0$) and $-\mathbf{1}$ ($\alpha_R = -1$, $\alpha_I = 0$, $\beta_R = 0$), with foci on the α_I–β_R plane. The hyperbolic subgroups will be represented by one branch of plane hyperbolæ with foci on the α_R axis. The parabolic subgroups appear as straight lines in the $\alpha_R = 2$ plane bounded by the ($P\pm$)-intercepts with the equilateral hyperboloid.

Conversely, any *SU(1,1)* $\simeq$ *Sp(2,ℜ)* group element (different from $\mathbf{1}$ or $-\mathbf{1}$) whose trace $T = \operatorname{tr}\mathbf{u}$ is in $(-2, 2)$, $\{2\}$, or $(2, \infty]$, may be placed on a one-parameter elliptic, parabolic, or hyperbolic subgroup, respectively. If $T \leq -2$, no such subgroup can be found, but one may write $\mathbf{u} = (-\mathbf{1})\mathbf{u}'$ and place $\mathbf{u}'$ on a one-parameter subgroup as before. In *SO(2,1)*, all elements $\mathbf{\Gamma}$ may be placed on one-parameter subgroups connected to the identity.

We also have the subset (*not* a subgroup) given by

$$\mathbf{g}_\gamma = \begin{pmatrix} \lambda + \operatorname{Re}\mu & \operatorname{Im}\mu \\ \operatorname{Im}\mu & \lambda - \operatorname{Re}\mu \end{pmatrix}, \qquad \mathbf{u}_\gamma = \begin{pmatrix} \lambda & \mu \\ \mu^* & \lambda \end{pmatrix},$$

$$\mathbf{\Gamma}_\gamma = \frac{1}{1 - |\gamma|^2}\begin{pmatrix} \lambda^2 + |\mu|^2 & 2\lambda\operatorname{Re}\mu & -2\lambda\operatorname{Im}\mu \\ 2\lambda\operatorname{Re}\mu & \lambda^2 + \operatorname{Re}\mu^2 & -\operatorname{Im}\mu^2 \\ -2\lambda\operatorname{Im}\mu & -\operatorname{Im}\mu^2 & \lambda^2 - \operatorname{Re}\mu^2 \end{pmatrix}, \qquad \overline{u}_\gamma\{0, \lambda, \mu\}. \tag{14f}$$

A.4 The general case of rank N

For rank $N = 1$, we saw, *Sp(2N,ℜ)* is homomorphic to the lowest-dimensional counterparts of two other Cartan-classified families, to a total of four groups. For $N \geq 2$, the only accidental homomorphism occurs for *Sp(4,ℜ)* $\simeq$ *SO(3,2)*, and is 2:1. We will now give Bargmann's treatment [2] of the covering of the *general* symplectic group *Sp(2N,ℜ)*.

A.4.1 *Sp(2N,ℜ)*

The group *Sp(2N,ℜ)* is defined as the set of real $2N \times 2N$ matrices $\mathbf{g}$ obeying the $2N$-dimensional version of (2):

$$\mathbf{g}\mathbf{M}_{Sp(2N)}\mathbf{g}^\top = \mathbf{M}_{Sp(2N)}, \qquad \mathbf{M}_{Sp(2N)}^\top = -\mathbf{M}_{Sp(2N)}, \qquad \det\mathbf{M}_{Sp(2N)} \neq 0. \tag{15a}$$

When we write the matrices involved in terms of $N \times N$ submatrices, we may choose

$$\mathbf{g} := \begin{pmatrix} \mathbf{A} & \mathbf{B} \\ \mathbf{C} & \mathbf{D} \end{pmatrix}, \qquad \mathbf{M}_{Sp(2N)} := \begin{pmatrix} \mathbf{0} & +\mathbf{1}_N \\ -\mathbf{1}_N & \mathbf{0} \end{pmatrix}, \tag{15b}$$

where $\mathbf{1}_N$ is the N-dimensional unit matrix. This leads to the following relations between the $N \times N$ submatrices:

$$\mathbf{AB}^\top = \mathbf{BA}^\top, \quad \mathbf{AC}^\top = \mathbf{CA}^\top, \quad \mathbf{BD}^\top = \mathbf{DB}^\top, \quad \mathbf{CD}^\top = \mathbf{DC}^\top, \tag{15c}$$

$$\mathbf{AD}^\top - \mathbf{BC}^\top = \mathbf{1}_N. \tag{15d}$$

The inverse of a symplectic matrix may thus be written as

$$\mathbf{g}^{-1} = \mathbf{M}_{Sp(2N)}\, \mathbf{g}^\top \mathbf{M}_{Sp(2N)} = \begin{pmatrix} \mathbf{D}^\top & -\mathbf{B}^\top \\ -\mathbf{C}^\top & \mathbf{A}^\top \end{pmatrix}. \tag{16}$$

The number of independent parameters of *Sp(2N,ℜ)* is $2N^2 + N$.

A.4.2 The Bargmann form for *Sp(2N,ℜ)*

In order to explore the connectivity properties of the *Sp(2N,ℜ)* manifold and parametrize its covering group, we shall present a generalization of the *Sp(2,ℜ)* = *SU(1,1)* isomorphism. Although clearly *Sp(2N,ℜ)* is not *isomorphic* to any pseudo-unitary group, its *inclusion* in *U(N,N)* will display the connectivity properties through its unitary *U(N)* maximal compact subgroup, generalizing the role of *U(1)* = *SO(2)* in *Sp(2,ℜ)*.

We construct first the $2N \times 2N$ matrix $\mathbf{W}_N = \mathbf{W} \otimes \mathbf{1}_N$, where $\mathbf{W} = \mathbf{W}_1$ is the 2×2 matrix (5) which gives the $N \times N$ block coefficients. Taking now $\mathbf{g}$ from (15), we write

$$\begin{aligned} \mathbf{u}(\mathbf{g}) :&= \mathbf{W}_N^{-1} \mathbf{g} \mathbf{W}_N \\ &= \frac{1}{2}\begin{pmatrix} [\mathbf{A}+\mathbf{D}] - i[\mathbf{B}-\mathbf{C}] & [\mathbf{A}-\mathbf{D}] + i[\mathbf{B}+\mathbf{C}] \\ [\mathbf{A}-\mathbf{D}] - i[\mathbf{B}+\mathbf{C}] & [\mathbf{A}+\mathbf{D}] + i[\mathbf{B}-\mathbf{C}] \end{pmatrix} =: \begin{pmatrix} \boldsymbol{\alpha} & \boldsymbol{\beta} \\ \boldsymbol{\beta}^* & \boldsymbol{\alpha}^* \end{pmatrix}. \end{aligned} \tag{17}$$

The symplecticity property of $\mathbf{g}$ becomes thus

$$\begin{aligned} \mathbf{u} \mathbf{M}_{E(N,N)} \mathbf{u}^\dagger &= \mathbf{M}_{E(N,N)}, \\ \mathbf{M}_{E(N,N)} :&= i \mathbf{W}_N^{-1} \mathbf{M}_{Sp(2N)} \mathbf{W}_N = \begin{pmatrix} +\mathbf{1}_N & \mathbf{0} \\ \mathbf{0} & -\mathbf{1}_N \end{pmatrix}. \end{aligned} \tag{18a}$$

This condition alone would define $\mathbf{u}$ as a pseudo-unitary *U(N,N)* matrix, but the restriction (17) stemming from the reality of $\mathbf{g}$, makes $\boldsymbol{\alpha}^*$ the complex conjugate of $\boldsymbol{\alpha}$, and $\boldsymbol{\beta}^*$ that of $\boldsymbol{\beta}$, restricting $\mathbf{u}$ to *Sp(2N,ℜ)* ⊂ *U(N,N)*.

The $N \times N$ submatrices of the Bargmann-form *Sp(2N,ℜ)* matrices obey

$$\boldsymbol{\alpha}\boldsymbol{\alpha}^\dagger - \boldsymbol{\beta}\boldsymbol{\beta}^\dagger = \mathbf{1}, \qquad \boldsymbol{\alpha}^\dagger\boldsymbol{\alpha} - \boldsymbol{\beta}^\top\boldsymbol{\beta}^* = \mathbf{1}, \tag{18b}$$

$$\boldsymbol{\alpha}\boldsymbol{\beta}^\top - \boldsymbol{\beta}\boldsymbol{\alpha}^\top = \mathbf{0}, \qquad \boldsymbol{\alpha}^\top\boldsymbol{\beta}^* - \boldsymbol{\beta}^\dagger\boldsymbol{\alpha} = \mathbf{0}. \tag{18c}$$

Since $\mathbf{1} + \boldsymbol{\beta}\boldsymbol{\beta}^\dagger$ is a positive definite matrix, $\boldsymbol{\alpha}$ has an inverse. From the last equations, $\boldsymbol{\alpha}^{-1}\boldsymbol{\beta}$ and $\boldsymbol{\beta}^*\boldsymbol{\alpha}^{-1}$ are shown to be symmetric. The inverse follows:

$$\mathbf{u}^{-1} = \mathbf{M}_{E(N,N)} \mathbf{u}^\dagger \mathbf{M}_{E(N,N)}^{-1} = \begin{pmatrix} \boldsymbol{\alpha}^\dagger & -\boldsymbol{\beta}^\top \\ -\boldsymbol{\beta}^\dagger & \boldsymbol{\alpha}^\top \end{pmatrix}. \tag{19}$$

Finally, corresponding to (6a), the mapping inverse to (17) is

$$\begin{pmatrix} \mathbf{A} & \mathbf{B} \\ \mathbf{C} & \mathbf{D} \end{pmatrix} = \mathbf{g}(\mathbf{u}) = \mathbf{W}_N \mathbf{u} \mathbf{W}_N^{-1} = \begin{pmatrix} \mathrm{Re}\,(\boldsymbol{\alpha}+\boldsymbol{\beta}) & -\mathrm{Im}\,(\boldsymbol{\alpha}-\boldsymbol{\beta}) \\ \mathrm{Im}\,(\boldsymbol{\alpha}+\boldsymbol{\beta}) & \mathrm{Re}\,(\boldsymbol{\alpha}-\boldsymbol{\beta}) \end{pmatrix}. \tag{20}$$

A.4.3 The subgroup *U(N)* ⊂ *Sp(2N,ℜ)*

The maximal compact subgroup of *Sp(2N,ℜ)* is *U(N)*. This fact may be seen knowing that the maximal compact subgroup of *GL(2N,ℭ)* —the group of complex $2N \times 2N$ matrices— is *U(2N)*; this is the weakest restriction which puts an upper bound to the norm of this row and column vectors.[4] The intersection of *U(2N)* with the Bargmann form of *Sp(2N,ℜ)* is the set of matrices satisfying both $\mathbf{u}_0\mathbf{u}_0^\dagger = \mathbf{1}$ and (18), which have therefore vanishing off-diagonal blocks and conjugate diagonal ones, *i.e.*,

$$\mathbf{u}_0 = \begin{pmatrix} \boldsymbol{\alpha} & \mathbf{0} \\ \mathbf{0} & \boldsymbol{\alpha}^* \end{pmatrix}, \qquad \boldsymbol{\alpha}\boldsymbol{\alpha}^\dagger = \mathbf{1}. \tag{21a}$$

The set of $\mathbf{u}_0$'s thus constitutes a *U(N)* group. In the real form (15) of *Sp(2N,ℜ)*, this N^2-parameter subgroup is the set of matrices

$$\mathbf{g}_0 = \begin{pmatrix} \mathrm{Re}\,\boldsymbol{\alpha} & -\mathrm{Im}\,\boldsymbol{\alpha} \\ \mathrm{Im}\,\boldsymbol{\alpha} & \mathrm{Re}\,\boldsymbol{\alpha} \end{pmatrix} = \mathbf{g}_0^{\top -1}, \qquad \boldsymbol{\alpha}\boldsymbol{\alpha}^\dagger = \mathbf{1}. \tag{21b}$$

All these matrices are *orthogonal* $2N \times 2N$ matrices, but not the most general ones, since the group *O(2N)* has $2N^2 - N$ parameters.

A.4.4 The *Sp(2N,ℜ)* manifold

A well known theorem in matrix theory states that any real matrix $\mathbf{R}$ may be decomposed into the product of an orthogonal $\mathbf{Q}$ and a symmetric positive definite matrix $\mathbf{S}$, uniquely, as $\mathbf{R} = \mathbf{QS}$. Additionally, Bargmann [2, §2.3] shows that if $\mathbf{R} \in$ *Sp(2N,ℜ)*, then also $\mathbf{Q}$ and $\mathbf{S}$ belong to this group. Through $\mathbf{W}_N$ [Eq. (17)] the matrices $\mathbf{Q}$ and $\mathbf{S}$ map onto unitary and hermitian positive definite ones. Restriction to the Bargmann form of *Sp(2N,ℜ)* in (17) details that $\mathbf{u}(\mathbf{Q})$ is given by $\boldsymbol{\alpha} \in$ *U(N)* and $\boldsymbol{\beta} = \mathbf{0}$ [*i.e.*, as in (21*a*), rather than simply a phase as in (14*a*)], and $\mathbf{u}(\mathbf{S})$ with $\boldsymbol{\alpha} = \boldsymbol{\alpha}^\dagger$ and $\boldsymbol{\beta} = \boldsymbol{\beta}^\top$, obeying (18). The former set of matrices is an N^2-dimensional real manifold with the topology of *U(N)*, while the real dimension of the latter is $N^2 + N$ with the euclidean topology of $\Re^{N^2+N}$. This last fact may be seen either through counting N^2 parameters for hermitian and N^2+N for symmetric complex matrices, minus N^2 conditions from the two independent equations in (18); or, succintly [2],

$$\mathbf{u}(\mathbf{Q}) = \begin{pmatrix} \boldsymbol{\alpha} & \mathbf{0} \\ \mathbf{0} & \boldsymbol{\alpha}^* \end{pmatrix}, \qquad \boldsymbol{\alpha}\boldsymbol{\alpha}^\dagger = \mathbf{1}, \tag{22a}$$

$$\mathbf{u}(\mathbf{S}) = \exp\begin{pmatrix} \mathbf{0} & \boldsymbol{\xi} \\ \boldsymbol{\xi}^* & \mathbf{0} \end{pmatrix}, \qquad \boldsymbol{\xi} = \boldsymbol{\xi}^\top. \tag{22b}$$

Since $\boldsymbol{\alpha} \in$ *U(N)* $\Rightarrow |\det\boldsymbol{\alpha}| = 1$, the group *U(N)* is the direct product of the compact group of *unimodular* unitary matrices *SU(N)*, times the *U(1)* group of determinant phases $e^{i\theta}$, $\theta \in \mathfrak{S}_1$ (the circle). Topologically, thus,

$$Sp(2N,\Re) \sim U(1) \times SU(N) \times \Re^{N^2+N}. \tag{23}$$

Since both *SU(N)* and $\Re^{N^2+N}$ are simply connected, the connectivity of *Sp(2N,ℜ)* is that of $U(1) \sim \mathfrak{S}_1$, *i.e.*, connected and infinitely connected. This is the generalization of the *Sp(2,ℜ)* case presented in **A.1.2**; there, the $SU(1) = \{1\}$ factor was absent.

[4]The fact that *U(N)* is the maximal compact subgroup of *Sp(2N,ℜ)* is quite clear, otherwise, if we recall that we realize the latter as generated by quadratic monomials in the N-dimensional oscillator raising and lowering operators; the symmetry group of the system is generated by the N^2 mixed products

A.4.5 The Bargmann parameters for *Sp(2N,ℜ)*

We shall now generalize the Bargmann parametrization (10) of *SU(1,1)* to *Sp(2N,ℜ)* in its *pseudo-unitary* form (17); then, through covering $\mathfrak{S}_1$ by ℜ, we shall parametrize the universal covering group $\overline{Sp(2N,R)}$ of *Sp(2N,R)*. We write $\mathbf{u}$ in (17) as

$$\mathbf{u}\{\omega,\boldsymbol{\lambda},\boldsymbol{\mu}\} = \begin{pmatrix} e^{i\omega}\mathbf{1} & \mathbf{0} \\ \mathbf{0} & e^{-i\omega}\mathbf{1} \end{pmatrix} \begin{pmatrix} \boldsymbol{\lambda} & \boldsymbol{\mu} \\ \boldsymbol{\mu}^* & \boldsymbol{\lambda}^* \end{pmatrix}, \qquad \omega \in \mathfrak{S}_1, \quad \det\boldsymbol{\lambda} > 0, \tag{24a}$$

where $\boldsymbol{\lambda}$ and $\boldsymbol{\mu}$ satisfy Eqs. (18*b*) and (18*c*) with $\boldsymbol{\alpha} \mapsto \boldsymbol{\lambda}$, $\boldsymbol{\beta} \mapsto \boldsymbol{\mu}$. These are the Bargmann parameters for *Sp(2N,ℜ)*. The crux of the matter is to separate the *U(1)* factor in (23) into a single phase parameter $\omega \in \mathfrak{S}_1$, so that (10*b*) is generalized to

$$\omega = \frac{1}{N}\arg\det\boldsymbol{\alpha}, \qquad e^{iN\omega} = \frac{\det\boldsymbol{\alpha}}{|\det\boldsymbol{\alpha}|}, \tag{24b}$$

$$\boldsymbol{\lambda} = e^{-i\omega}\boldsymbol{\alpha}, \qquad \det\boldsymbol{\lambda} = |\det\boldsymbol{\alpha}| > 0, \tag{24c}$$

$$\boldsymbol{\mu} = e^{-i\omega}\boldsymbol{\beta} \tag{24d}$$

Here, *unlike* the $(N=1)$-dimensional case, $\boldsymbol{\lambda}$ is *not* a redundant parameter.

The product for the *Sp(2N,ℜ)* Bargmann parameters, $\mathbf{u}\{\omega,\boldsymbol{\lambda},\boldsymbol{\mu}\} = \mathbf{u}\{\omega_1,\boldsymbol{\lambda}_1,\boldsymbol{\mu}_1\}\,\mathbf{u}\{\omega_2,\boldsymbol{\lambda}_2,\boldsymbol{\mu}_2\}$, is obtained straightforwardly and yields

$$\begin{aligned}\omega &= \frac{1}{N}\arg\det[e^{i(\omega_1+\omega_2)}\boldsymbol{\lambda}_1\boldsymbol{\lambda}_2 + e^{i(\omega_1-\omega_2)}\boldsymbol{\mu}_1\boldsymbol{\mu}_2^*] \\ &= \frac{1}{N}\arg[e^{iN(\omega_1+\omega_2)}\det\boldsymbol{\lambda}_1\ \det\boldsymbol{\nu}\ \det\boldsymbol{\lambda}_2] \\ &= \omega_1+\omega_2+\omega_\nu,\end{aligned} \tag{25a}$$

$$\boldsymbol{\lambda} = e^{-i\omega_\nu}\boldsymbol{\lambda}_1\,\boldsymbol{\nu}\,\boldsymbol{\lambda}_2, \tag{25b}$$

$$\boldsymbol{\mu} = e^{-i\omega_\nu}(\boldsymbol{\lambda}_1\boldsymbol{\mu}_2 - e^{-2i\omega_2}\boldsymbol{\mu}_1\boldsymbol{\lambda}_2^*), \tag{25c}$$

where the role of ν in (12*d*) is taken by the nonsingular matrix $\boldsymbol{\nu}$:

$$\boldsymbol{\nu} := \mathbf{1} + e^{-2i\omega_2}\boldsymbol{\lambda}_1^{-1}\,\boldsymbol{\mu}_1\,\boldsymbol{\mu}_2^*\,\boldsymbol{\lambda}_2^{-1}, \tag{26a}$$

$$\omega_\nu := \frac{1}{N}\arg\det\boldsymbol{\nu}. \tag{26b}$$

The nonsingularity of the matrix $\boldsymbol{\nu}$, necessary for a proper definition of the argument ω_ν, can be proven through noting that the operator norms $[\mathbf{v}^\dagger\mathbf{A}^\dagger\mathbf{A}\mathbf{v} \le |\mathbf{A}|^2\mathbf{v}^\dagger\mathbf{v}$ for an arbitrary vector $\mathbf{v}]$ of the symmetric matrices $\boldsymbol{\lambda}_i^{-1}\boldsymbol{\mu}_i$ and $\boldsymbol{\mu}_i^*\boldsymbol{\lambda}_i^{-1}$ are bounded by $0 \le 1-|\boldsymbol{\lambda}_i|^2 < 1$, $i=1,2$. Consequently, $|\boldsymbol{\nu}-\mathbf{1}| < 1$.

The *Sp(2N,ℜ)* matrices (15), written through (20) in terms of the Bargmann parameters, read

$$\mathbf{g}\{\omega,\boldsymbol{\lambda},\boldsymbol{\mu}\} = \begin{pmatrix} \cos\omega\mathbf{1} & -\sin\omega\mathbf{1} \\ \sin\omega\mathbf{1} & \cos\omega\mathbf{1} \end{pmatrix} \begin{pmatrix} \mathrm{Re}\,(\boldsymbol{\lambda}+\boldsymbol{\mu}) & -\mathrm{Im}\,(\boldsymbol{\lambda}-\boldsymbol{\mu}) \\ \mathrm{Im}\,(\boldsymbol{\lambda}+\boldsymbol{\mu}) & \mathrm{Re}\,(\boldsymbol{\lambda}-\boldsymbol{\mu}) \end{pmatrix}. \tag{27}$$

This generalizes (11).

A.4.6 $\overline{Sp(2N,ℜ)}$ and *Mp(2N,ℜ)*

For the matrix realizations of *Sp(2N,ℜ)* in (24) and (27), only $\omega \equiv \omega (\mathrm{mod}\ 2\pi)$ makes sense. As in the $(N=1)$-dimensional case, however, the composition law (25), taken form $\omega \in ℜ$, suffices to define the universal covering group $\overline{Sp(2N,\Re)}$ of $Sp(2N,\Re)$. We shall denote its elements by $\overline{u}\{\omega,\boldsymbol{\lambda},\boldsymbol{\mu}\}$. The group unit is given by $\overline{u}\{0,\mathbf{1},\mathbf{0}\}$, and $\overline{u}\{\omega,\boldsymbol{\lambda},\boldsymbol{\mu}\}^{-1} = \overline{u}\{-\omega,\boldsymbol{\lambda}^{\dagger},-e^{2i\omega}\boldsymbol{\mu}^{\top}\}$. The center of $\overline{Sp(2N,\Re)}$ is the set $\overline{u}\{n\pi,\mathbf{1},\mathbf{0}\}, n \in \mathfrak{Z}$, and the symplectic group is given by $Sp(2N,\Re) = \overline{Sp(2N,\Re)}/\{\overline{u}\{2n\pi,\mathbf{1},\mathbf{0}\},\ n \in \mathfrak{Z}\}$.

The N-dimensional *metaplectic* group, defined by

$$Mp(2N,\Re) := \overline{Sp(2N,\Re)}/\{\overline{u}\{4\pi n,\mathbf{1},\mathbf{0}\},\ n \in \mathfrak{Z}\}, \tag{28}$$

is the two-fold cover of $Sp(2N,\Re)$, and its elements may be denoted $\tilde{g}(\omega,\boldsymbol{\lambda},\boldsymbol{\mu})$, with $\omega \equiv \omega(\text{mod } 4\pi)$. Again, we have no representation through finite matrices.

References

[1] V. Bargmann, Irreducible unitary representations of the Lorentz group. *Ann. Math.* **48**, 568–640 (1947).

[2] V. Bargmann, Group representations in Hilbert spaces of analytic functions. In: *Analytical Methods in Mathematical Physics*, P. Gilbert and R. G. Newton, Eds. (Gordon and Breach, New York, 1970), pp. 27–63.

Appendix B

Representations of the algebra sp(2,R)

ABSTRACT: Finite-dimensional representations of the Lie algebra *sp(2,ℜ)* are used in geometric aberration optics, and the self-adjoint ones are relevant for wave optics, especially in the paraxial approximation. In this appendix we gather some information about this Lie algebra, its self-adjoint, indecomposable, and finite-dimensional representations.

B.1 The Lie algebra *sp(2,ℜ) = su(1,1) = so(2,1)*

We dedicate this section to obtain the generators of the Lie group *Sp(2,ℜ)* summarized in Appendix A; these constitute a basis for the Lie algebra *sp(2,ℜ)*, isomorphic to *sl(2,ℜ)*, *su(1,1)*, and *so(2,1)*, the Lie algebras of the corresponding homomorphic groups. It helps intuition to work with the 'relativistic' *so(2,1)*, justifying the notation and relating its structure to that of the compact rotation algebra *so(3)* more readily, since the properties and conventions of the latter are well known and established.

B.1.1 The cartesian basis

From equations $(A.14a,b,c)$[1] we may find the basic matrix representatives of the one-parameter subgroup generators J_k, $k=1,2,0$, $g_k(\tau)=\exp(i\tau J_k)$. We indicate their correspondence by "↔":

$$\begin{array}{ccccccc} & & sp(2,\Re) & & su(1,1) & & so(2,1) \\ J_1 & \leftrightarrow & \frac{-1}{2i}\begin{pmatrix}0&1\\1&0\end{pmatrix} & \leftrightarrow & \frac{1}{2i}\begin{pmatrix}0&-i\\i&0\end{pmatrix} & \leftrightarrow & \frac{1}{i}\begin{pmatrix}0&0&1\\0&0&0\\1&0&0\end{pmatrix}, \end{array} \tag{1a}$$

$$J_2 \leftrightarrow \frac{-1}{2i}\begin{pmatrix}1&0\\0&-1\end{pmatrix} \leftrightarrow \frac{-1}{2i}\begin{pmatrix}0&1\\1&0\end{pmatrix} \leftrightarrow \frac{1}{i}\begin{pmatrix}0&-1&0\\-1&0&0\\0&0&0\end{pmatrix}, \tag{1b}$$

$$J_0 \leftrightarrow \frac{-1}{2i}\begin{pmatrix}0&1\\-1&0\end{pmatrix} \leftrightarrow \frac{1}{2}\begin{pmatrix}1&0\\0&-1\end{pmatrix} \leftrightarrow \frac{1}{i}\begin{pmatrix}0&0&0\\0&0&1\\0&-1&0\end{pmatrix}. \tag{1c}$$

These operators satisfy the commutation relations

$$[J_0,J_1]=iJ_2, \qquad [J_1,J_2]=-iJ_0, \qquad [J_2,J_0]=iJ_1. \tag{2a,b,c}$$

[1] That is, equations (14*a*), (14*b*), and (14*c*) of Appendix A.

The minus sign in (2b) distinguishes *so(2,1)* from *so(3)*, the compact rotation algebra.

We may now abstract the above commutation relations for any particular representation of the quantities $\{J_i\}_{i=0}^{2}$.

Most of the representation theory of semisimple Lie algebras (and groups) deals with representations that are self-adjoint (or unitary) and irreducible. These provide a deep understanding of many of the properties such as muliplet classification by the compact subgroup generator [1]. For noncompact groups, there are also integral transforms on the continuous spectrum of the parabolic [2] or the hyperbolic generator [3]. Generally one requires the existence of a complex Hilbert space $\mathcal{H}$ with a sesquilinear inner product $(\cdot,\cdot)$. Analytic continuation of the expressions from the self-adjoint representations provides the conventions for the finite dimensional, *non*-self-adjoint representations.

B.1.2 Raising, lowering, and Casimir operators

We follow Bargmann [1] in the definition of the raising and lowering operators as the *complex*[2] linear combinations of the *sp(2,ℜ)* generators J_k, $k = 0, 1, 2$:

$$J_\uparrow := J_1 + iJ_2, \qquad J_\downarrow := J_1 - iJ_2. \tag{3a, b}$$

Their commutation relations with the ***weight*** operator J_0 are

$$[J_0, J_{\uparrow\downarrow}] = \pm J_{\uparrow\downarrow}, \qquad [J_\uparrow, J_\downarrow] = -2J_0. \tag{5}$$

The Casimir operator —commuting with all J_k— is

$$\begin{aligned} C &:= J_1^2 + J_2^2 - J_0^2 \\ &= J_\uparrow J_\downarrow - J_0^2 + J_0 \\ &= J_\downarrow J_\uparrow - J_0^2 - J_0. \end{aligned} \tag{5}$$

B.1.3 A Hilbert space and a basis

A ***representation*** of *sp(2,ℜ)* is a mapping of *sp(2,ℜ)* onto a linear space of operators on $\mathcal{H}$, with a vector basis $\mathbb{J}_k$, $k = 1, 2, 0$, whose commutators in $\mathcal{H}$ follows equations (2). We shall require the $\mathbb{J}_k$ to be *self-adjoint* in $\mathcal{H}$, *i.e.*,

$$(\mathbb{J}_k^\dagger f, g) := (f, \mathbb{J}_k g) = (\mathbb{J}_k f, g), \qquad f, g \in \mathcal{H}, \quad k = 0, 1, 2; \tag{6}$$

the domain of $\mathbb{J}_k$ and of $\mathbb{J}_k^\dagger$ are assumed to be the same. The raising and lowering operators (3) will then be one the adjoint of the other:

$$(\mathbb{J}_\uparrow f, g) = (f, \mathbb{J}_\downarrow g), \qquad (\mathbb{J}_\downarrow f, g) = (f, \mathbb{J}_\uparrow g). \tag{7a, b}$$

---

[2] These should not be confused with the *real* linear combinations $J_\pm := J_0 \pm J_1$ that generate the parabolic subgroups of *sp(2,ℜ)*.

The operator C representing the Casimir operator through $(C.3)$ will thus be also self-adjoint in $\mathcal{H}$.

We consider a complete basis for $\mathcal{H}$ given by the simultaneous eigenfunctions of J_0 and C with (real) eigenvalues μ and q, respectively:

$$\mathsf{C} f^q_\mu = q f^q_\mu, \qquad q \in \Sigma(\mathsf{C}, \mathcal{H}), \tag{8a}$$
$$\mathsf{J}_0 f^q_\mu = \mu f^q_\mu, \qquad \mu \in \Sigma(\mathsf{J}_0, \mathcal{H}), \tag{8b}$$

where we use the eigenvalues as eigenfunction labels, and denote by $\Sigma(\mathsf{A}, \mathcal{H})$ the spectrum of A in $\mathcal{H}$.

B.1.4 Normalization coefficients

When we appply $\mathsf{J}_\uparrow$ and $\mathsf{J}_\downarrow$ to the above pair of equations and use $(4a)$, we see that $\mathsf{J}_\uparrow f^q_\mu$ and $\mathsf{J}_\downarrow f^q_\mu$, if not null, are eigenfunctions of C with the same eigenvalue q, and of J_0 with eigenvalues $\mu + 1$ and $\mu - 1$, respectively. The spectrum of J_0 in $\mathcal{H}$ must be thus a collection of *equally spaced* points. If the classification (8) resolves the eigenfunctions uniquely, then

$$\mathsf{J}_\uparrow f^q_\mu = c^q_{\uparrow\mu} f^q_{\mu+1}, \tag{9a}$$
$$\mathsf{J}_\downarrow f^q_\mu = c^q_{\downarrow\mu} f^q_{\mu-1}. \tag{9b}$$

The constant $c^q_\uparrow\mu$ will be zero if $\mu + 1$ is *not* a point in the spectrum of J_0 in $\mathcal{H}$, and analogously for $c^q_\downarrow\mu$, that is, $\mu \pm 1 \notin \Sigma(\mathsf{J}_0, \mathcal{H}) \Rightarrow c^q_{\uparrow\downarrow\mu} = 0$. The eigenfunctions may be normalized to unity: $(f^q_\mu, f^q_\mu) = 1$ for all $\mu \in \Sigma(\mathsf{J}_0, \mathcal{H})$. That inner product of (9) with itself may be written

$$\begin{aligned}
|c^q_{\uparrow\downarrow\mu}|^2 (f^q_{\mu\pm1}, f^q_{\mu\pm1}) &= (\mathsf{J}_{\uparrow\downarrow} f^q_\mu, \mathsf{J}_{\uparrow\downarrow} f^q_\mu) && \text{[by (9)]}\\
&= (f^q_\mu, \mathsf{J}_{\downarrow\uparrow} \mathsf{J}_{\uparrow\downarrow} f^q_\mu) && \text{[by (7)]}\\
&= \big(f^q_\mu, [\mathsf{C} + \mathsf{J}_0^2 \pm \mathsf{J}_0] f^q_\mu\big) && \text{[by (5)]}\\
&= (q + \mu^2 \pm \mu)(f^q_\mu, f^q_\mu). && \text{[by (8)]}
\end{aligned} \tag{10}$$

Hence, $|c^q_{\uparrow\downarrow\mu}|^2 = 0$ when $q + \mu^2 \pm \mu = 0$, and otherwise they must be *positive*. Any eigenvalue pair q, μ must therefore satisfy

$$|c^q_{\uparrow\downarrow\mu}|^2 = q + \mu^2 \pm \mu \geq 0. \tag{11}$$

The regions of positivity of $|c^q_{\uparrow\mu}|^2$ and of $|c^q_{\downarrow\mu}|^2$ are depicted in Figure B.1, next page.

B.2 The self-adjoint irreducible representations

B.2.1 Bounds on multiplets

Consider an eigenfunction f^q_μ such that q and μ *do* satisfy (11). They will determine a point *outside* the striped regions of Figure B.1, or on the boundaries. Through successive application of $\mathsf{J}_\uparrow$ and $\mathsf{J}_\downarrow$ we can produce the sequence of eigenfunctions $f^q_{\mu\pm1}, f^q_{\mu\pm2}, \dots$ which should also fall outside the striped regions. For fixed q, the ***multiplet*** of eigenvalues $\{\mu\}$ forms thus a vertical lattice of points. If any of these points falls on the *forbidden* regions, (11) is violated and $\mu \notin \Sigma(\mathsf{J}_0, \mathcal{H})$ for that value of q, since further application of $\mathsf{J}_\downarrow$ or of $\mathsf{J}_\uparrow$ will yield zero; if the point falls *on* the boundary, this value of μ will be a *bound* —lower or upper— of the multiplet. In this regard we have the following distinct intervals for q:

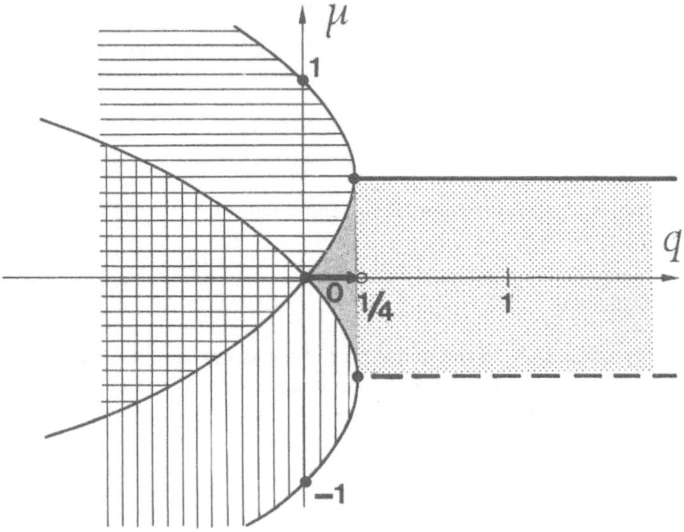

Figure B.1 The *forbidden* regions $q + \mu^2 \mp \mu < 0$ are marked with horizontal and vertical stripes, respectively. No eigenfunction f^q_μ of $\mathbb{C}$ and J_0 within a self-adjoint representation space may correspond to points q, μ inside these striped regions.

$q > \frac{1}{4}$ Multiplets are unbounded.

$q \leq 0$ Multiplets are upper- or lower-bound; to describe them it is convenient to introduce the *Bargmann index* k, related to q:

$$q(k) := k(1-k),$$
$$k(q) := \tfrac{1}{2} + \sqrt{\tfrac{1}{4} - q}, \qquad \mathrm{Re}\, k \geq \tfrac{1}{2},\ \mathrm{Im}\, k \geq 0. \tag{12}$$

Then:

\+ If a point q, μ falls on the *upper* branch of the *upper* parabola, $\mu_{\min} := k$ is the *lower* bound of its multiplet.

– If a point q, μ falls on the *lower* branch of the *lower* parabola, $\mu_{\max} := -k$ is the *upper* bound of the multiplet.

0 For $q = 0$ we have the ***sp(2,ℜ) trivial*** representation by zero: $J_n f^0_0 = 0$ for $n = 0, \uparrow, \downarrow, 1, 2$.

$0 < q \leq \frac{1}{4}$ Both half-bounded *and* unbounded multiplets coexist in this ***exceptional*** interval: the former ones happen when some μ falls *on* the parabolæ, the latter when the unit spacing of the μ's allow them to *jump over* the forbidden regions. (+) Lower-bound multiplets are obtained when ***either*** $\mu_{\min} = k$ as before (falling on the upper branch of the upper parabola), *or* when $\mu_{\min} = 1 - k$ (falling on the lower branch of the upper parabola). (–) Upper-bounded multiplets occur when *either* $\mu_{\max} = -k$ (falling on the lower branch of the lower parabola) *or* $\mu_{\max} = k - 1$ (falling on the upper branch of the lower parabola). When $q = \frac{1}{4}$, then $k = \frac{1}{2}$, and the two branches coalesce on the same point $\mu_{\min} = \frac{1}{2}$ and $\mu_{\max} = -\frac{1}{2}$. Unbounded multiplets occur when $\mu = \epsilon + n$, $|\epsilon| < k$, $n \in \mathcal{Z}$ (the set of integers).

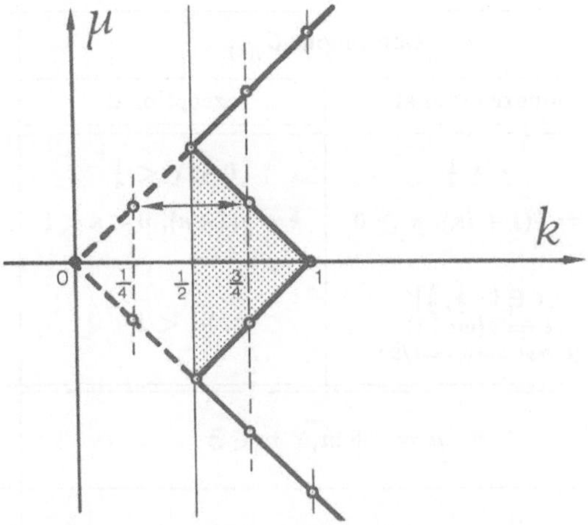

Figure B.2 The Bargmann index k is used as coordinate axis to show the *sp(2,ℜ)* representation structure in the exceptional interval. The bold lines correspond to the lower- and upper-bound representations. The open gray region indicates the exceptional continous representations. The dotted lines prolong the index k to the origin, unfolding thereby the discrete series doubling due to $q(k) = q(1-k)$.

B.2.2 Resolution of irreducible subspaces

In contradistinction to the familiar *su(2)* case, where the Casimir operator eigenvalue $j(j+1)$ uniquely specifies the spectrum of the compact generator, the *sp(2,ℜ)* Casimir operator eigenvalue does not. This is so because, as we have seen in the preceding subsection, $\mathcal{H}$ is *not* irreducible under the action of the algebra. In addition to the direct integral decomposition $\mathcal{H} = \int_{\Re}^{\oplus} \mathcal{H}^{(q)}$ into eigenspaces of the Casimir operator $\mathbb{C}$, within each $\mathcal{H}^{(q)}$, J_0 exhibits *more than one self-adjoint extension* (*two* for $q < 0$, *three* for $q = 0$, and a *one-parameter family* for $q > 0$). Since we have identified the multiplets themselves, the reduction of $\mathcal{H}^{(q)}$ to its irreducible components proceeds as follows: *(i)* we build the ***linear span*** of all functions $\{f_\mu^q\}$ belonging to a given multiplet (q fixed), and *(ii)* the ***completion*** of this space with respect to the original inner product $(\cdot,\cdot)$ will *define* the Hilbert space $\mathcal{H}_E^{(q)}$ —labelled by E— which is irreducible under *sp(2,ℜ)* and where $\Sigma(J_0, \mathcal{H}_E^{(q)})$ is *unique*. We shall now specify what E is. See Figure B.2, above.

B.2.2.1 The continuous series $C_{q>0}^{\epsilon}$

For the case of *unbounded multiplets* in $q > 0$, the label E that fully specifies the self-adjoint irreducible representations is denoted ϵ, and is

$$\epsilon \equiv \mu \,(\mathrm{mod}\ 1) \in (-\tfrac{1}{2}, \tfrac{1}{2}]. \tag{13}$$

The spectrum of J_0 is $\mu = \epsilon + n$, $n \in \mathfrak{Z}$. There are two subintervals of interest:

Nonexceptional In $q > \frac{1}{4}$, $[k = \frac{1}{2}(1 + i\kappa),\ \kappa \in \Re^+]$, the full range $(-\frac{1}{2}, \frac{1}{2}]$ is available for ϵ. This is the *continuous nonexceptional* series of representations.

Exceptional In $0 < q \leq \frac{1}{4}$, $[\frac{1}{2} \leq k < 1]$, ϵ is constrained by $|\epsilon| < 1 - k$. This is the *continuous exceptional* series of representations. See the Table in the following page.

Representation series:	Continuous $C^{\epsilon}_{q(k)}$		Discrete $D^{\pm}_{k}$
	nonexceptional	exceptional	
$q \in \Sigma(\mathbb{C})$ Bargmann label k: $q = k(1-k)$	$q \geq \frac{1}{4}$ $k = \frac{1}{2}(1+i\kappa),\ \kappa \geq 0$	$0 < q < \frac{1}{4}$ $k = \frac{1}{2}(1+\kappa),\ 0<\kappa<1$	$q \leq \frac{1}{4}$ $k > 0$
'multivaluation' index: ϵ	$\epsilon \in (-\frac{1}{2}, \frac{1}{2}]$ $\epsilon \equiv \epsilon\,(\mathrm{mod}\ 1)$ (except $\kappa = 0, \epsilon = 1/2$)	$\lvert\epsilon\rvert < k$	$\epsilon = \pm k$
$\mu \in \Sigma(\mathsf{J}_0)$	$\mu = \epsilon + m, \quad m \in \mathfrak{Z}$		$\pm\mu = \epsilon + m$ $m \in \mathfrak{Z}^{0+}$
$\nu \in \Sigma(\mathsf{J}_1)$ $\sigma \in \Sigma(\mathsf{A})$	$\nu \in \Re, \quad \sigma \in \{-1, 1\}$		$\nu \in \Re$ $\pm\sigma = 1$
$\xi \in \Sigma(\mathsf{J}_-)$	$\xi \in \Re$		$\pm\xi \in \Re^+$

Table. Casimir operator eigenvalues and spectra of the elliptic (J_0), hyperbolic (J_1), and parabolic (J_-) subalgebra representatives for all representation series. (The outer algebra automorphism A is described in reference [3].)

B.2.2.2 The discrete series $D^{\pm}_{k}$

The lower- and upper-bound multiplets belong to the so-called ***discrete***[3] representation series. There, it is $\mu_{\min}$ or $\mu_{\max}$ which becomes the label E specifying the representation bound. The main division concerns the direction in which the multiplet extends, while the bound itself is given in terms of Bargmann's label. We thus take E to be + or −, and we write, following the established convention:

Positive $\quad D^{+}_{k}: \quad k > 0, \quad \mu = k + n,\ n \in \{0, 1, 2, \ldots\} = \mathfrak{Z}^{0+}.$

Negative $\quad D^{-}_{k}: \quad k > 0, \quad \mu = -k + n,\ n \in \{0, -1, -2, \ldots\} = \mathfrak{Z}^{0-}.$

We may uphold the choice of the parameter ϵ, simply setting $\epsilon = \pm k$ and abandon the modulo 1 condition since now it is a lower bound for $\{\mu\}$. It is also known [3] that by means of an (outer) automorphism A of the algebra, $\mathsf{A} : \{J_+, J_0, J_-\} = \{J_-, -J_0, J_+\}$ we may intertwine the positive and the negative discrete series.

As we can see in Figures B.1 and B.2, the number of discrete-series multiplets corresponding to a given value of the Casimir eigenvalue q is *two* for $q < 0$ ($k > 1$ *i.e.*, D^{+}_{k} and D^{-}_{k}), *three* for $q = 0$ ($k = 1$ *i.e.*, D^{+}_{1}, D^{-}_{1} and the trivial D^{0}), *four* for $0 < q < \frac{1}{4}$ ($\frac{1}{2} < k < 1$ *i.e.*, D^{+}_{k}, D^{+}_{1-k}, D^{-}_{k}, and D^{-}_{1-k})[4]

[3]The name *discrete* for these series was given by V. Bargmann [1], who considered the single-valued *group* representations, rather than the algebra representations as here. In that case, the spectrum $\{\mu\}$ is restricted to *integers* for *SO(2,1)*, and the half-integers for *Sp(2,R)*. For the group, the allowed values of Bargmann's label k are *discrete*. For the algebra, they are continuous.

[4]The well-known *oscillator* representation falls on $q = 3/16$; there we have $k = 3/4$, and so $D^{+}_{3/4}$ and $D^{+}_{1/4}$ are the irreducible representations spanned by the odd and even states. The $D^{-}_{3/4}$ and $D^{-}_{1/4}$ representation spaces would contain negative unbounded energies. They are disregarded as unphysical.

and finally *two* again for $q = \frac{1}{4}$ ($k = \frac{1}{2}$ *i.e.*, $D^+_{1/2}$ and $D^-_{1/2}$). In all but the first case, these representations coexist with continuous-series ones. The plethora of cases is conveniently reduced by disregarding the Casimir operator eigenvalue q in favor of Bargmann's label $k > 0$ and $\pm$ for the discrete series, and q —or equally well k— and ϵ for the continuous series[5] In the table of last page we abstract this information.

For all self-adjoint representations series, using (11) and the Bargmann label (12), we may write

$$J_\uparrow f^k_\mu = \sqrt{(\mu + k)(\mu - k + 1)}\, f^k_{\mu+1}, \tag{14a}$$

$$J_\downarrow f^k_\mu = \sqrt{(\mu - k)(\mu + k - 1)}\, f^k_{\mu-1}, \tag{14b}$$

where the radicands are positive.

B.3 Indecomposable and finite-dimensional representations

We now relax the condition (7) that the raising operator be the adjoint of the lowering operator. We should keep the condition of self-adjointness for J_0 and $\mathbb{C}$, however, if in (8) we still want real eigenvalue labels μ and q for the multiplet members. This means that both $J_\uparrow J_\downarrow$ and $J_\downarrow J_\uparrow$ are self-adjoint, as may be seen from (5). We may thus allow the matrix elements of $J_{\uparrow\downarrow}$ to keep the absolute values of (14), but to differ by conjugate *phases* from those of $(J_{\downarrow\uparrow})^\dagger$, so that $c^k_{\uparrow\mu} c^k_{\downarrow\mu+1}$ be real, *i.e.*,

$$c^{q(k)}_{\uparrow\ \mu} = e^{i\phi(k,\mu)} \sqrt{|(\mu + k)(\mu - k + 1)|}, \tag{15a}$$

$$c^{q(k)}_{\downarrow\ \mu} = e^{-i\phi(k,\mu-1)} \sqrt{|(\mu - k)(\mu + k - 1)|}. \tag{15b}$$

B.3.1 Indecomposable representations

The Bargmann label k will prove to describe the class of indecomposable representations better than the Casimir operator eigenvalue q.

Due to (12), $q(k) = q(1 - k)$, but k itself, allowed to range over $k > 0$ above, may be extended here to the real line[6] As for the discrete series, we see from (14b) that $\mu_{\min} = k$ is a lower bound for multiplets extending up, and from (14a) that $\mu_{\max} = -k$ is an upper bound for multiplets extending down, since $J_\downarrow f^k_{\mu_{\min}} = 0$ and $J_\uparrow f^k_{\mu_{\max}} = 0$ (here we have switched from q to k to denote the representation). Yet, these are *one-way* barriers, because we may arrive at the lower bound from *below*, *i.e.*, $J_\uparrow f^k_{\mu_{\min}-1} = e^{i\phi}\sqrt{2k} f^k_{\mu_{\min}}$ and correspondingly $J_\downarrow f^k_{\mu_{\max}+1} = e^{-i\phi'}\sqrt{2k} f^k_{\mu_{\max}}$. See Figure B.3, next page.

Suppose now we have k integer and we start with $f^k_{\mu=0}$. In Figure B.3 this falls within the 'non-self-adjoint' region to the right of the origin. We may now raise μ with $J_\uparrow$ past the $\mu_{\min} = k$ barrier, *into* the D^+_k region, or we may lower it with $J_\downarrow$ into the D^-_k region. Once there, however, we cannot go back to f^k_0 because of the one-way barriers. It follows that f^k_μ, $\mu \in \mathfrak{Z}$ is a basis for an *indecomposable* representation of $sp(2,\Re)$ [and of $Sp(2,\Re)$], with two irreducible, self-adjoint pieces D^+_k and D^-_k. The block form of the algebra (and group) representation is thus $\begin{pmatrix} D^+ & \times & 0 \\ 0 & \times & 0 \\ 0 & \times & D^- \end{pmatrix}$.

[5]Further study of the *exceptional interval* $0 < k < 1$ in terms of quantum mechanical eigenfunctions of a harmonic oscillator with a *weakly* attractive or repulsive x^{-2}–core, is undertaken in reference [4].

[6]The Bargmann label is, in fact, complex for the continuous representation series: $k = \frac{1}{2}(1 + i\kappa)$, $\kappa \geq 0$.

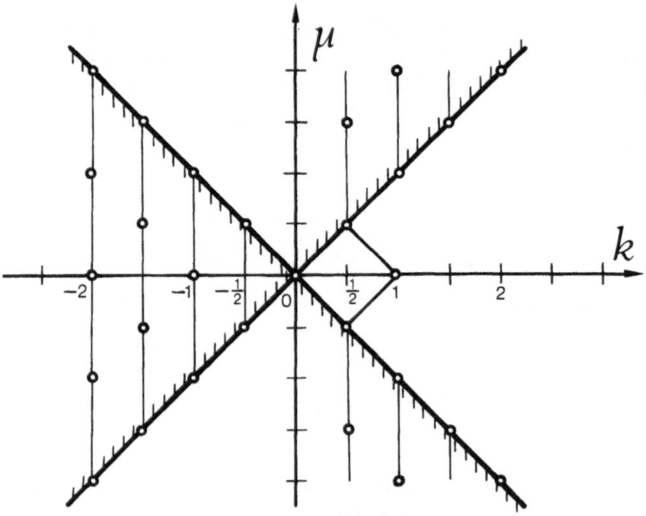

Figure B.3 The Bargmann index k is prolongued to negative values, showing the one-way barriers that hold the irreducible spaces within the indecomposable ones. The finite-dimensional representation multiplets f^j_μ are to the left.

For noninteger k, when $\mu \equiv k \pmod 1$ we obtain upper-triangular indecomposable representations of 2×2 block form, with $\mathcal{D}^+$ in the 1–1 position. Similarly, when $\mu \equiv 1 - k \pmod 1$, the lower-triangular indecomposable representations contain $\mathcal{D}^-$ in the 2–2 position.

We consider now the region $k < 0$ as continuation of the label k in (14) to negative values. Set $j := -k$, positive. Then, the Casimir eigenvalue is $q = k(1-k) = -j(j+1)$ [*cf.* Eq. (5)]; $\mu = -j$ remains a lower bound barrier and $\mu = j$ and upper bound barrier. Hence, for $j = \frac{1}{2}, 1, \frac{3}{2}, 2 \ldots$, the set of vectors $\{f^j_\mu\}^j_{\mu=-j}$ forms a finite-dimensional basis for an irreducible $sp(2,\Re)$ representation [also valid for the group $Sp(2,\Re)$], which we may call $\mathcal{D}^j$. The block form of the representation for all values $\mu \equiv j \pmod 1$ is $\begin{pmatrix} \times & 0 & 0 \\ \times & \mathcal{D}^j & \times \\ 0 & 0 & \times \end{pmatrix}$. If j is not in the above range of values but some μ falls on a boundary, the block form reduces to 2×2 block triangular cases.

B.3.2 The finite-dimensional representations of $sp(2,\Re)$

We now consider specifically the finite dimensional (*non*-self-adjoint) representations $\mathcal{D}^j$, where $-k = j = \frac{1}{2}, 1, \frac{3}{2}, 2, \ldots$. The phase of the normalization constants in (14) may be chosen to be unity, so that

$$\mathrm{J}_\uparrow f^j_\mu = \sqrt{(j-\mu)(j+\mu+1)}\, f^j_{\mu+1}, \tag{16a}$$

$$\mathrm{J}_\downarrow f^j_\mu = \sqrt{(j+\mu)(j-\mu+1)}\, f^j_{\mu-1}. \tag{16b}$$

These equations have now exactly the same form as the familiar raising and lowering operator action of the rotation algebra $so(3)$ [5, Eqs.(3.20)]. What we have done is part of the inverse Weyl trick: replacing $J_1 \mapsto iJ_1$, $J_2 \mapsto iJ_2$, whereby the minus sign in (2b) is now a plus, the Casimir operator C in (5) is now $-J^2$, and the i's have been brought into the radicands of (14) to yield those of (16). The other part of the trick, applied to groups, is the analytic continuation of the group parameters. This we need not do here; instead, we remain within $sp(2,\Re)$.

B.3.3 The finite-dimensional representations of the group

The relation between the general element of the $sp(2,\Re)$ algebra and its exponentiation to $g\begin{pmatrix} a & b \\ c & d \end{pmatrix} \in Sp(2,\Re)$ is the following:

$$\exp i(\theta_0 J_0 + \theta_1 J_1 + \theta_2 J_2) = g\begin{pmatrix} \cos\frac{\theta}{2} - \frac{\theta_2}{\theta}\sin\frac{\theta}{2} & -\frac{\theta_0+\theta_1}{\theta}\sin\frac{\theta}{2} \\ \frac{\theta_0-\theta_1}{\theta}\sin\frac{\theta}{2} & \cos\frac{\theta}{2} + \frac{\theta_2}{\theta}\sin\frac{\theta}{2} \end{pmatrix}, \tag{17a}$$

where

$$\theta = \sqrt{\theta_0^2 - \theta_1^2 - \theta_2^2}. \tag{17b}$$

This can be verified to be consistent with (A.14) and (1) in the basic representation.

The self-adjoint representations of the algebra exponentiate to unitary representations of the group. These were found by Bargmann [1,§10] in 1947, and can be seen summarized for the symplectic group parameters $\begin{pmatrix} a & b \\ c & d \end{pmatrix}$ in reference [2]. The finite-dimensional D^j representation matrix elements in the basis $\{f_\mu^j\}_{\mu=-j}^j$ were also given by Bargmann [1, §10g] and written in terms of hypergeometric functions. In polynomial form, we find

$$D^j_{m,m'}\begin{pmatrix} a & b \\ c & d \end{pmatrix} = \sqrt{\frac{(j-m')!\,(j+m')!}{(j-m)!\,(j+m)!}} \sum_n \binom{j-m}{j+m'-n}\binom{j+m}{n} a^n b^{j+m-n} c^{j+m'-n} d^{n-m-m'}. \tag{18}$$

This is a polynomial of degree $2j$ in the symplectic matrix group parameters.

References

[1] V. Bargmann, Irreducible unitary representations of the Lorentz group, *Ann. Math.* **48**, 568–640 (1947).

[2] D. Basu and K.B. Wolf, The unitary irreducible representations of SL(2,R) in all subgroup reductions, *J. Math. Phys.* **23**, 189–205 (1982).

[3] N. Mukunda, Unitary representations of the group O(2,1) in an O(1,1) basis, *J. Math. Phys.* **8**, 2210–2220 (1967); *ib.* Matrices of finite Lorentz transformations in a noncompact basis. I and II, *J. Math. Phys.* **10**, 2086–2092, 2092–2098 (1969).

[4] K.B. Wolf, Equally-spaced energy spectra: the harmonic oscillator with a centrifugal barrier or with a centripetal well, *Kinam* **3**, 323–346 (1981).

[5] L.C. Biedenharn and J.D. Louck, *Angular Momentum in Quantum Physics*, Encyclopedia of Mathematics, Vol. 8. Ed. by G.–C. Rota (Addison-Wesley, Reading Mass., 1981).

THIS
VOLUME
WAS COMPILED
IN JANUARY OF 1986
AT IIMAS–UNAM
MEXICO
CITY

The matrix for this volume was prepared entirely at the computer instalations of the **Instituto de Investigaciones en Matemáticas Aplicadas y en Sistemas**, Universidad Nacional Autónoma de México. I would like to thank my colleagues in sharing some of the burden of being editor. Special recognition is due to **Gilberto Becerril** and **Ricardo Espriella** for the nuts-and-bolts action at the computer systems and the cranky printer. **Miguel Navarro Saad** helped a lot with typography and back-up computer assistance while I escaped to work in delicious Cuernavaca. Also special thanks are due to **Enrique Pérez**, Dirección General de Servicios de Cómputo Académico UNAM, for graciously lending us a better printer. Last but not least, **José Luis Abreu**, Director of IIMAS, for his faith in the **Tipografía Científica Automatizada** project and a decent laser printer.

—Bernardo Wolf

Lecture Notes in Physics

For information about Vols. 1–172, please contact your bookseller or Springer-Verlag.

Vol. 173: Stochastic Processes in Quantum Theory and Statistical Physics. Proceedings, 1981. Edited by S. Albeverio, Ph. Combe, and M. Sirugue-Collin. VIII, 337 pages. 1982.

Vol. 174: A. Kadić, D.G.B. Edelen, A Gauge Theory of Dislocations and Disclinations. VII, 290 pages. 1983.

Vol. 175: Defect Complexes in Semiconductor Structures. Proceedings, 1982. Edited by J. Giber, F. Beleznay, J. C. Szép, and J. László. VI, 308 pages. 1983.

Vol. 176: Gauge Theory and Gravitation. Proceedings, 1982. Edited by K. Kikkawa, N. Nakanishi, and H. Nariai. X, 316 pages. 1983.

Vol. 177: Application of High Magnetic Fields in Semiconductor Physics. Proceedings, 1982. Edited by G. Landwehr. XII, 552 pages. 1983.

Vol. 178: Detectors in Heavy-Ion Reactions. Proceedings, 1982. Edited by W. von Oertzen. VIII, 258 pages. 1983.

Vol. 179: Dynamical Systems and Chaos. Proceedings, 1982. Edited by L. Garrido. XIV, 298 pages. 1983.

Vol. 180: Group Theoretical Methods in Physics. Proceedings, 1982. Edited by M. Serdaroğlu and E. İnönü. XI, 569 pages. 1983.

Vol. 181: Gauge Theories of the Eighties. Proceedings, 1982. Edited by R. Raitio and J. Lindfors. V, 644 pages. 1983.

Vol. 182: Laser Physics. Proceedings, 1983. Edited by J. D. Harvey and D. F. Walls. V, 263 pages. 1983.

Vol. 183: J. D. Gunton, M. Droz, Introduction to the Theory of Metastable and Unstable States. VI, 140 pages. 1983.

Vol. 184: Stochastic Processes – Formalism and Applications. Proceedings, 1982. Edited by G. S. Agarwal and S. Dattagupta. VI, 324 pages. 1983.

Vol. 185: H. N. Shirer, R. Wells, Mathematical Structure of the Singularities at the Transitions between Steady States in Hydrodynamic Systems. XI, 276 pages. 1983.

Vol. 186: Critical Phenomena. Proceedings, 1982. Edited by F.J.W. Hahne. VII, 353 pages. 1983.

Vol. 187: Density Functional Theory. Edited by J. Keller and J.L. Gázquez. V, 301 pages. 1983.

Vol. 188: A. P. Balachandran, G. Marmo, B.-S. Skagerstam, A. Stern, Gauge Symmetries and Fibre Bundles. IV, 140 pages. 1983.

Vol. 189: Nonlinear Phenomena. Proceedings, 1982. Edited by K. B. Wolf. XII, 453 pages. 1983.

Vol. 190: K. Kraus, States, Effects, and Operations. Edited by A. Böhm, J. W. Dollard and W. H. Wootters. IX, 151 pages. 1983.

Vol. 191: Photon Photon Collisions. Proceedings, 1983. Edited by Ch. Berger. V, 417 pages. 1983.

Vol. 192: Heidelberg Colloquium on Spin Glasses. Proceedings, 1983. Edited by J. L. van Hemmen and I. Morgenstern. VII, 356 pages. 1983.

Vol. 193: Cool Stars, Stellar Systems, and the Sun. Proceedings, 1983. Edited by S. L. Balliunas and L. Hartmann. VII, 364 pages. 1984.

Vol. 194: P. Pascual, R. Tarrach, QCD: Renormalization for the Practitioner. V, 277 pages. 1984.

Vol. 195: Trends and Applications of Pure Mathematics to Mechanics. Proceedings, 1983. Edited by P. G. Ciarlet and M. Roseau. V, 422 pages. 1984.

Vol. 196: WOPPLOT 83. Parallel Processing: Logic, Organization and Technology. Proceedings, 1983. Edited by J. Becker and I. Eisele. V, 189 pages. 1984.

Vol. 197: Quarks and Nuclear Structure. Proceedings, 1983. Edited by K. Bleuler. VIII, 414 pages. 1984.

Vol. 198: Recent Progress in Many-Body Theories. Proceedings, 1983. Edited by H. Kümmel and M. L. Ristig. IX, 422 pages. 1984.

Vol. 199: Recent Developments in Nonequilibrium Thermodynamics. Proceedings, 1983. Edited by J. Casas-Vázquez, D. Jou and G. Lebon. XIII, 485 pages. 1984.

Vol. 200: H. D. Zeh, Die Physik der Zeitrichtung. V, 86 Seiten. 1984.

Vol. 201: Group Theoretical Methods in Physics. Proceedings, 1983. Edited by G. Denardo, G. Ghirardi and T. Weber. XXXVII, 518 pages. 1984.

Vol. 202: Asymptotic Behavior of Mass and Spacetime Geometry. Proceedings, 1983. Edited by F. J. Flaherty. VI, 213 pages. 1984.

Vol. 203: C. Marchioro, M. Pulvirenti, Vortex Methods in Two-Dimensional Fluid Dynamics. III, 137 pages. 1984.

Vol. 204: Y. Waseda, Novel Application of Anomalous (Resonance) X-Ray Scattering for Structural Characterization of Disordered Materials. VI, 183 pages. 1984.

Vol. 205: Solutions of Einstein's Equations: Techniques and Results. Proceedings, 1983. Edited by C. Hoenselaers and W. Dietz. VI, 439 pages. 1984.

Vol. 206: Static Critical Phenomena in Inhomogeneous Systems. Edited by A. Pękalski and J. Sznajd. Proceedings, 1984. VIII, 358 pages. 1984.

Vol. 207: S. W. Koch, Dynamics of First-Order Phase Transitions in Equilibrium and Nonequilibrium Systems. III, 148 pages. 1984.

Vol. 208: Supersymmetry and Supergravity/Nonperturbative QCD. Proceedings, 1984. Edited by P. Roy and V. Singh. V, 389 pages. 1984.

Vol. 209: Mathematical and Computational Methods in Nuclear Physics. Proceedings, 1983. Edited by J. S. Dehesa, J. M. G. Gomez and A. Polls. V, 276 pages. 1984.

Vol. 210: Cellular Structures in Instabilities. Proceedings, 1983. Edited by J. E. Wesfreid and S. Zaleski. VI, 389 pages. 1984.

Vol. 211: Resonances – Models and Phenomena. Proceedings, 1984. Edited by S. Albeverio, L. S. Ferreira and L. Streit. VI, 369 pages. 1984.

Vol. 212: Gravitation, Geometry and Relativistic Physics. Proceedings, 1984. Edited by Laboratoire "Gravitation et Cosmologie Relativistes", Université Pierre et Marie Curie et C.N.R.S., Institut Henri Poincaré, Paris. VI, 336 pages. 1984.

Vol. 213: Forward Electron Ejection in Ion Collisions. Proceedings, 1984. Edited by K. O. Groeneveld, W. Meckbach and I. A. Sellin. VII, 165 pages. 1984.

Lecture Notes in Physics

Vol. 214: H. Moraal, Classical, Discrete Spin Models. VII, 251 pages. 1984.

Vol. 215: Computing in Accelerator Design and Operation. Proceedings, 1983. Edited by W. Busse and R. Zelazny. XII, 574 pages. 1984.

Vol. 216: Applications of Field Theory to Statistical Mechanics. Proceedings, 1984. Edited by L. Garrido. VIII, 352 pages. 1985.

Vol. 217: Charge Density Waves in Solids. Proceedings, 1984. Edited by Gy. Hutiray and J. Sólyom. XIV, 541 pages. 1985.

Vol. 218: Ninth International Conference on Numerical Methods in Fluid Dynamics. Edited by Soubbaramayer and J. P. Boujot. X, 612 pages. 1985.

Vol. 219: Fusion Reactions Below the Coulomb Barrier. Proceedings, 1984. Edited by S. G. Steadman. VII, 351 pages. 1985.

Vol. 220: W. Dittrich, M. Reuter, Effective Lagrangians in Quantum Electrodynamics. V, 244 pages. 1985.

Vol. 221: Quark Matter '84. Proceedings, 1984. Edited by K. Kajantie. VI, 305 pages. 1985.

Vol. 222: A. García, P. Kielanowski, The Beta Decay of Hyperons. Edited by A. Bohm. VIII, 173 pages. 1985.

Vol. 223: H. Saller, Vereinheitlichte Feldtheorien der Elementarteilchen. IX, 157 Seiten. 1985.

Vol. 224: Supernovae as Distance Indicators. Proceedings, 1984. Edited by N. Bartel. VI, 226 pages. 1985.

Vol. 225: B. Müller, The Physics of the Quark-Gluon Plasma. VII, 142 pages. 1985.

Vol. 226: Non-Linear Equations in Classical and Quantum Field Theory. Proceedings, 1983/84. Edited by N. Sanchez. VII, 400 pages. 1985.

Vol. 227: J.-P. Eckmann, P. Wittwer, Computer Methods and Borel Summability Applied to Feigenbaum's Equation. XIV, 297 pages. 1985.

Vol. 228: Thermodynamics and Constitutive Equations. Proceedings, 1982. Edited by G. Grioli. V, 257 pages. 1985.

Vol. 229: Fundamentals of Laser Interactions. Proceedings, 1985. Edited by F. Ehlotzky. IX, 314 pages. 1985.

Vol. 230: Macroscopic Modelling of Turbulent FLows. Proceedings, 1984. Edited by U. Frisch, J. B. Keller, G. Papanicolaou and O. Pironneau. X, 360 pages. 1985.

Vol. 231: Hadrons and Heavy Ions. Proceedings, 1984. Edited by W. D. Heiss. VII, 458 pages. 1985.

Vol. 232: New Aspects of Galaxy Photometry. Proceedings, 1984. Edited by J.-L. Nieto. XIII, 350 pages. 1985.

Vol. 233: High Resolution in Solar Physics. Proceedings, 1984. Edited by R. Muller. VII, 320 pages. 1985.

Vol. 234: Electron and Photon Interactions at Intermediate Energies. Proceedings, 1984. Edited by D. Menze, W. Pfeil and W. J. Schwille. VII, 481 pages. 1985.

Vol. 235: G.E.A. Meier, F. Obermeier (Eds.), Flow of Real Fluids. VIII, 348 pages. 1985.

Vol. 236: Advanced Methods in the Evaluation of Nuclear Scattering Data. Proceedings, 1985. Edited by H. J. Krappe and R. Lipperheide. VI, 364 pages. 1985.

Vol. 237: Nearby Molecular Clouds. Proceedings, 1984. Edited by G. Serra. IX, 242 pages. 1985.

Vol. 238: The Free-Lagrange Method. Proceedings, 1985. Edited by M. J. Fritts, W. P. Crowley and H. Trease. IX, 313 pages. 1985.

Vol. 239: Geometrics Aspects of the Einstein Equations and Integrable Systems. Proceedings, 1984. Edited by R. Martini. V, 344 pages. 1985.

Vol. 240: Monte-Carlo Methods and Applications in Neutronics, Photonics and Statistical Physics. Proceedings, 1985. Edited by R. Alcouffe, R. Dautray, A. Forster, G. Ledanois and B. Mercier. VIII, 483 pages. 1985.

Vol. 241: Numerical Simulation of Combustion Phenomena. Proceedings, 1985. Edited by R. Glowinski, B. Larrouturou and R. Temam. IX, 404 pages. 1985.

Vol. 242: Exactly Solvable Problems in Condensed Matter and Relativistic Field Theory. Proceedings, 1985. Edited by B. S. Shastry, S. S. Jha and V. Singh. V, 318 pages. 1985.

Vol. 243: Medium Energy Nucleon and Antinucleon Scattering. Proceedings, 1985. Edited by H. V. von Geramb. IX, 576 pages. 1985.

Vol. 244: W. Dittrich, M. Reuter, Selected Topics in Gauge Theories. V, 315 pages. 1986.

Vol. 245: R. Kh. Zeytounian, Les Modèles Asymptotiques de la Mécanique des Fluides I. IX, 260 pages. 1986.

Vol. 246: Field Theory, Quantum Gravity and Strings. Proceedings, 1984/85. Edited by H. J. de Vega and N. Sánchez. VI, 381 pages. 1986.

Vol. 247: Nonlinear Dynamics Aspects of Particle Accelerators. Proceedings, 1985. Edited by J. M. Jowett, M. Month and S. Turner. VIII, 583 pages. 1986.

Vol. 248: Quarks and Leptons. Proceedings, 1985. Edited by C. A. Engelbrecht. X, 417 pages. 1986.

Vol. 249: Trends in Applications of Pure Mathematics to Mechanics. Proceedings, 1985. Edited by E. Kröner and K. Kirchgässner. VIII, 523 pages. 1986.

Vol. 250: Lie Methods in Optics. Proceedings 1985. Edited by J. Sánchez Mondragón and K. B. Wolf. XIV, 249 pages. 1986.

Selected Issues from

Lecture Notes in Mathematics